Thermochemistry and Energy Transformations

Thermochemistry and Energy Transformations

Prof. Dr. Mukesh Pandey

www.arclerpress.com

Thermochemistry and Energy Transformations

Prof. Dr. Mukesh Pandey

Arcler Press
4164 Lakeshore Road
Burlington, ON L7L 1A4
Canada
www.arclerpress.com
Email: orders@arclereducation.com

ISBN: 978-1-77956-960-8 (Hardcover)

ABOUT THE AUTHOR

Prof. Dr. Mukesh Pandey served as Rector(Pro Vice Chancellor), Director SoEEM & Dean of Faculty of Energy Technology. He has looked after as Coordinator for establishing IIIT, Bhopal (under PPP Mode by MHRD). He held various administrative positions as Director (R&D), Coordinator of RGPV- NBA Nodal Centre, Dy. Registrar, Member of Engineering Accreditation Evaluation Committee [EAEC] and Moderation Committee of NBA(National Board of Accreditation).

He is multi-skilled professional with around 30 years of rich exposure and versatile experience in Industries, Administration, Academics , Research domains and institution building. He received his B.E. from SATI Vidisha, M.Tech. from N.I.T. Bhopal and Ph.D. from RGPV, Bhopal. He has always had a holistic approach to education to impart knowledge. He is not just a visionary in his field but also a praiseworthy administrator. He played an instrumental role in the establishment of Rajiv Gandhi Technical University Bhopal, Madhya Pradesh since it's inception in 1999.

Professor Pandey works in the area of Renewable Energy and Environment having focus on Solar Thermal Energy and he has developed an Energy Technology Park at RGPV,Bhopal by making use of the renewable energy resources; Solar Roof Top Plant on University Building, Energy Systems, Solar Thin Film systems, Solar wind Hybrid System, , Bio-Diesel Reactor, CO2 Carbon Sequestration Unit, Dual Rotor Wind Turbine etc.

Professor Pandey has an iconic track record of quality teaching, innovation and research he has been Principal Investigator of many Govt. Funded Projects by DST, MNRE, MPCST, AICTE etc including Principal Investigator of Innovative & Breakthrough Technology based 30KW CL-CSP SOLAR international R&D PROJECT (India-Japan Joint Venture Project) installed in RGPV campus. He has also facilitated as Reviewer for many national and international journals of repute and associated with many academic and research organizations at various levels.

Over the past several years he has taught M.Tech. and Ph.D. Research Scholars in the areas of Energy, Environment, Direct Energy Conversion and Integrated Energy

Systems and has been key-note speaker and resource person at several International and National Conferences and programmes.

On International level for academic & collaborative research purposes, he has visited several countries such as Australia,Turkey, Japan, Hongkong, Thailand, and Singapore etc. Prof. Pandey has been an active member of governing bodies, Steering Committees of institutions and number of professional academy and societies also member of expert committees of AICTE, NBA,NAAC , UGC etc.

He has a keen Interest in quality and outcome based research and has already supervised 18 quality Ph.Ds, currently under him 07 Ph.D. Research Scholars are working , many M.Tech. Projects. He has published more than 120 Research papers in International/ National Journals and more than 100 papers in International/National Conferences of repute and Authored 13 books to his credit,awarded 01 patent and he has received 03 Awards .

TABLE OF CONTENTS

List of Figures .... *xi*

Preface .... *xvii*

Chapter 1 Introduction to Thermochemistry 1

Introduction 1

1.1. Historical Background 6

1.2. Fundamental Concepts 11

1.3. Types of Energy Changes 16

1.4. Measurement Techniques 20

1.5. Applications of Thermochemistry 25

1.6. Thermochemical Equations 30

Summary 35

References 37

Chapter 2 Energy, Work, and Heat 43

Introduction 43

2.1. Types of Energy 48

2.2. The Concept of Work 53

2.3. The Relationship Between Energy and Work 57

2.4. Heat Transfer Mechanisms 61

2.5. The Laws of Thermodynamics 66

2.6. Practical Applications of Energy, Work, and Heat 71

Summary 76

References 78

Chapter 3 Enthalpy Changes in Chemical Reactions 83

Introduction 83

3.1. The Concept of Enthalpy 88

3.2. Types of Enthalpy Changes 92
3.3. Measuring Enthalpy Changes 97
3.4. Factors Affecting Enthalpy Changes 101
3.5. Applications of Enthalpy Changes 106
3.6. Case Studies of Enthalpy Changes 111
Summary 115
References 118

Chapter 4 Calorimetry and Experimental Methods 121
Introduction to Calorimetry 121
4.1. Basic Principles of Calorimetry 126
4.2. Types of Calorimeters 131
4.3. Experimental Methods in Calorimetry 136
4.4. Factors Affecting Calorimetric Measurements 141
4.5. Case Studies in Calorimetry 146
4.6. Advances in Calorimetric Techniques 151
Summary 155
References 158

Chapter 5 Hess's Law and Applications 163
Introduction 163
5.1. Fundamental Principles of Hess's Law 167
5.2. Mathematical Representation 172
5.3. Experimental Verification 176
5.4. Applications in Chemistry 181
5.5. Industrial Applications 186
5.6. Educational Importance 190
Summary 194
References 197

Chapter 6 Entropy and the Second Law of Thermodynamics 201
Introduction 201
6.1. Historical Background 206
6.2. Fundamental Concepts of Thermodynamics 210
6.3. The Second Law of Thermodynamics 215
6.4. Entropy in Isolated Systems 220

6.5. Applications of Entropy....225
6.6. Misconceptions about Entropy....230
Summary....234
References....237

Chapter 7 Gibbs Free Energy and Equilibrium....241
Introduction....241
7.1. The Concept of Gibbs Free Energy....246
7.2. Thermodynamic Principles....252
7.3. Gibbs Free Energy and Chemical Reactions....256
7.4. Equilibrium in Chemical Systems....261
7.5. The Role of Gibbs Free Energy in Equilibrium....266
7.6. Applications of Gibbs Free Energy and Equilibrium....270
Summary....274
References....277

Chapter 8 Energy Transformations in Physical and Chemical Processes....281
Introduction....281
8.1. Types of Energy....286
8.2. Energy Transformations in Physical Processes....291
8.3. Energy Transformations in Chemical Processes....296
8.4. The Law of Conservation of Energy....301
8.5. Real-World Examples of Energy Transformations....305
8.6. Implications of Energy Transformations....310
Summary....314
References....316

Chapter 9 Thermochemistry of Solutions and Mixtures....321
Introduction....321
9.1. Fundamental Concepts of Thermochemistry....326
9.2. Types of Solutions....331
9.3. Thermodynamic Properties of Solutions....336
9.4. Mixtures and Their Thermochemical Behavior....341
9.5. Calorimetry in Thermochemistry....346
9.6. Real-World Applications of Thermochemistry....351

Summary ... 355
References ... 358

Chapter 10 Thermochemistry in Industrial and Biological Systems ... 361
Introduction ... 361
10.1. Fundamental Principles of Thermochemistry ... 365
10.2. Thermochemical Processes in Industrial Applications ... 370
10.3. Thermochemistry in Biological Systems ... 375
10.4. Thermochemistry and Environmental Impact ... 381
10.5. Advances in Thermochemical Research ... 386
10.6. Case Studies of Thermochemistry in Action ... 391
Summary ... 395
References ... 397

Index ... 401

LIST OF FIGURES

Figure 1.1. Introduction to thermochemistry
Figure 1.2. Importance of energy changes in chemical reactions
Figure 1.3. Development of thermochemistry in the 18th century
Figure 1.4. Key figures in thermochemistry, such as Antoine Lavoisier
Figure 1.5. Definition of heat and temperature
Figure 1.6. The concept of enthalpy
Figure 1.7. Exothermic reactions
Figure 1.8. Endothermic reactions
Figure 1.9. Calorimeter
Figure 1.10. Use of bomb calorimeters for measuring heat
Figure 1.11. Applications of thermochemistry
Figure 1.12. Thermochemistry applications in metabolic processes
Figure 1.13. Thermochemistry equations
Figure 1.14. Standard enthalpy of formation
Figure 2.1. Energy, work, and heat
Figure 2.2. Heat as a form of energy transfer
Figure 2.3. Kinetic energy and its applications
Figure 2.4. Gravitational potential energy
Figure 2.5. Work done in physics
Figure 2.6. Work-energy theorem
Figure 2.7. Principle of conservation of energy
Figure 2.8. Work done by a constant force
Figure 2.9. Heat transfer mechanisms
Figure 2.10. Convection
Figure 2.11. First law: Conservation of energy
Figure 2.12. Entropy and energy quality
Figure 2.13. Energy efficiency in machines

Figure 2.14. Heat engines and thermodynamic cycles
Figure 3.1. Enthalpy change of reaction
Figure 3.2. Enthalpy in chemistry
Figure 3.3. Explanation of enthalpy as a thermodynamic quantity
Figure 3.4. Relationship between enthalpy and internal energy
Figure 3.5. Enthalpy of formation
Figure 3.6. Standard enthalpy of combustion
Figure 3.7. Measure enthalpy changes
Figure 3.8. Hess's law and its applications measuring enthalpy changes
Figure 3.9. Temperature and pressure effects
Figure 3.10. Physical state of reactants and products
Figure 3.11. Role in predicting reaction spontaneity
Figure 3.12. Enthalpy changes in use in biochemical reactions
Figure 3.13. Example of a combustion reaction
Figure 3.14. Example of a phase change reaction
Figure 4.1. Understanding calorimetry in chemistry
Figure 4.2. Importance of calorimetry
Figure 4.3. Basic principle of calorimetry
Figure 4.4. Heat transfer mechanisms
Figure 4.5. Constant pressure calorimeters
Figure 4.6. Bomb calorimeters
Figure 4.7. Differential scanning calorimeters
Figure 4.8. Experimental methods in calorimetry
Figure 4.9. Calorimeter calibration
Figure 4.10. Factors affecting calorimetric measurements
Figure 4.11. Sample purity and composition
Figure 4.12. Food calorimetry
Figure 4.13. Calorimetric measurements of biological interactions
Figure 5.1. Hess's law
Figure 5.2. Historical background of Hess's law
Figure 5.3. Conservation of energy
Figure 5.4. Relationship between Hess's law and thermodynamics
Figure 5.5. Calculating enthalpy changes

Figure 5.6. Enthalpy of formation
Figure 5.7. Calorimetry and its role in measuring enthalpy
Figure 5.8. An experimental demonstration of Hess's law
Figure 5.9. Using Thermodynamics to predict the outcomes
Figure 5.10. Calculating enthalpy changes for non-standard conditions
Figure 5.11. Hess's law use in chemical manufacturing processes
Figure 5.12. Teaching Hess's law in academic settings
Figure 6.1. Entropy and the second law of thermodynamics
Figure 6.2. Entropy in physical sciences
Figure 6.3. Development of thermodynamic principles
Figure 6.4. Key figures in thermodynamics (e.g., Carnot, Clausius, and Kelvin)
Figure 6.5. Fundamental concepts of thermodynamics
Figure 6.6. Types of thermodynamic processes
Figure 6.7. The second law of thermodynamics
Figure 6.8. Statement of the second law of thermodynamics
Figure 6.9. Examples of the second law in everyday life of thermodynamics
Figure 6.10. Entropy in isolated systems
Figure 6.11. Entropy changes in isolated systems
Figure 6.12. Entropy in chemical reactions
Figure 6.13. Entropy in information theory
Figure 6.14. Misconceptions about entropy
Figure 7.1. Gibbs free energy and equilibrium
Figure 7.2. Gibbs free energy in thermodynamics
Figure 7.3. The concept of Gibbs free energy
Figure 7.4. Mathematical expression of Gibbs free energy
Figure 7.5. Components of Gibbs free energy
Figure 7.6. First law of thermodynamics
Figure 7.7. Relationship between thermodynamic principles and Gibbs free energy
Figure 7.8. Gibbs free energy and chemical reactions
Figure 7.9. Calculation of Gibbs free energy changes in reactions
Figure 7.10. Equilibrium in chemical systems
Figure 7.11. Dynamic equilibrium
Figure 7.12. The role of Gibbs free energy in equilibrium
Figure 7.13. Relationship between Gibbs free energy and the equilibrium constant

Figure 8.1. Energy transformations
Figure 8.2. Importance of studying energy transformations
Figure 8.3. Kinetic energy
Figure 8.4. Potential energy
Figure 8.5. Thermal (Heat) energy
Figure 8.6. Phase changes
Figure 8.7. Energy transfer in systems (e.g., conduction and convection)
Figure 8.8. Exothermic reactions
Figure 8.9. Endothermic reactions
Figure 8.10. The law of conservation of energy
Figure 8.11. Energy transformations applications in physical processes
Figure 8.12. Photosynthesis in plants of energy transformations
Figure 8.13. Battery operation in electronic devices of energy transformations
Figure 9.1. Thermochemistry of solutions and mixtures
Figure 9.2. Studying solutions and mixtures
Figure 9.3. What is enthalpy?
Figure 9.4. Heat capacity and its role in solutions
Figure 9.5. Types of solutions in chemistry
Figure 9.6. Solvent-solute interactions
Figure 9.7. Factors affecting solubility
Figure 9.8. Changes in enthalpy during dissolution
Figure 9.9. Effect of temperature on solubility
Figure 9.10. Mixtures and their thermochemical behavior
Figure 9.11. Ideal solution vs non ideal solution
Figure 9.12. Raoult's law
Figure 9.13. Calorimetry in thermochemistry
Figure 9.14. Measuring heat changes in solutions
Figure 9.15. Industrial processes involving solutions in thermochemistry
Figure 10.1. Thermochemistry in industrial and biological systems
Figure 10.2. Thermochemistry in various fields
Figure 10.3. Laws of thermodynamics
Figure 10.4. Heat capacity and calorimetry
Figure 10.5. Role of thermochemistry in chemical manufacturing
Figure 10.6. Energy efficiency in industrial processes in chemical

Figure 10.7. Thermochemistry in biological systems
Figure 10.8. Role of enzymes in thermochemical reactions
Figure 10.9. Thermoregulation in living organisms
Figure 10.10. Green chemistry and sustainability
Figure 10.11. Impact of thermochemistry on climate change
Figure 10.12. Innovative thermochemical energy storage technology
Figure 10.13. Role of computational chemistry in thermochemistry

PREFACE

The study of energy conversion is central to the comprehension of physical and chemical transformations that govern the natural world. Any physical process and all chemical reactions entail energy changes in the form of heat evolved, work accomplished, or potential energy. These energy transactions not only explain why reactions occur but also determine process direction and possibility. Thermochemistry, being a branch of chemistry, provides the technique to measure, analyze, and forecast these energy changes. It provides the bridge between theoretical chemistry and practical use and is therefore a cornerstone of scientific research and industrial application.

Energy has been compared to nature's force. In engine combustion of fuels, to metabolic reactions sustaining life, energy changes mold our lives. Heat, work, and free energy determine the matter behavior when subjected to certain conditions. An understanding of these is not only of academic interest but is essential to problem-solving in daily life, such as efficient energy use, alternative fuels, and high-performance materials synthesis. Thermochemistry prepares students with tools to measure energy change, predict reaction spontaneity, and utilize the knowledge in a vast array of scientific and engineering fields.

Thermochemistry and Energy Transformations has been prepared to provide a structured and comprehensive account of the subject. The work outlines fundamental principles, experimental techniques, and theoretical models along with their applications to industrial and biological processes. Each chapter builds sequentially upon the earlier one such that readers progress from basic concepts to advanced applications with ease and continuity.

Chapter 1 acquaints the reader with Thermochemistry. The introductory chapter explains what the term "thermochemistry" is, presents its scope, and places it in physical chemistry as a whole. It follows the importance of studying energy change during reactions and processes, noting how thermochemistry lies at the root of areas such as material science, energy technology, and biological metabolism.

Chapter 2 covers Energy, Work, and Heat. Here, the basic concepts of energy in chemical systems are taken into account. The chapter describes the distinctions between work and heat as processes for the transmission of energy and gives an introduction to the first law of thermodynamics. Instances of mechanical work, heat transmission, and alterations in internal energy provide a solid basis for what is to come.

Chapter 3 explains Enthalpy Changes in Chemical Reactions. Enthalpy, a basic theory of thermochemistry, is explained and elaborated upon. The chapter explains standard enthalpies of formation, combustion, and reaction. Real-life examples, such as fuel burning and neutralization reactions, illustrate how enthalpy values are calculated and applied practically.

Chapter 4 discusses Calorimetry and Experimental Methods. Experimental methods are required for accurate measurement of heat changes, and in this chapter, calorimetry is presented as the primary method. Constant-pressure and constant-volume calorimeters are defined together with their construction and application in measuring enthalpy changes. Food calorimetry and bomb calorimetry are illustrations.

Chapter 5 covers Hess's Law and Applications. The law, which states that enthalpy changes are independent of the path, is shown with examples. The chapter shows how Hess's Law enables calculation for enthalpy changes which are difficult to calculate directly, and how it is used in calculating enthalpies of reaction, formation, and combustion.

Chapter 6 addresses Entropy and the Second Law of Thermodynamics. Entropy is defined as a probability and disorder measure, and its use in spontaneous processes. The second law of thermodynamics is explained through examples of everyday life such as melting, diffusion, and heat flow, giving students the rationale as to why some processes are natural and others are not.

Chapter 7 explains Gibbs Free Energy and Equilibrium. Gibbs free energy is introduced as the criterion for spontaneity at constant pressure and temperature. The chapter relates Gibbs free energy to chemical equilibrium, and it explains how constants of equilibrium are derived and why energy factors determine reaction direction. Such a discussion connects thermodynamics to chemical kinetics and reaction feasibility.

Chapter 8 discusses Energy Transformations in Physical and Chemical Processes. Examples ofhow energy transformations dictate natural processes as well as technological developments are phase changes, combustion reactions, and electrochemical systems. The chapter emphasizes the importance of energy efficiency and efficient use of energy in today's environment.

Chapter 9 covers the Thermochemistry of Solutions and Mixtures. Dissolution, mixing, and dilution energy changes are the topics covered here. Entthalpies of solution, hydration, and mixing are explained in the chapter with applications in chemistry, biology, and industry. Salt dissolution and solution enthalpy in drugs are used as examples of its application.

Chapter 10 concludes with Thermochemistry in Industrial and Biological Systems. The final chapter spans the gap between theory and practice by discussing how thermochemical principles are applied in extensive-scale industrial reactions, such as ammonia synthesis, and biological energy conversions, such as cellular respiration. It emphasizes the universal applicability of thermochemistry to engineered and natural systems.

The text is intended for undergraduate and graduate students of chemistry, chemical engineering, and allied sciences who desire a strong foundation in thermochemistry. It will serve as a reference also for industrial chemists, energy researchers, and biologists where considerations of energy are of vital importance. For students, the logical order of presentation from basic to advanced material provides depth and clarity. The blend of practice and theory will render it an invaluable source for practitioners and researchers.

The book will benefit curriculum planning and classroom teaching for teachers, and the students will be able to benefit from its clear explanations, practical examples, and sensible structure.

CHAPTER 1

INTRODUCTION TO THERMOCHEMISTRY

Contents

Introduction 1
1.1. Historical Background 6
1.2. Fundamental Concepts 11
1.3. Types of Energy Changes 16
1.4. Measurement Techniques 20
1.5. Applications of Thermochemistry 25
1.6. Thermochemical Equations 30
Summary 35
References 37

INTRODUCTION

Definition of Thermochemistry

Thermochemistry, a central branch of physical chemistry, deals with the relationship between heat, work, and chemical reaction. It is broadly interested in energy changes of chemical reactions, which are crucial to the study of reaction dynamics and thermodynamic principles. It not only helps in predicting the extent and direction of the reaction but also in improving the number of chemical processes. One of the fundamental concepts of thermochemistry is the measurement of enthalpy change, which also decides whether the reaction will be exothermic or endothermic. By quantitative analysis of such energy changes, chemists can make rational decisions regarding reactants and products, resulting in efficient formulations in many industrial applications. To study the intricate dynamics of thermochemistry, its role should be understood in broader chemical contexts, particularly in combustion and reaction mechanisms. Research has demonstrated how mechanistic chemical reaction mechanisms create a better understanding of practical combustion systems, as put forward by Braun-Unkhoff et al. (2014).

Figure 1.1. *Introduction to thermochemistry.*

Source: https://i.ytimg.com/vi/Iz_Q6bbXdfE/hq720.jpg?sqp=-oaymwE8CK4FEI-IDSFryq4qpAy4IARUAAAAAGAElAADIQj0AgKJD8AEB-AHUBoAC4AOKAg0IABABGIwBIFQoFTAP&rs=AOn4CLDR7hpq7O_vp3QnnnZnMwIkO42iKQ.

These advanced mechanisms not only assist in designing and optimizing gas turbine combustion chambers but also help illustrate the underlying thermodynamic principles. They optimize our ability to manage fuel and load flexibility at the expense of reducing pollutant discharge, a significant hurdle in the environmentally conscious world of today. Thermochemistry therefore stands at the crossroads of energy production and environmental science, illustrating its important place in driving technology as well as ecological objectives. Practical applications of thermochemistry extend beyond theory into a variety of applications, impacting industrial processes and leading-edge research on new materials. Of interest is the synthesis of molecules and assessing their stability using the application of thermochemical principles. Recent studies focus on advancements in the synthesis of virtual molecules that obey the octet rule, again underscoring how thermochemistry can help identify acceptable chemical structures with desired properties, as discussed by Hamaekers et al. (2017). This study offers new evidence in molecule designing, particularly in the case of intricate elements such as boron chemistry. As the field evolves, integrating thermochemical concepts with computational approaches is likely to yield new ideas and solutions, shaping the future of chemistry and materials science (Figure 1.1).

Explanation of Thermochemistry as a Branch of Chemistry

Thermochemistry is an elementary branch of chemistry that connects thermodynamics to chemical reactions. It is the branch of chemistry that deals with the transformation of energy in chemical reactions, primarily as heat transfer. It assists in the comprehension of how energy is transferred in chemical reactions and provides insight into the reactivity and stability of compounds. The concepts of thermochemistry are the building blocks for many applications, including the kinetics of reactions and design of chemical processes. By using calorimetry, the technique of measuring heat transfer, thermochemists are able to determine enthalpy changes in reactions. Not only does this quantitative approach advance theoretical knowledge, but also it finds use in materials science and energy production. Besides, experimental approaches control the course of study of thermochemistry. A wonderful technique is through the use of computational models imitating chemical reactions under varying conditions. Such models can forecast the ways that energy changes influence product formation and reaction routes. Such as the manner in which heat interacts with materials, such as carbon-based charring ablators on space vehicles, highlights the engineering significance of thermochemistry. Advanced computational methods, including the Two-dimensional Implicit Thermal Response and Ablation Program (TITAN), demonstrate how thermal response codes link with fluid mechanics in order to handle sophisticated issues in material science and chemistry (Chen et al., 2012). Therefore, thermochemistry bridges the divide between empirical experiment and theoretical calculation to offer an integrated understanding of energy dynamics in chemical reactions. Furthermore, thermochemistry is not restricted to academia but is also applicable in industrial and environmental contexts where energy management is paramount. The ability to forecast thermodynamic characteristics is responsible for designing efficient energy systems, e.g., optimizing burning processes and reducing poisonous emissions. Coupling state-of-the-art simulations, e.g., computational tools tailored for fluid dynamics research, allows for the examination of reacting equilibrium chemistry under a wide variety of conditions (Anderson et al., 2025). The synchronization of computational chemistry with experimental data presents a solid foundation for establishing new sustainable practices. With increased global energy needs, the application of thermochemistry in environmental chemistry as well as its use in green technologies emphasizes its significance in addressing the challenges of current energy usage and management.

Importance of Energy Changes in Chemical Reactions

The application of energy changes in chemical reactions is at the core of our understanding of thermochemistry because these changes dictate the feasibility and reaction rate. Energy conversions, predominantly heat, form an integral part of the definition of exothermic and endothermic reactions, which release and absorb heat, respectively. This allows chemists to predict the spontaneity of a reaction and identify conditions under which reactions are optimally performed. For example, in studying the dissolution of complex minerals such as jarosite, new thermochemical data indicate that it is stable on Mars due to some energy needs under varied pressure and temperature conditions. The stability not only suggests for planetary science but also indicates that energy dynamics are of utmost concern in thinking about chemical stability away from the Earth (Drouet et al., 2005). In combustion chemistry, energy changes are particularly important as they display the inherent processes of release of energy when fuels undergo combustion. Combustion chemistry has a rich history with some foundational work contributed by great pioneers like Lavoisier and Bunsen, whose work can be used even today. The interdependence between the generation of heat and combustion of various fuels is evidence of the importance of studying these energy changes. Moreover, the contributions of Hinshelwood and Semenov regarding chain reactions significantly enhanced the knowledge of combustion processes that also involve the measurement of explosion limits for premixed gases. Explanation of such mechanisms not only enhances our knowledge about practical application such as engine efficiency and safety but also highlights the intricate relationship between the transformation of energy and chemical kinetics in real conditions (Penner et al., 1963). Understand the importance of energy change in chemical reactions is not just about scientific curiosity; it also has practical applications across a wide range of scientific and industrial processes. For instance, energy control of chemical processes can lead to better catalysts, optimum reaction pathways, and sustainable procedures with minimized energy loss and waste. Predictive abilities derived from thermochemical data allow scientists to tailor conditions for predicted reaction outcomes. Thus, the understanding of these considerations is imperative for further advancement in fields like materials science, environmental chemistry, and pharmacy. With energy questions sitting squarely at the center of scientific inquiry, embracing the principles of thermochemistry will continue to be at the forefront of our capability to innovate and understand chemical systems that are extremely complex (Figure 1.2).

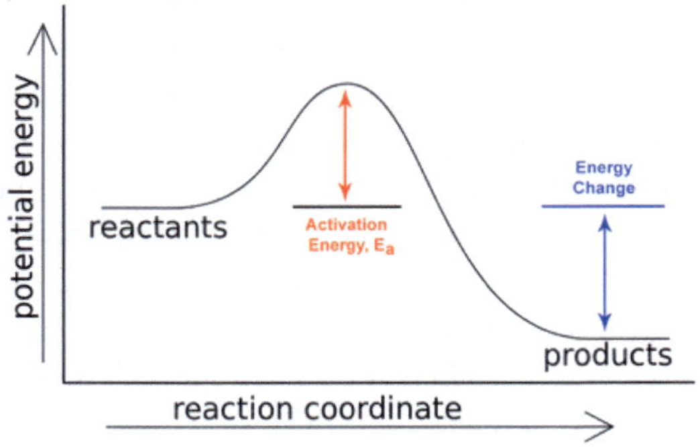

Figure 1.2. *Importance of energy changes in chemical reactions.*

Source: https://homework.study.com/cimages/multimages/16/ac_com_exo8028647243577574320.png.

Overview of Key Concepts in Thermochemistry

Thermochemistry is a basic field in the study of chemical reaction energy changes, specifically via heat transfer. One of the most important concepts is the differentiation between exothermic and endothermic reactions; exothermic releases heat, increasing the temperature of the environment, whereas endothermic takes in heat, decreasing the temperature. This heat work is significant not just in the lab but also in industry, where effectiveness in the utilization of energy is significant. The principles of thermochemistry enable prediction of spontaneity of reactions and are fundamental to applications across various fields, including materials science and environmental chemistry. The significance of the discipline is captured in Lavoisier's contributions, who laid early groundwork that has persisted in influencing contemporary research and practice, pointing to the discipline's historical roots and contemporary relevance (Penner et al., 1963). Beyond this kind of reaction, enthalpy is the central idea of thermochemistry. Enthalpy, symbolized as H, is a quantification of the internal energy of a thermodynamic system plus the pressure-volume product. Enthalpy changes (ΔH) during chemical reactions give us an idea of the energy process involved, something that is essential for the computation of reaction equilibrium and efficiency. The correlation of enthalpy with other thermodynamic functions, such as Gibbs free energy, aids in the explanation of reaction viability and their barriers to energy. Most evident is the case when one reflects upon the thermochemical nature of solid-state reactions, i.e., dehydrogenation reactions of sodium hydride and sodium hydroxide, where complex phase interactions can have a profound influence on dehydrogenation behavior (Gregory et al., 2015). By understanding these principles, chemists can develop more efficient energy storage and conversion systems. A second critical area of thermochemistry is the application of calorimetry in determining heat changes of chemical reactions. Calorimetry provides experimental data that are crucial to the verification of theoretical models and the optimization of reaction kinetics. Scientists can measure accurately heat absorbed or released through the use of calorimeters, thereby verifying the enthalpy change associated with specific chemical conversions. This empirical approach increases the predictive power of thermochemical concepts to allow researchers to engineer reactions for precise outcomes. The linkage between calorimetric measurements and theoretical models propels new designs in chemical engineering, particularly in the optimization of combustion systems, as seen in the ongoing advancement of combustion chemistry. These papers emphasize the interaction of experimental techniques and theoretical understanding, subsidizing the discipline of thermochemistry in the general framework of chemical science (Penner et al., 1963).

1.1. HISTORICAL BACKGROUND

The history of thermochemistry is marked by breakthroughs that transformed our understanding of energy transformation in chemical reactions. Early scientific investigations into heat and energy are attributed to the work of philosophers like Aristotle, who speculated about the nature of matter. It wasn't until the 18th century that Antonie Lavoisier's work set the stage for contemporary chemistry. His accurate experiments with burning revealed the concept of conservation of mass and began the idea that heat is a form of energy released or taken up in chemical alterations. This revolutionary finding provided access to deeper investigation of the connection between heat and chemical reactions, which culminated ultimately in the codification of thermochemistry as we know it today. As the science went on, the 19th century witnessed tremendous advances that extended the laws of thermochemistry. Scientists such as Pierre and Marie Curie, and later Johannes van der Waals, began studying the interaction between work and heat in chemical systems. The development of the first law of thermodynamics—the statement of conservation of energy—was a significant achievement shaping further studies. Concurrently with this, new calorimetry methods emerged that allowed researchers to quantify heat changes in reactions and, in the process, provide the discipline with greater empiricism. Study of gas-phase reactions, particularly radical ones, became central, as illustrated by studies of radical reactions relevant to atmospheric and combustion chemistry. This basis established conditions required in more systematic comprehension of reaction mechanisms and energetics, as evidenced by recent research into reaction rates and thermochemical calculations (Rissanen et al., 2012). In applied thermochemistry, the advancement of spectroscopy and enhanced analytical techniques further immersed the discipline's achievements. These techniques enabled scientists to examine molecular interaction complexities and energy transfer with previously unattainable accuracy. For instance, with the emergence of spectroscopic methods, scientists became capable of acquiring information regarding reaction mechanisms at a molecular level, such as how specific molecular structures could potentially influence energy change during reactions. The importance of temperature and pressure conditions that had significant effects on reaction kinetics and thermochemical characteristics became apparent to scientists. By comprehensive research into radical reactivity and the effect of environmental factors such as pressure and temperature on reaction rates, a more general framework was developed that would be used to direct models applied to numerous fields, such as climate science and energy production (Woods et al., 2025). Thus, the historical evolution of thermochemistry not only reflects advances in theoretical science but also a growing fascination with applications in other scientific disciplines.

1.1.1. Development of Thermochemistry in the 18th Century

The 18th century was a pivotal time in the history of thermochemistry, fueled primarily by the Enlightenments era of inquiry and innovation. This period witnessed the convergence of various scientific disciplines, including chemistry, physics, and engineering, which had collectively enhanced the understanding of heat and chemical reactions. Most notable was the work of Antoine Lavoisier with basic principles that would give birth to thermochemistry. Lavoisier's strict experiments led to the discovery of combustion as a chemical reaction with emphasis on the role of heat in the reaction. His new technique of measuring mass before and after reactions determined that heat was a measurable entity inextricably bonded to the chemical makeup of materials. This scientific investigation of heat flow between chemical reactions not only helped advance a better understanding of energy transfer by the scientific community, but also established thermochemistry as a significant sub-discipline to chemistry itself. From the ground laid by Lavoisier, the 18th century saw increasing interest in the specific heat capacities of materials, which had a profound effect on the study of thermochemistry. Researchers began to realize that materials required varying amounts of heat to change temperature, leading to the discovery of specific heat—a notion of tremendous value for the understanding of energy change in the course of chemical reactions. Researchers such as Joseph Black also played a critical role in this research by inventing equipment to measure heat transfer and conducting experiments in order to establish certain heat ratios. The investigation of latent heat also developed the field by demonstrating the heat transferred during phase changes. This quiet appreciation of the specific and latent heats provided the basis for the advances of thermodynamic concepts in the future, opening the way for thermochemistry to further become intertwined with applied uses, such as calorimetry and industrial processes. Throughout the course of the century, the use of thermochemical concepts in applied uses began to materialize, such as in metallurgy and the production of gases. The discovery of instruments like the calorimeter, driven by the demand for accurate measurement of chemical reactions, proved the practical use of thermochemistry. At the same time, the study of gas behavior at various temperatures and pressures also gained momentum. Heat's role in the process was highlighted during combustion and gas expansion experiments, which led to the concept of energy production and conservation. Also, the emerging science of gas chemistry, which dealt with air and its constituents, began to emerge with a greater emphasis placed on thermodynamic characteristics. This stepwise application of theory to practice not only served to emphasize the increasingly important role that thermochemistry played in scientific inquiry but also paved the way for future advancements in energy conversion technologies, such as those identified in Fabry et al. (2013) and Special Libraries Association (1934), which depend on an integrated knowledge of heat and chemical reactions (Figure 1.3).

Figure 1.3. *Development of thermochemistry in the 18th century.*

Source: https://cdn.britannica.com/68/115468-050-91355391/Antoine-Laurent-Lavoisier-French-work-laboratory.jpg?w=400&h=225&c=crop.

1.1.2. Key Figures in Thermochemistry, Such as Antoine Lavoisier

Thermochemistry's early movers are owed a lot by the science, and one such cornerstone is Antoine Lavoisier. Lavoisier's experiments on chemical reactions and the conservation of mass law totally transformed the science community. His scrupulous experimentation laid the groundwork for the emergence of calorimetry, the core method of thermochemistry to measure heat transfer. By establishing that substances exchange heat in chemical reactions, Lavoisier illuminated the contribution of thermal energy to reactions, beginning the shift from a singular interest in chemical makeup to energetic consideration. His research presented a systematic chemistry, with empirical observation balanced by theoretical constructs that laid the foundation for later scientists to explore further the relationship between heat and chemical processes. Following in the footsteps of Lavoisier, scientists like Adolphe Wurtz expanded thermochemistry research in substantial ways. Wurtz's Paris-based research school was a hub of innovation, where theoretical and empirical approaches to chemistry were given equal weight, including components of thermochemical studies. His research, particularly in organic and biological chemistry, emphasized the importance of incorporating thermodynamic aspects into the study of complex mixtures, such as those that exist in biological systems. Wurtz's research showed how thermochemical modeling is not only useful in mainstream chemistry but also with the intricacies of biological interactions and food chemistry, thereby proving the widespread utility of thermochemistry in many scientific fields. His integration of thermodynamic models into experimental chemistry also cemented the fundamental concepts established by Lavoisier. Lavoisier's expression of conservation of mass and energy continues to this day in modern thermochemical research, significantly influencing current scientific research. These principles are not just relevant in classical chemical investigation but also essential in interpreting complex biological systems that are typically multicomponent mixtures in need of predictive models for equilibrium behavior. Development of models such as COSMO-RS highlights the ongoing relevance of basic thermochemistry principles in addressing current challenges in chemistry. Through the integration of higher-order concepts in thermodynamics and statistical physics, modern scientists are able to comprehensively predict physico-chemical properties in complex environments. This investigation reflects Lavoisier's legacy, demonstrating how early conceptual understanding can develop and adapt with the latest scientific breakthroughs to reveal more about chemical and biological interactions under thermochemistry, highlighting the continued influence of early titans such as Lavoisier on this vibrant discipline (Figure 1.4).

Figure 1.4. *Key figures in thermochemistry, such as Antoine Lavoisier.*

Source: https://metamind.academy/wp-content/uploads/2024/03/father-of-chemistry.png.

1.1.3. Evolution of Thermochemical Principles Over Time

The evolution of thermochemical concepts has been a step-by-step process with dramatic leaps in the understanding of the relationship between heat and chemical reactions. Early research established the foundation of concepts, e.g., laws of thermodynamics, that govern energy conservation and direction of spontaneous processes. The transition from classical to modern thermochemistry involved sophisticated experiments that revealed the subtle relationships between reaction spontaneity and enthalpy changes. Major breakthroughs, such as Hess Law and the standard enthalpy concept, made the scientific exploration of reaction energies and prediction of product formation possible. This advancement is not merely academic; these concepts are generally applicable, and fields such as materials science and environmental chemistry are included, thereby emphasizing the significance of thermochemical understanding in an increasingly complex world. As the mid-20th century approached, the interactions between planetary sciences and thermochemistry began to gain attention, particularly on the knowledge of the composition of meteorites as a record of the early Solar System. Condensation calculation development, in return, that simulates mineral formation under variable temperature and pressure is a critical milestone in this evolution. Integration of laboratory data and theoretical models introduced refinement in the understanding of planetary formation chemistry and meteorite analysis. For instance, the establishment of how elements and isotopes partition between the condensed and vapor phases clarified geological and chemical evolution of planetary bodies. The combination of thermochemical principles with astrophysical processes highlights the relevance of this science in unraveling celestial mysteries, thus documenting a profound integration of thermochemistry in planetary science and geology. Moreover, recent research indicates the usability of thermochemical principles in astrobiology and search for life beyond the Earth. By employing experimental microbiology along with thermochemical modeling, researchers have been successful in proposing potential bio-signatures for extraterrestrial life, particularly in extreme environments like early Mars. Experiments with terrestrial basalt as an analog demonstrated the effect of microbial activity on mineral dissolution and provided evidence into the biotic and abiotic weathering process. The significance of the research is that specific minerals formed through microbial interaction could be employed to identify past life and therefore redefine the extent of life's possibility in the universe. This then development of thermochemical principles is not only specific to its earth origin; it stretches into planetary exploration and origins of life contexts that showcase the open-ended and adaptive character of thermochemical inquiry in an evolving scientific environment.

1.2. FUNDAMENTAL CONCEPTS

In thermochemistry, the energy conversion principle is one that helps to understand many chemical processes. Energy, defined as the capacity to perform work, plays a central role in determining the feasibility and direction of reactions. According to the first law of thermodynamics, energy cannot be created or destroyed but can be converted from one form to another, with significant implications in chemical reactions. For instance, during combustion, energy is easily released when chemical bonds in reactants are broken and new bonds in products are formed. This not only liberates energy as heat but also portrays the precarious balance between reactants and products, emphasizing the conservation of energy in chemical systems. Understanding these energy transformations enables chemists to predict the result of reactions and develop best conditions for a given process, thus laying the basis for development in theoretical as well as practical chemistry. A significant aspect of thermochemistry is distinguishing between exothermic and endothermic reactions, which are characterized by heat exchange with or from the environment. Exothermic reactions, such as combustion, release energy, causing temperature increase in the surroundings. This property is crucial in many applications, ranging from the energy used to drive engines to the generation of heat for industrial processes. In contrast, endothermic reactions remove heat, causing the temperature of their environments to fall, as occurs in the process of photosynthesis. Familiarity with such thermodynamic concepts not only helps one understand the basic behavior of chemical reactions but also the skill to control conditions to produce a desired outcome. For instance, their application in solid phase reactions, for instance, sodium hydroxide and sodium hydride, serves to illustrate the character of such interactions and the function of thermochemical analysis in material discovery (Gregory et al., 2015). In thermochemistry, enthalpy change recording and analysis are a vital tool to quantify the energy trends in chemical reactions. Enthalpy (*H*) is a state function defined as internal energy plus pressure and volume, so chemists can quantify heat transfer at constant pressure. Scientists can easily measure heat taken up or released during reactions using techniques such as calorimetry. These values are essential to designing thermochemical cycles, which allow for the enthalpy change of reactions difficult to experiment on. Besides, the history of combustion, articulated in terms of the early pioneers' works like those of Lavoisier and Bunsen, reveals the inherent natures of thermochemistry in chemistry. The early research by the pioneers demonstrates the need to learn the fundamental concepts of energy, heat transfer, and enthalpy in an attempt to further develop the field and solve modern scientific issues (Penner et al., 1963).

1.2.1. Definition of Heat and Temperature

The laws of heat and temperature are at the heart of the study of thermochemistry and serve as a foundation for the description of energy exchange in chemical reactions. Heat is energy transferred between a system and its surroundings due to a temperature difference. It is not a property of a system but a measurable energy change resulting from molecular interaction and kinetic energy differences between particles. On the other hand, temperature is a measure of the average kinetic energy of molecules within a system and is a thermal energy measure. The distinction is significant; while heat is energy in transit, temperature quantifies the state of matter as thermal, which controls reaction rates and equilibria. Identification of this correlation is critical in establishing energy changes in chemical reactions and has numerous uses in fields such as thermodynamics and material science. Heat is often expressed in thermochemistry using the specific heat capacity, and it refers to the amount of heat required to raise a unit mass of a substance's temperature by 1°C. Different substances have different heat capacities, based on the structure and molecular bonding. This variation has significant implications in reaction because it dictates how energy is absorbed or released during a chemical process. If, for instance, a reaction occurs within a solution, the net energy change is determined significantly by the specific heat capacity of the solvent. In addition, heat exchange can occur by conduction, convection, or radiation, each of which is characterized by its own set of thermodynamic conditions that in turn influence temperature and heat behavior within a chemical system. Hence, a comprehension of the interrelationship between heat and temperature is most significant in understanding reaction behaviors and efficiencies thermodynamic. Critical examination of heat and temperature not only sheds light on fundamental principles in thermochemistry but also gives insights into practical applications in various scientific fields. As an example, in combustion reactions, temperature has a critical effect on reaction rates and product formation. Design of proper combustion engines is heavily reliant on heat transfer behavior and temperature optimization for better efficiency and emission management. Detailed processes of chemical reactions, as highlighted in existing research, demonstrate the relevance of a profound understanding of such concepts towards improving technologies like gas turbine engines and alternative fuel systems (Braun-Unkhoff et al., 2014). Additionally, the techniques founded on this understanding allow researchers and engineers to predict behavior in complex systems and thus are central to the enhancement of energy efficiency and sustainability in the field of thermochemistry (Figure 1.5).

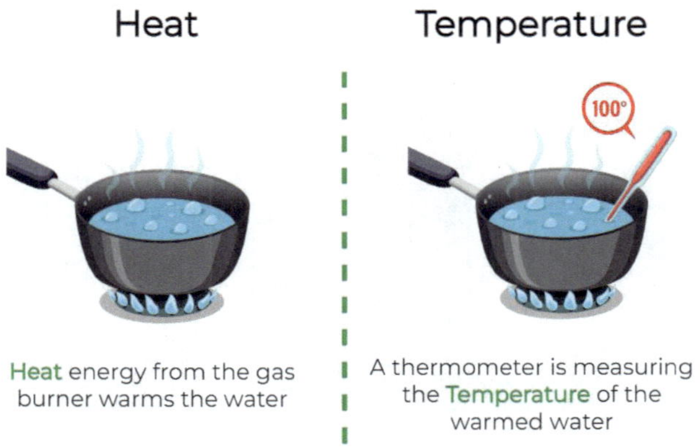

Figure 1.5. *Definition of heat and temperature.*

Source: https://media.geeksforgeeks.org/wp-content/uploads/20220907121301/DifferencebetweenHeatandTemperature.jpg.

1.2.2. The Concept of Enthalpy and Its Significance

In thermochemistry, the concept of enthalpy is a basis for studying energy change in chemical reactions and phase changes. Enthalpy, H, is a thermodynamic quantity that incorporates the internal energy of a system and the product of the volume and pressure of the system. The relationship supplies the correct evaluation of heat energy transfer during constant-pressure processes, particularly useful in calorimetry. Scientists can determine whether reactions are exothermic, releasing energy, or endothermic, absorbing energy from the surroundings, from changes in enthalpy. Not only does this deepen our understanding of chemical behavior but also inform the design of many handy applications, such as effective chemical reactors and energy systems. The significance of enthalpy cannot be limited to theory alone; it plays a crucial part in practical applications in fields including material science and chemical engineering. For instance, enthalpy changes in combustion reactions are very important for the development of new energy systems. Combustion of fuels is one method of producing energy, and by determining the enthalpy changes for these reactions, scientists can limit the fuel loss and decrease emissions of pollutants. Other than combustion, the enthalpic properties of electrochemical systems, such as the calorimetry of nickel cadmium batteries used in spacecraft, display its application in the development of energy storage devices. An understanding of the enthalpic processes in these systems allows engineers to create more effective technologies for the management of energy (Foley et al., 2025).

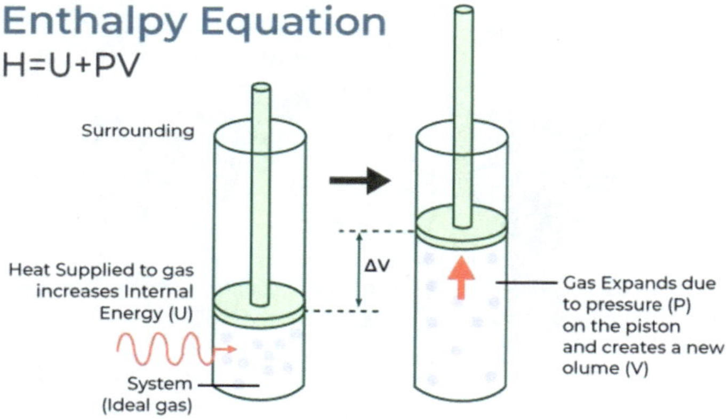

Figure 1.6. *The concept of enthalpy.*

Source: https://media.geeksforgeeks.org/wp-content/uploads/20230502124906/Enthalpy-1.webp.

Converting such knowledge into useful engineering applications requires good understanding of the underlying thermodynamic principles, particularly in combustion. Aerothermochemistry studies focus on the intricate interplay between enthalpy changes and combustion processes, as mentioned in contemporary research studies. The coupling of theoretical models with experimentation involves significant challenges, notably in coping with temperature and composition gradients in flame fronts. As discussed in literature that accounts for the theory of laminar flames, establishing a strong connection between theory of combustion and its applications is still a research-intensive area (Marble et al., 1956). In the end, overcoming these challenges emphasizes the importance of enthalpy as not only a theoretical aspect but as an essential tool in achieving technological advancements in energy production and consumption (Figure 1.6).

1.2.3. Understanding the Laws of Thermodynamics

The laws of thermodynamics are fundamental principles which govern energy interactions within chemical reactions and are a major component of thermochemistry. The first law, sometimes stated as the law of conservation of energy, states that energy cannot be created nor destroyed but only changed. This principle implies that energy in an enclosed system remains constant. For example, in a chemical reaction, reactant energy is converted to products with energy lost to the environment accounted for. This basic principle is critical to predicting reaction behavior and efficiency, especially in biochemical systems where the law is useful in modeling reaction mechanisms and metabolic pathways. By applying this law, scientists can find methods to maximize the utilization of energy in various chemical processes and promote sustainable consumption of energy resources. The second law of thermodynamics together with the first law leads to the formulation of the concept of entropy as a measure of disorder within a system. This law supposes that in any spontaneous process, the entropy of an isolated system will increase with time and compel natural processes to a state of equilibrium. This tendency toward greater disorder is a sign of the intrinsic constraints on energy efficiency and implies energy conversions to be less than 100% efficient. An example is where, during metabolic reactions, energy is dissipated as heat, resulting in an increase in entropy within the system. It is significant for chemists and biologists to understand entropy since it accounts for why certain reactions are thermodynamically favored. The consequences of the second law therefore extend into purely practical matters such as making better catalysts and improving energy conversion technology. The third law of thermodynamics, which asserts that as temperature is reduced to absolute zero, the entropy of a perfect crystal will go to some minimum value, further establishes the context of thermodynamic knowledge. This law states that it is not possible to attain absolute zero by any physical process, and hence it is a theoretical concept and not a realizable goal. In thermochemistry, the third law plays an important role in explaining reaction mechanisms and low-temperature behavior where kinetic energy becomes very small. The third law is applied as a criterion for calculating absolute entropies of substances with data that is critical in thermodynamic cycles and calculations. The combination of these three laws gives a complete framework that goes deeper in our understanding of energy processes in chemical systems, hence creating the groundwork for progress in theoretical and applied thermochemistry, eventually creating innovations in fields ranging from material science to bioengineering.

1.3. TYPES OF ENERGY CHANGES

In the field of thermochemistry, the character of the various types of energy changes is key to comprehending how energy is transferred and transformed within chemical reactions. One of the big energy conversions is the internal energy of a system, which includes the total of kinetic and potential energy of its molecules. Such an energy can be altered by a number of processes, such as heating, cooling, or chemical reaction. For instance, the energy gained in an endothermic process results in a rise in the internal energy, whereas an exothermic process results in energy release and therefore a decrease in the internal energy. The associated changes in internal energy are quantifiable according to the first law of thermodynamics, which states that energy cannot be created or destroyed but only transformed from one type to another. The law serves to form a foundation for the study of more sophisticated energy transformations in a thermochemical context. Another crucial factor of energy alteration in thermochemistry is enthalpy, a measurement of total energy consisting of internal energy and energy to move the surrounding environment. Change in enthalpy, represented by ΔH, is a crucial factor in classifying the predictability and spontaneity of reactions.

For example, when comparing halomethane cations such as CH_3Cl and their derivatives, researchers identified specific dissociation channels that reflect the disruption of chemical bonds resulting in unique energy changes. These changes can be systematically studied using devices such as imaging photoelectron photoion coincidence spectroscopy. As indicated in current studies, the enthalpy changes associated with dissociative photoionization mechanisms reveal intricate thermochemical networks that enlighten our understanding of molecular reactivity and stability, crucial in the construction of knowledge in energy conversions (Bodi et al., 2012). Energy changes are also influenced by external factors such as magnetic fields and ionization conditions. For instance, in astrophysical contexts, magnetized protoplanetary disk outflows play a significant role in the process of planet formation, and the coupling between magnetic and chemical processes is illustrated. Major energy changes are created by pressure changes in toroidal magnetic fields by strong magnetization, and it has been shown by full magnetohydrodynamic simulations. The coupling between thermal and magnetic energies illustrates how energy transformations are complex within various systems. Moreover, these simulations also point to the importance of understanding how changes in energy in a thermochemical framework can enlighten more universal questions in science, including those related to the evolution of planets and the assembly of celestial objects (Bai et al., 2018).

1.3.1. Exothermic Reactions and Their Characteristics

Exothermic processes are defined by their natural quality of dispersing heat energy into the surroundings, which reduces the overall enthalpy of the system. This type of chemical reaction is fundamental to natural and industrial processes because it is a fundamental process of energy transfer. Examples include combustion reactions such as the burning of hydrocarbons that not only provide heat but are utilized to produce power. With these reactions, bonds in the reactant atoms are broken and bonds are formed in the products, often at the expense of releasing energy as heat. The enthalpic changes involved with exothermic processes can be quantified, and so scientists can ascertain the quantity of energy released, a significant factor in determining reaction efficiency and viability in various uses.

The kinetics of exothermic reactions also heavily rely on conditions such as temperature, pressure, and reactant concentration. For instance, an increase in temperature is commonly observed to increase the rate of reaction and drive the system towards more complete conversion of reactants to products. Interestingly, a series of studies, including for instance radical reactions in the gas phase, have revealed reaction kinetics to exhibit various types of behaviors under certain experimental conditions. As studies have revealed, the rates of reactions are affected not only by changes in temperature, but also can change the courses of the reactions themselves, thus the energy profile of exothermic reactions (Rissanen et al., 2012). This knowledge is important for exothermic reaction control for purposeful intentions, e.g., in combustion engines, or even in biochemical processes. Apart from their industrial relevance, exothermic reactions have a crucial role to play in atmospheric chemistry and the environment. Radical species-molecule reactions, such as the reaction of oxygen with hydrocarbons, can lead to release of substantial amounts of energy, indirectly affecting global climatic regimes. The dynamics of such interactions are critical in the formation of air pollutants and greenhouse gases, which are exemplified in studies of atmospheric reactions where some products can control factors like mass-independent isotopic fractionation (Blake et al., 2004). Thus, carefully examining exothermic reactions not only enhances our fundamental understanding of thermochemistry but is also central in the design of strategies towards minimizing environmental impacts while maximizing energy utilization in various industries (Figure 1.7).

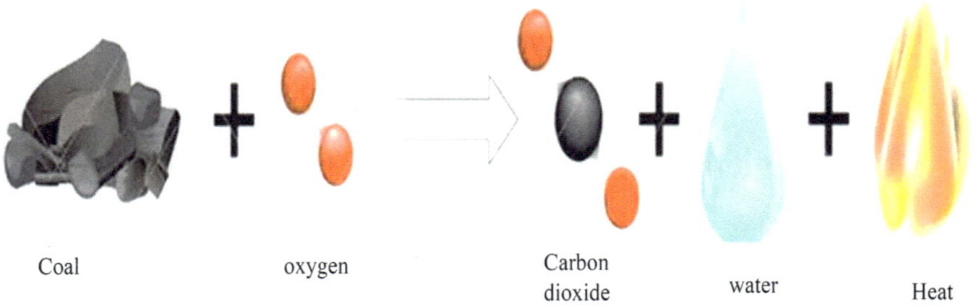

Figure 1.7. *Exothermic reactions.*

Source: https://blogger.googleusercontent.com/img/b/R29vZ2xl/AVvXsEj6Wnm47_HdsFvVbGVBvZkAIpxd7Yske66zA_xvk06NVO8v866eKrPOwfGBQp9TMsl-CpMMboUvUCEbBtmcmFios8SwdSVghdsSUaCbUFAnCKTuR0rVKqdtmgPYg-Gly6flx0oYpj7uXys57qdHSpwUrF9I6yaZVRjDK5IH5l3dB7DJazhHIHinrb3OuD/s16000-rw/exothermic%20reaction.webp.

1.3.2. Endothermic Reactions and Their Implications

Endothermic reactions, characterized by the absorption of heat from their surroundings, are critical in numerous chemical operations and applications. In such reactions, the change in enthalpy is positive, signifying that the energy input needed to dissociate bonds is greater than the energy released upon bond formation. This aspect is not only of basic theoretical chemistry but also carries a major implication in practical contexts as well. For example, the foundation of endothermic reactions is crucial to understanding how certain biological processes survive and even prosper in conditions of varying temperatures. With the heat absorption, such reactions can propel vital metabolic processes, indicating the interplay between biological efficiency and thermochemical principles. The significance of endothermic reactions also seeps into industrial applications and energy solutions. Certain processes, such as those happening during refrigeration and air conditioning, include endothermic reactions to pull heat from a target area, hence lowering temperature effectively. Moreover, innovations in the process of creating materials that are based on endothermic reactions have also been witnessed, especially with respect to energy storage and conversion technologies. For instance, recent thermal energy storage advances tend to use materials that react endothermically to absorb and store excess heat temporarily, promoting more efficient energy redistribution at the time of demand. Such uses highlight the calculated manipulations of the thermochemical principles toward solving critical energy issues in sustainable technologies. Learning approaches also highlight the importance of conceptual understanding of endothermic reactions. Through the integration of innovative pedagogical methods such as STEAM projects in learning environments, students can gain access to these concepts of thermochemical reactions at a deeper level. The projects can include hands-on projects that reinforce endothermic and exothermic reaction concepts, bridging theoretical knowledge with real-world applicability. For instance, experiential learning activities involving reactions with the aim of cultivating reactions in varying situations can assist students in comprehending the enthalpy complexity and reaction categories, thereby rectifying common misconceptions encountered in the learning of thermochemistry. As demonstrated in recent studies, cultivating such interactive learning situations acts robustly to improve comprehension of major concepts in thermochemistry, including those of endothermic reactions (Afrizal et al., 2021). Finally, a good grasp of these ideas is required in order to apply academic as well as practical uses of thermochemistry in the future (Figure 1.8).

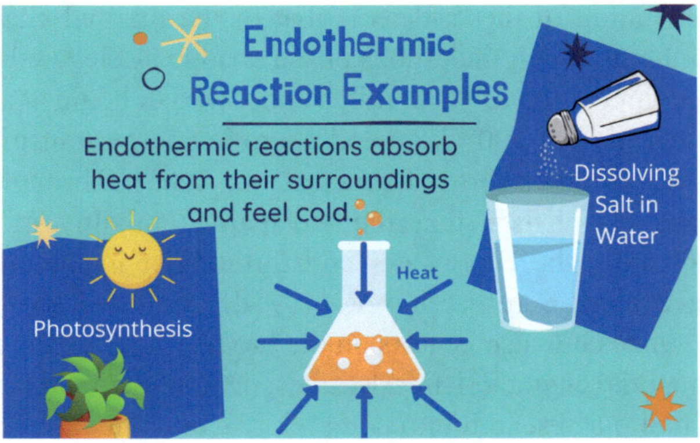

Figure 1.8. *Endothermic reactions.*

Source: https://sciencenotes.org/wp-content/uploads/2020/12/endothermic-reaction-examples.png.

1.3.3. Phase Changes and Associated Energy Changes

While studying thermochemistry, it is essential to know about phase changes and associated energy changes to know about the behavior of matter under various conditions. A phase change is when a substance transitions between states—solid, liquid, or gas—due to changes in temperature or pressure. The changes are characterized by dramatic energy changes in the form of absorbed or released heat energy. For instance, during melting, energy is taken in by a solid to enable particles to overcome intermolecular forces and transition into a liquid state. Alternatively, when a substance freezes, energy release makes particles be structured in a solid state. The built-in energy needs for such transitions reflect the material's natural properties and lead to variations in melting and boiling temperatures in different substances. Thermodynamics provides a platform on which the energetic aspect of phase changes can be measured by such basic principles as entropy and enthalpy. Enthalpy, represented symbolically as ΔH, is heat at constant pressure in a system and is important for calculating the energy change in phase changes. For instance, enthalpy of fusion is utilized to calculate the quantity of heat absorbed by a solid when it melts and the enthalpy of vaporization to calculate heat absorbed when a liquid evaporates. The relationship between enthalpy and phase change is often illustrated with calorimetry experiments where some temperature change is monitored. Entropy, a system disorder, is also a key consideration at phase change. For example, solids turning to liquids or gases have increasing entropy due to greater particle mobility, illustrating the energy change and molecular structure relationship. A practical application of these ideas is that of the thermal characteristics of yavapaiite, a mineral which undergoes phase change at elevated temperatures. The thermal decomposition of yavapaiite is observed to begin at around 500°C, with full disintegration at 700°C, an endothermic energy transition process in the transition into another form (Drouet et al., 2005). Through calorimetry, researchers have established that the yavapaiite enthalpy of formation is significant, indicating the energy demand throughout its decomposition and synthesis. Phase change research not only deepens our understanding of individual substances but also centers on the broader implications for chemical reactions and material characteristics. Besides, the foundational concepts behind these changes fuel our research into complex reactions in varied surroundings, such as dark clouds in space, where phase interactions of molecules with carbon show the intricate balance between structure and energy under some conditions (Bergeat et al., 2013).

1.4. MEASUREMENT TECHNIQUES

A crucial aspect of thermochemistry involves the accurate measurement of thermal properties using a variety of measurement techniques. Perhaps the simplest measurement carried out within thermochemical study is an assessment of heat capacity, a marker that is critical to an understanding of how materials respond to changes in thermal energy. Differential Scanning Calorimetry (DSC) is used typically to achieve this, allowing researchers to distinguish heat flow associated with material changes, for example, phase changes or chemical reactions. Quantitative methods like DSC, for transition metals and their compounds, can provide data for the thermal stability and phase energetics of the Hf-Zr-C-N-O system, according to literature on these compounds. Through the careful investigation of phase equilibria and with the aid of robust experimental methods paired with computational predictions, these techniques aim to validate thermodynamic models through base thermal properties required for advancing materials science and thermochemical understanding (Hong et al., 2019). Aside from heat capacity, enthalpy measurements are another essential part of thermochemical research. Techniques such as bomb calorimetry are used to determine the heat of combustion, a significant factor in determining the energy content of fuels and the thermodynamic feasibility of reactions. Accurate measurement of enthalpy is particularly crucial when examining compounds such as zirconium and hafnium carbides or nitrides with variable thermal stability and reaction mechanisms. The precision of calorimetric methods allows scientists to derive beneficial thermochemical data that correlates with experiment results and calculation results. This combination of experimental data and calculation models not only offers a clearer insight into phase stability and transformation behaviors, but also guides the synthesis of materials with desirable thermal properties, enhancing their usefulness in various technological processes (Hu et al., 2015). Moreover, newer developments in measurement technology, i.e., *in situ* spectroscopy and nano-calorimetry, are actually revolutionizing thermochemical analysis by allowing the observation of material behavior in real time when subjected to varying thermal conditions. Sophisticated techniques enable the investigation of complex phenomena hard to measure earlier, creating a more sophisticated understanding of thermodynamic properties. For instance, using advanced spectroscopic equipment can facilitate how one is able to comprehend structural transformations under oxidative or reductive conditions, which is critical in materials science. With scientists continually developing these measurement techniques, probing the delicate interaction of thermal properties and chemical behavior becomes even simpler, and this provides new dimensions in thermochemistry. Lastly, these advancements not only support theoretical frameworks but also assist in revealing new materials and reaction pathways, enhancing our overall comprehension of thermochemical dynamics (Hong et al., 2019).

1.4.1. Calorimetry and Its Applications in Thermochemistry

Calorimetry is a crucial branch of study in thermochemistry, providing a method for the measurement of heat transfer involving chemical reactions and physical change. By measuring the heat absorbed or evolved, calorimetry gives a full picture of reaction energetics. This is particularly relevant to the measurement of enthalpy changes that serve to estimate the thermodynamic stability of various compounds. For instance, recent advances in calorimetric techniques have enabled researchers to investigate phase equilibria in complex systems such as the Hf-Zr-C-N-O system. The study of stable transition metal carbides and nitrides, such as zirconium and hafnium compounds, indicates the contribution calorimetry has to our understanding of materials properties under varying conditions, increasing our capacity to forecast their performance in industrial processes (Hong et al., 2019). Further, the application of calorimetry is employed when studying spinel structures, particularly iron manganese oxides. Enthalpy of samples such as $(MnxFe_1{-}x)_3O_4$ highlights the delicate balance of oxidation-reduction reactions and cation distribution, giving an insight into their thermodynamic behavior.

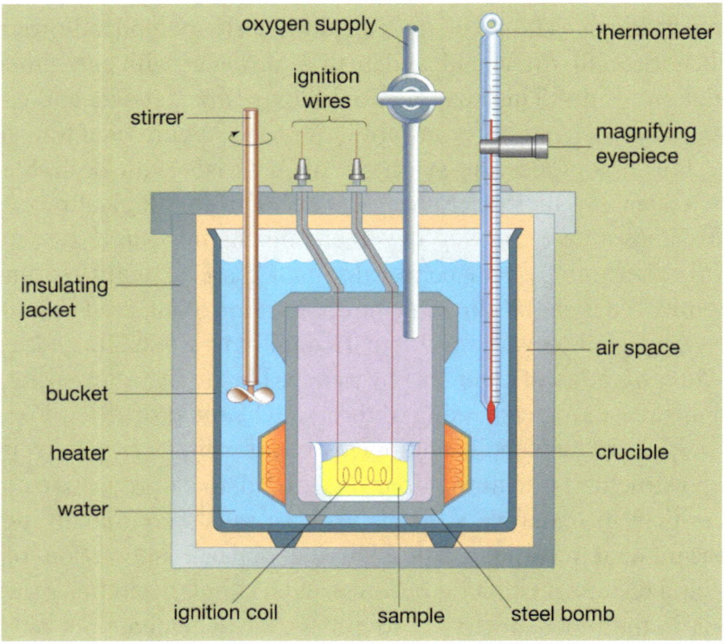

Figure 1.9. *Calorimeter.*

Source: https://cdn.britannica.com/54/7054-050-7B02ED39/Cross-section-bomb-calorimeter.jpg.

These investigations highlight the role of entropy in chemical reactions and suggest that entropy of mixing in certain compositions may be a deviation from theoretical predictions. Through the measurement of enthalpy-entropy correlation, calorimetry not only investigates the thermodynamic stability of such substances but also guides synthesis of advanced materials designed for specific functions (Coradin et al., 2004). The accumulated information from calorimetric studies is a cornerstone in thermochemistry, where the complex relationships among heat, energy, and chemical structure are revealed. As techniques become increasingly refined with advances in calorimetry by researchers, the combination of computational methods, including ab initio computations, will further provide greater predictive power to thermochemical data. Through reconciliations of experimental results and theoretical models, great opportunities will be found to examine new materials and optimize their functions. Hence, calorimetry remains an essential instrument to enhance thermochemical knowledge, facilitating a deeper insight into the energetic profiles that control chemical reactivity and material performance in a host of scientific and industrial contexts (Figure 1.9).

1.4.2. Use of Bomb Calorimeters for Measuring Heat

Bomb calorimeter use is an important method in thermochemistry for precise measurement of the heat of combustion of materials. This sophisticated device allows scientists to measure the energy released in chemical reactions, such as in fuel and food analysis. A bomb calorimeter is a robust metal container, or a bomb, in which the sample is combusted under controlled conditions, in a known volume of water. The energy liberated as heat upon combustion raises the temperature of the surrounding water, and from calorimetry principles, scientists can measure the energy content of the sample analyzed. Such accuracy is significant in uses ranging from energy production to nutrition since it gives insights into caloric values required for food and fuel efficiency testing, which are part of an energy budget of nutrition research (Ponniah et al., 1987). In nutrition science, bomb calorimeters are used to define the calorific value of various feed components and flesh. The accurate measurement of such caloric values enables scientists to compute energy budgets and absorption and conversion efficiencies of various diets. Its use in the preparation of isocaloric test diets is required, and this enables researchers to make comparative studies comparing energy utilization in humans and animals. The combined understanding of caloric content derived through bomb calorimetry thus has a critical place in nutrition requirement studies, guiding dietary recommendations and maximizing energy use in food production processes. Immediate measurement of energy content through this technique emphasizes its use in diet formulation maximization and improving health performance (Ponniah et al., 1987). Outside of nutritional uses, bomb calorimeters also have an important function in developing the chemical sciences with basic knowledge of thermodynamic behavior of compounds. Experimental determination of combustion heat by researchers allows them to calculate standard molar enthalpies of formation in gaseous phases, which aids theoretical chemistry. For instance, studies of imidazolidine derivatives illustrate how bomb calorimetry may be used in conjunction with techniques like high-temperature Calvet microcalorimetry to more accurately inform our understanding of molecular energies and properties. Such sophisticated measurements allow for the interpolation of information on enthalpy changes, proton affinities, and basicities and therefore allow us to inform our knowledge of molecular stability and interactions. Thus, the bomb calorimeter is a must-have device in the elaboration of practical and theoretical thermochemical research (Figure 1.10).

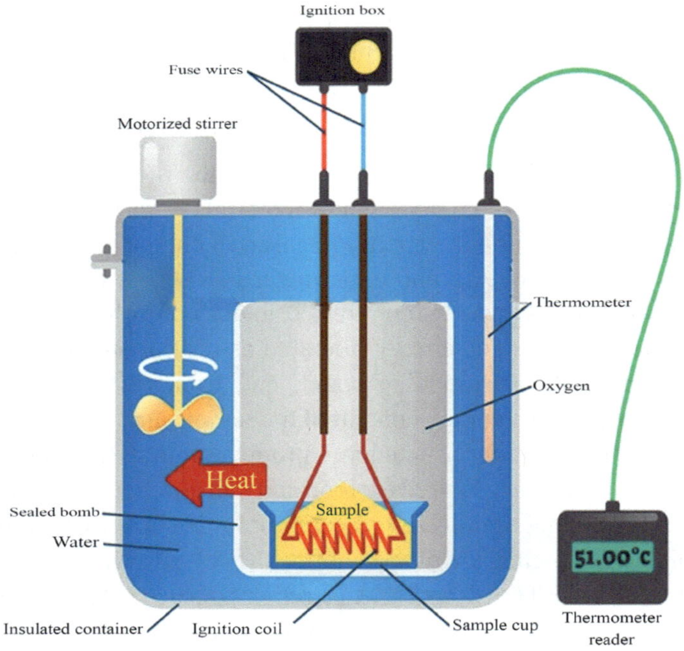

Figure 1.10. *Use of bomb calorimeters for measuring heat.*

Source: https://www.shutterstock.com/image-vector/bomb-calorimeter-vector-illustration-labeled-600w-1607802400.jpg.

1.4.3. Importance of Accurate Measurements in Thermochemical Studies

Precise measurements constitute the basis of the study of thermochemistry because they are the basis for reproducible and dependable results. Within thermochemistry, precision in results from experimentation directly impacts the validity of conclusions drawn regarding enthalpy changes, reaction kinetics, and modes of energy transfer. Deviation from agreed standards for measurements of concentration, pressure, and temperature has the potential to invalidate the entire thermodynamic system. Consideration of the impact of experimental conditions highlights the importance of precise calibration and adherence to procedures with low error margins. Without such rigor, there are uncertainties involved in the results that can lead to misinterpreted understanding of thermochemical behaviors that are vital to theoretical and practical science. Moreover, the need for accurate measurements is more than mere academic interest; it has profound implications to practical applications in industries. For instance, in combustion chemistry, the accurate thermochemical properties of reactants and products can affect efficiency and emissions. In situations where hydrogen isocyanide (HNC) plays a central role in the combustion chemistry of nitrogen, precise measurements not only input into models of chemical reactions but also inform the development of cleaner, more environmentally sustainable technologies (Marshall et al., 2017). Such innovations are based on the ground-level information that precise measurements provide, and so how closely the details of lab experimentation are connected to broader societal goals like sustainability and innovation. Moreover, the creation of innovative techniques and technologies continues to further enhance the quality of accuracy for thermochemical measurements. The introduction of high-quality gear, such as high-resolution calorimetry and spectrometry, enables researchers to define complex thermal properties with unprecedented levels of precision. Not only does this gear streamline the process of acquiring thermochemical data, but it also enhances the capacity to simulate and model thermodynamic processes. As indicated in magnetized molecular outflow research, the role of accurate measurement cannot be overstated. Here, understanding the thermochemistry of such systems facilitates a clearer insight into stellar evolution and the formation of planets (Bai et al., 2018). Overall, excellence in measurement techniques is not necessarily necessary but inevitable in bridging the loopholes between experimental science and application, solidifying the pivotal role of thermochemistry across various scientific fields.

1.5. APPLICATIONS OF THERMOCHEMISTRY

Applications of thermochemistry are diverse and are significant in various fields of science, particularly with reference to the investigation of chemical reactions and processes. Through the determination of changes in heat in reactions, thermochemistry comes in handy when used to predict reaction spontaneity, an aspect that is vital in research work as well as in industry. For instance, thermochemical calculations help determine the reaction enthalpy, which in turn influences planning and optimization of chemical reactors. In the science of materials, correct knowledge of the thermochemical behavior of materials aids in the synthesis of compounds with desired stabilities and performance at elevated temperatures. Hence, thermochemistry not only assists scientists in gaining insight into the rudimentary chemical kinetics but also assists them in designing new materials with increased functionality in application ranging from catalysis to energy storage. In energy storage, thermochemistry has been critical in the exploration of effective hydrogen storage that forms a central component in the development of clean energy systems. A case in point is research on ammonia borane, which is a possible hydrogen storage material that has come under unprecedented scrutiny since it contains large hydrogen contents. The study of ammonia borane dehydrogenation clearly elucidates the complexity of solid-state chemistry, which is not understandable intuitively with standard chemical methods. Scientists employ techniques such as \\\\textit{ab-initio} molecular dynamics and metadynamics to decipher the mechanisms of the reaction that occurs during the release of hydrogen to assist the design of effective hydrogen storage materials.

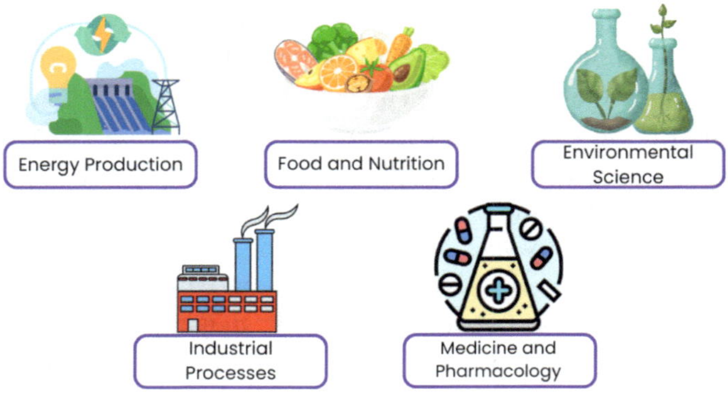

Figure 1.11. *Applications of thermochemistry.*

Source: https://images.examples.com/wp-content/uploads/2024/10/Applications-of-Thermochemistry.png.

These studies highlight the crucial place that the thermochemical theories hold in addressing key challenges in the realm of clean energy, and thus establishing its relevance to modern scientific research. Alongside this, the combination of advanced computational methods and thermochemical analysis has significantly revolutionized our understanding of molecular interactions as well as reaction kinetics. For instance, the development of sophisticated density functional theory (DFT) approaches, such as XYG3, enhances prediction accuracy for reaction barriers as well as for nonbonding interaction predictions for main group molecules. This holds very profound impacts on thermochemical application, where computational precision correlates with successful design and optimization of materials. By employing such computational methods, scientists can forecast reaction routes and establish suitable conditions for chemical transformations, ranging from organic synthesis to energy conversion technologies. Therefore, the combination of thermochemistry and novel computational methods stimulates innovation in various fields, emphasizing its significant importance in theoretical and applied chemistry (Figure 1.11).

1.5.1. Role in Chemical Engineering and Industrial Processes

In industrial processes and chemical engineering, the science of thermochemistry is important in evaluating the feasibility and efficiency of various reactions. Understanding the energetic and molecular behavior under chemical conversions allows engineers to design processes for optimal yield and least resource utilization. A case in point is the introduction of thermodynamic data into model systems to enable the optimization of reactor and separator design, linking energy requirements to economic viability. This is particularly applicable when polymer use occurs at high temperature, with thermochemical investigation determining the material and process selection that can withstand the hostile environment, ultimately leading to aerospace and electronics industry innovation. As performance and sustainability needs increase, the role of thermochemistry remains at the forefront of establishing and continuing to improve these valuable industrial processes. Chemical engineers also utilize the concepts of thermochemistry for use in energy transfer systems such as fluorescent lamps and electronic ballasts. With intimate knowledge of plasma ionization dynamics, engineers can simulate the lamp's behavior under various conditions so that lamps with improved efficiency and longevity can be designed. The feasibility of user-defined parameters like lamp size and gas pressure is directly related to thermochemical insights about reaction dynamics and heat transfer efficiency in such systems. By modeling time-dependent conductance of the devices, designers gain a greater understanding of the physical processes governing performance, leading to improved product design and energy efficiency. In this regard, thermochemistry is not merely a theoretical problem but a fundamental part that directs relevant industrial development. Furthermore, the applicability of thermochemistry extends to the innovation of materials. High-temperature polyimides, for instance, demonstrate the application of thermochemical principles in creating materials with enhanced mechanical performance under harsh environments. Thermal stability and oxidative degradation are connected; with this knowledge, engineers can make time projections on material behavior and enhance the life of composites and adhesives in various applications.

Models of research, like one related to viscoelasticity of polyimides, emphasize the importance of thermochemical information for maintaining material integrity in real-world conditions. As chemical engineers extend the boundaries of thermochemical properties of materials, they will be more and more capable of responding to evolving industry needs and environmental demands with accelerating technological innovation in most fields.

1.5.2. Importance in Environmental Science and Energy Production

The intricate interactions between thermochemistry and environmental research are essential in the calculation of energy production and utilization. Thermochemistry, the exploration of energy transformations during chemical reactions, is essential in evaluating the environmental impact of various energy sources. Processes of producing energy are directly proportional to their chemical properties, particularly with reference to greenhouse gas emissions and resource sustainability. Optimization and simplification of these processes through thermochemical analysis allows scientists to produce cleaner alternatives with smaller environmental impacts. In an effort to fight global warming, the demand for environmentally friendly sources of energy increases, catapulting thermochemistry onto the limelight of research aimed at minimizing environmental impact. Understanding the thermochemical factors that guide energy systems is crucial in spearheading solutions that reconcile energy generation and environmental protection. Moreover, advances in technology for recycling, particularly in rare earth elements, demonstrate the relevance of thermochemistry to environmental science. Recycling and refining rare earth elements is energy consuming, and improving the processes can lead to dramatic increases in sustainability and resource efficiency. Plasma mass separation, for instance, is one novel recycling process that minimizes environmental damage generally associated with standard recycling methods. As illustrated by recent studies, this approach aims to mitigate existing inefficiencies by providing a more environmentally friendly means of reclaiming rare earth elements, which are critical to many modern technologies. The opportunities for integrating such thermochemical breakthroughs within existing waste treatment infrastructure provide the significant points of intersection among energy production, resource recovery, and environmental preservation, thereby emphasizing the significance of thermochemical principles in enabling an enduring future (Fisch et al., 2017). In addition to recycling technologies, the finding of ionic liquids (ILs) is another instance of thermochemistry's role in innovations of green energy technologies. Amino acid-based ionic liquids are one of the ILs that have been of interest since they share certain characteristics that are associated with energy consumption. These ILs have favorable thermodynamic properties that could make them useful in various chemical processes including energy generation and storage. Their general compatibility with mainstream molecular dynamics models guarantees their widespread use, encouraging greater computational investigations into their possible applications. Through the use of thermochemical information, scientists have the ability to realize new uses for ILs that will potentially give rise to more efficient and less wasteful energy production devices. Thus, the incorporation of thermochemical studies into environmental science and energy studies is a call to action towards fostering innovative solutions to address sustainability problems on a global scale (Chaban et al., 2014).

1.5.3. Applications in Biochemistry and Metabolic Processes

The intersection of thermochemistry and biochemistry provides incredibly useful information about the energetic organization of metabolic processes. Thermodynamics, particularly Gibbs free energy, play a key role in predicting the favorability of biochemical reactions. Quantum chemistry has enabled scientists to build computational models to approximate standard Gibbs reaction energies for complex metabolism reactions. This is not merely enhancing the quality of thermodynamic data but also filling in the gaps of our understanding of metabolism. Having the capability to quantify the energies of reaction allows scientists to model metabolic pathways more realistically, resulting in breakthroughs in areas such as enzyme engineering and synthetic biology. Surprisingly, studies have shown that quantum chemical approaches are just as accurate as the traditional group contribution methods in quantifying Gibbs energies, particularly in central metabolic reactions, where their relevance to biochemical research is underscored (Aspuru-Guzik et al., 2014). Within enzymatic catalysis, the C–H bond conversion indicates the intricate interplay between thermochemistry and biochemical conversion. Thymidylate synthase (TSase), for instance, facilitates the key C–H to C and O transformation while providing thymidylate, a key building block of DNA synthesis. Such intricacies in the reaction give rise to the significance of thermodynamic calculations, specifically that of transition states. Quantum mechanics/molecular mechanics (QM/MM) free energy calculations have revealed a number of energetically similar transition states with different structures, exposing the reaction mechanisms at play. These findings not only address enzyme behavior but also require a re-interpretation of historical experimental data. To this degree, this information attests to the significance of including thermochemistry principles in the acquisition and regulation of enzymatic activities (Castillo et al., 2013). The understanding of metabolic processes through thermochemistry extends beyond rudimentary scientific examination to drug discovery and treatment of disease. The data obtained from thermodynamic models is strong in guiding targeted therapies, particularly in cancer therapy, where thymidylate inhibition can stop tumor growth. By way of transition states and potential intermediates in enzymatic reactions, scientists are able to design new inhibitors that can effectively disrupt cancer cell proliferation metabolic processes. Moreover, predictive capability provided by quantum chemical methods facilitates the exploration of novel intermediates in reactions, thereby expanding the pool of potential drugs used in therapeutics. In general, the application of thermochemical concepts in biochemistry not only aids a better understanding of metabolic reactions but also paves the way for emerging healthcare applications (Castillo et al., 2013) (Figure 1.12).

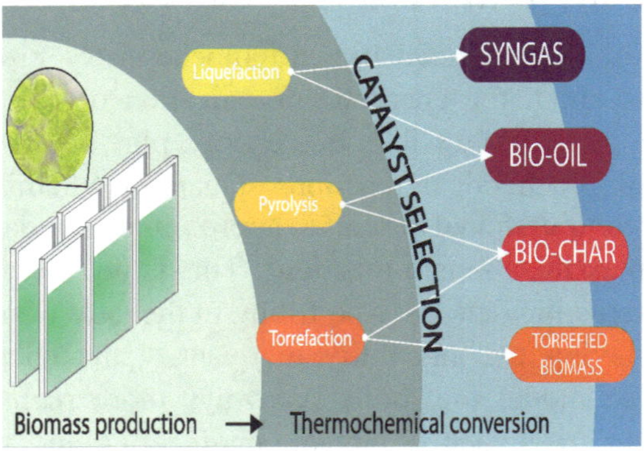

Figure 1.12. *Thermochemistry applications in metabolic processes.*

Source: https://pub.mdpi-res.com/catalysts/catalysts-11-00022/article_deploy/html/images/catalysts-11-00022-ag.png?1619688477.

1.6. THERMOCHEMICAL EQUATIONS

The study of thermochemical equations forms the core of studying the energy change associated with chemical reactions. Apart from quantifying the heat evolved or absorbed in a reaction, the equations also show the relationship between temperature and the changes in enthalpy that go along with it. Through the application of thermodynamics concepts, thermochemical equations form the basis of many scientific and engineering applications. For instance, they are crucial in defining reaction progress and equilibrium, as well as in helping predict the products and their stability under different conditions. This predictive function becomes very critical in material science to environmental chemistry applications where the construction of chemical processes that are efficient and safe is driven by thermochemical principles. In practice, the thermochemical equations become critical resources for chemists and engineers who need to optimize chemical reactions and energy efficiency. Improvement in chemical vapor deposition (CVD) processes is one of the best illustrations of the need to transfer thermochemical knowledge to reality. Models based on thermochemical equations play a crucial role in determining the correlation between process parameters and material characteristics, allowing performance improvement of high-temperature structural materials.

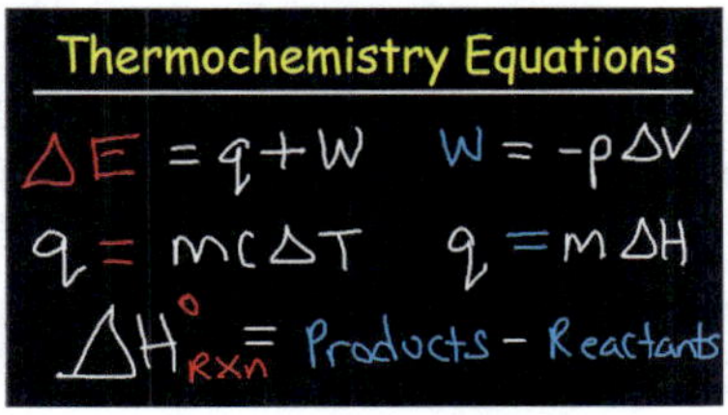

Figure 1.13. *Thermochemistry equations.*

Source: https://i.ytimg.com/vi/LsqKL3pBVMA/sddefault.jpg.

According to recent studies, traditional modeling techniques based on equilibrium thermochemistry have been found useful even when complexities and uncertainties in chemical kinetics arise. Thus, the integration of thermochemical equations into engineering practices not only improves material performance but also provides insight into reaction mechanisms and energy exchange (Goekoglu et al., 2025). Advances in computational models further enhance the role of thermochemical equations in research and application. The development of advanced radiation-hydrodynamics codes, featuring thermochemical considerations, demonstrates the usage of these equations on new frontiers, such as astrophysics and materials science. By approximating encounters between radiation and matter, scientists are able to deconstruct complex dynamics within different systems. These programs assist in investigating phenomenology such as radiation pressure and heat transfer, expanding the application horizon of thermochemical equations outside standard chemistry. Last but not least, the ongoing discovery and application of these equations demonstrate their significant role in advancing scientific knowledge and technological progress in academic and practical applications (Rosdahl et al., 2015) (Figure 1.13).

1.6.1. Writing and Interpreting Thermochemical Equations

The writing of thermochemical equations serves as the foundation in calculating energy changes of chemical reactions. These equations not only indicate the reactants and products involved but also quantify energy changes, in terms of enthalpy changes in most instances. The case of examining the formation of compounds like yavapaiite, $KFe(SO_4)_2$, from elements would lead one to view a thermochemical equation listing the enthalpy of formation, ΔH°f, as the key information. The true enthalpy value, determined by methods such as calorimetry, enhances our knowledge of the reactions energy pattern, which is vital for the forecasting of the stability of compounds in varying conditions. Because the thermochemical equation encircles both the character and structure of energy in a reaction, it is an essential tool for chemists seeking to study energy transfers occurring with chemical changes in a broad variety of conditions. Thermochemical equations are understood on the basis of a nuanced understanding of thermodynamics, namely the laws of conservation of energy. They usually use Gibbs free energy and entropy to predict the spontaneity of a reaction. As an example, explaining the decomposition of jarosite to yavapaiite, Fe_2O_3, and H_2O requires knowledge of the equilibrium conditions as well as how variables like temperature and pressure affect the thermodynamic stability of the product. Chemists can tell whether a reaction will be favorable under some conditions if they know the standard entropy, S°, and Gibbs free energy of formation, $\Delta G°$. This observation is critical in applications ranging from geochemical reactions to the creation of synthetic products, in which reaction and energy conditions can be optimized to produce desired outcomes in chemical processing and materials science, as demonstrated by the study of thermochemical phenomena in planetary atmospheres (Gilbert et al., 2025). Lastly, the composition and interpretation of thermochemical equations are vital not only for scientists but also for technical purposes in various scientific fields. Quantitative values succeeding these equations provide us with concepts regarding the landscape of energy in reactions, directing methods in chemical engineering, materials science, and environmental chemistry. By thermodynamically defining reactions, such as entrapment of transition states and energy barriers in thermal decomposition reactions, scientists are able to predict the behaviors of reactions and make conditions optimized for desired outcomes. The equations serve as a bridge between experimental findings and theory predictions, facilitating innovation in chemical synthesis and analysis. Improved understanding of such equations also aids in the presentation of complex information in a clear manner among researchers so that understanding of the thermochemical properties continues to develop both in theoretical frameworks and application (Drouet et al., 2005).

1.6.2. Standard Enthalpy of Formation and Its Significance

Standard enthalpy of formation is a fundamental thermochemical concept, that is, heat of reaction that accompanies the synthesis of one mole of a compound from its elements in standard states. This thermodynamic factor aids in the determination of energy profiles of chemical reactions in that it allows chemists to predict whether a reaction will be exothermic or endothermic. Through tabulation of the standard enthalpies of formation of compounds, scientists can use Hess's law to calculate the enthalpies of change for complex reactions, thus optimizing more effective chemical processes. Moreover, by looking at the values of the standard enthalpies of formation, one appreciates how stability in molecules and bond strengths influence reaction kinetics, thereby the relevance of this concept in theoretical and practical chemistry. The significance of standard enthalpy of formation extends beyond calculation in theory; it is also fundamental in practical applications in industry. For instance, in metallurgical operations, standard Gibbs energies of formation serve as the foundation for phase equilibria calculations and metal thermodynamic behavior investigation of metals like nickel and copper in extraction and refining procedures.

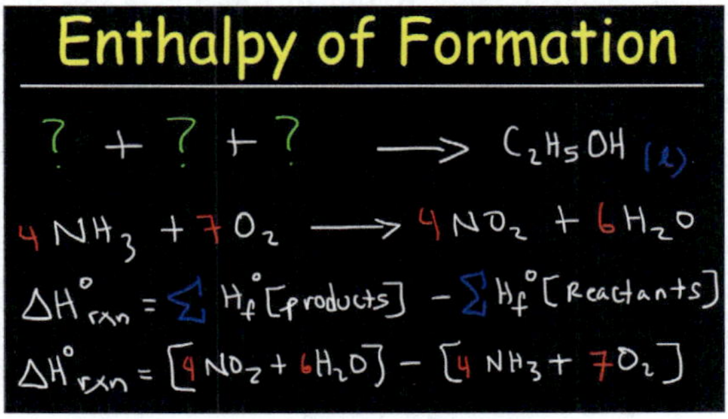

Figure 1.14. *Standard enthalpy of formation.*

Source: https://i.ytimg.com/vi/FP-VYfTYi8Y/maxresdefault.jpg.

Understanding such energy transformations becomes imperative to optimize pyrometallurgical conditions, wherein heat efficiencies directly affect production costs and environmental footprint. Thermodynamic data for nickel-copper sulfide systems, for example, describe the stability of various phases and their solubility limits, thus dictating factory operating conditions (Taskinen et al., 2012). The use of standard enthalpy of formation is therefore one of optimizing economic and environmental circumstances in chemical manufacture. The standard enthalpy of formation is also useful in environmental chemistry as well as materials science. The calorimetry of nickel-cadmium cells, for example, which are extensively used in rechargeable battery technology, relies on having knowledge of the thermodynamic properties of the involved materials. By defining the standard enthalpy of formation of the constituents in these electrochemical systems, researchers can predict the performance and lifespan of batteries under varying conditions (Foley et al., 2025). In this manner, studying the enthalpy change occurring in reaction within batteries gives vital information on energy efficiency, stability of cycles, and potential environmental impacts as a result of material degradation. Standard enthalpy of formation is not only a theoretical basis but also a pragmatic tool for creating technology and solving ecological issues (Figure 1.14).

1.6.3. Hess's Law and Its Applications in Calculating Enthalpy Changes

In thermochemistry, Hess's Law is a central idea, allowing the enthalpy change of a chemical process to be determined regardless of the path of the pathway. The law arises from the fundamental principle of state functions, which has it that the net enthalpy change is only a function of the initial and final states, and not upon the specific process used in altering between the states. By utilizing Hess Law, researchers are able to decompose complex reactions into a series of steps whose enthalpies are familiar. For instance, if a series of steps in a reaction series can be known or calculated, then the overall change in enthalpy can be obtained by summing the individual changes, providing a convenient means of examining thermodynamic characteristics. This method is particularly valuable when the measurement of direct enthalpy changes is impossible or not feasible. Applications of Hess's Law extend beyond theoretical calculations to industrial processes and routine laboratory practice. For example, in production product synthesis, engineers can calculate the energy inputs or outputs needed, optimizing energy efficiency and cost savings. In addition, this law assists us in establishing whether a reaction will occur spontaneously under particular conditions because enthalpy changes are simple predictors of thermodynamic favorability. Calorimetric experiments can also be carried out by chemists to make predictions of reaction enthalpies from measured heat changes when a reaction or phase change takes place (Dr. Arora, 2024). This adaptability shows the way in which Hess Law emerges not merely as an theoretical technique, but as an inherent utilitarian device that broadens our understanding of chemical reactions and enthalpic examination. In addition, the use of Hess's Law reaches beyond simple calculations of enthalpy to explain how various thermodynamic values interrelate. With the use of this law, scientists can conclude and understand such pivotal ideas as standard enthalpy of combustion and formation. The enthalpy values of standards allow comparison between different substances and reactions and making databases that chemists tend to refer to. The standardization not only applies to academic purposes but also provides input to real-life applications such as the estimation of fuel-based energy production or measuring reaction mechanisms (Dr. Arora, 2024). So,

Hess Law is not merely a mathematical approach; it is a broader conceptual framework that encourages thermochemical study, ultimately encouraging theoretical comprehension as well as practical potential in chemistry.

SUMMARY

One has to comprehend the significance of the science in making a connection between chemical reactions and energy alterations while studying thermochemistry. Thermochemical studies encompass several phenomena, primarily the energy alteration in chemical reactions. Such a comprehension is not only a requirement for academic work but also for real-world implementations, i.e., the development of more efficient energy systems. With growing demand for clean energy technologies, the use of thermochemical theories becomes even more critical. Breakthroughs in research fields such as electrocatalysts reveal how thermochemical understanding will propel innovation in energy conversion technologies. The catalytic activity of reactions such as the oxygen evolution reaction (OER) demonstrates the promise of thermochemistry to power revolutionary energy efficiency and sustainability in upcoming advancements (Li, 2022). Another significant aspect of thermochemistry is its compatibility with modern computational methods, which have revolutionized the field. The use of density functional theory (DFT) allows the elucidation of composition-structure-activity relationships of catalysts, which can contribute to enhancing the predictive capability in electrocatalysis. Specifically, determining composition-structure-activity relationships using DFT can accelerate the discovery of cost-effective and efficient catalysts for energy conversion reactions. The focus on low-cost, highly efficient electrocatalysts is part of a broader direction in thermochemistry to more sustainable approaches more compatible with the requirements of the world's energy. Through establishing a fuller picture of catalyst mechanism, scientists are able to overcome existing difficulties, ultimately paving the way for the development of paradigm-changing energy technologies (Liao et al., 2021). Lastly, the prospects of thermochemistry are good but fraught with challenges that require continued research and innovation. The ongoing advances in catalyst engineering and electrocatalysis that are able to facilitate reactions such as water splitting put thermochemistry on the center stage of green energy options. Effective research will hinge on interdisciplinarity, with a balance between theory and experiment to construct pathways towards cleaner energy systems. Speaking of the complicated scaling relationships that exist within catalysts, particularly within the OER, serves to further underscore the importance of deepening our understanding of thermochemical reactions. Ultimately, a thorough comprehension of thermochemical principles will not only serve to benefit the world of academic research but also propel innovation in green technologies that have a profound influence on energy consumption and conservation in the years to come (Li, 2022; Liao et al., 2021).

The area of thermochemistry encompasses a range of significant concepts necessary to explain chemical reaction and energy transformation. Analysis of non-covalent interactions is perhaps most crucial among these, as they are absolutely crucial in determining molecular stability and dynamics. Recent advancements in computational methods, specifically the advancement of dispersion-correcting potentials (DCPs), have greatly enhanced the accuracy of modeling these interactions. As DiLabio et al. (2013) have noted, the inclusion of new C-DCPs in standard computational models such as B3LYP provides precise predictions for complex reaction mechanisms, particularly under conditions such as the beta-scission of cumyloxyl radicals. This indicates the necessity of applying sophisticated theoretical tools to further expound on our comprehension of thermochemical phenomena, thereby enhancing theoretical wisdom and applications in chemistry. Another noteworthy feature of thermochemistry highlighted in the essay is the effectiveness of various computational methods to calculate thermochemical data, i.e., ionization energies and heat of formation. The W1 and W2 theoretical models significantly better than standard methods such as G2 and CBS-QB3, as indicated in the evidence cited in Martin et al. (2001). The ability of W1 theory to achieve benchmark-quality accuracy speaks to the merit of further advancing computational thermochemistry. Significantly, such methodologies contain enhancements that stabilize the calculation and improve outcomes for systems with slow basis set convergence, acknowledging their quality in application to thermochemistry. The consideration of such methodologies justifies the progressive enhancement of computational techniques, underscoring the significance of their role in more precisely defining our understanding of energetic change in chemical systems. Moreover, the relevance of thermochemistry extends beyond theoretical analysis and contributes to shaping real-world applications in many fields, including materials science and drug design. The calculated energy profiles from thermochemical calculations inform the design of new materials with targeted properties and the formulation of pharmaceuticals that optimize molecular level contacts. The new computational approaches detailed in the essay, specifically those derived through DCPs and upper-level theory, exhibit the convergence of theory with practice. As substantiated in the advances outlined, such as the enhanced accuracy attributed to the W1 theory for heats of formation, it is clear that continued investigation and employment of computational thermochemistry are fundamental to ongoing chemical study and advancement.

REFERENCES

1. Abramov, O., Andeer, P., Arvidson, R. E., Banfield, J. F., Banfield, K., Barker, D. A., Bennett, J. P., et al. (2017). Determination of geochemical bio-signatures in Mars-like basaltic environments. https://core.ac.uk/download/131317448.pdf (accessed on 4th October 2025).
2. Afrizal, A., Afrizal, A., Mardiah, M., Alin, Y., Puspitasari, D., Mela, N., Rahmawati, F., et al. (2021). Development of students' conceptual understanding through STEAM project integration in thermochemistry. https://core.ac.uk/download/478173574.pdf (accessed on 4th October 2025).
3. Anderson, P. G., Mayne, A. W., Rawlinson, E. G., Robinson, J. C., et al. (2025). Development and application of the GIM code for the Cyber 203 computer. https://core.ac.uk/download/pdf/42853745.pdf (accessed on 4th October 2025).
4. Argaman, N., & Makov, G. (1998). Density functional theory–an introduction. https://core.ac.uk/download/25241258.pdf (accessed on 4th October 2025).
5. Arora, S. (2024). The Chemistry Connection: From Atoms to Applications. Libros Publication. https://play.google.com/store/books/details?id=uo1cEQAAQBAJ&source=gbs_api (accessed on 4th October 2025).
6. Aspuru-Guzik, A., Dunn, I., Even, A. B., Jinich, A., et al. (2014). Quantum chemical approach to estimating the thermodynamics of metabolic reactions. https://core.ac.uk/download/28952413.pdf (accessed on 4th October 2025).
7. Bai, X. N., Goodman, J., & Wang, L. (2018). Global simulations of protoplanetary disk outflows with coupled non-ideal magnetohydrodynamics and consistent thermochemistry. http://arxiv.org/abs/1810.12330 (accessed on 4th October 2025).
8. Bergeat, A., Hickson, K. M., Loison, J. C., Mereau, R., et al. (2013). The gas-phase chemistry of carbon chains in dark cloud chemical models. http://arxiv.org/abs/1311.3865 (accessed on 4th October 2025).
9. Blake, G. A., Liang, M. C., Miller, C. E., Muller, J. P., et al. (2004). Evidence for O-atom exchange in the O(^1D) + N_2O reaction as the source of mass-independent isotopic fractionation in atmospheric N_2O. https://core.ac.uk/download/10211203.pdf (accessed on 4th October 2025).
10. Bodi, A., Baboul, A. G., Baer, T., Berry, R. J., et al. (2012). A halomethane thermochemical network from iPEPICO experiments and quantum chemical calculations. https://core.ac.uk/download/pdf/9427899.pdf (accessed on 4th October 2025).

11. Braun-Unkhoff, M., Goos, E., Kathrotia, T., Kick, T., et al. (2014). The importance of detailed chemical mechanisms in gas turbine combustion simulations. https://core.ac.uk/download/31013967.pdf (accessed on 4th October 2025).

12. Carneiro, A. (2021). The Research School of Chemistry of Adolphe Wurtz, Paris, 1853–1884. https://core.ac.uk/download/384444417.pdf (accessed on 4th October 2025).

13. Chaban, V. V., & Fileti, E. E. (2014). The scaled-charge additive force field for amino acid based ionic liquids. http://arxiv.org/abs/1407.5852 (accessed on 4th October 2025).

14. Chen, Y. K., & Gokcen, T. (2012). Effect of surface nonequilibrium thermochemistry in simulation of carbon-based ablators. https://core.ac.uk/download/pdf/10571160.pdf (accessed on 4th October 2025).

15. Cimas, A., Silva, A. L. R., Monte, M. J. S., Ribeiro da Silva, M. D. M. C. R., Vale, N., & Gomes, P. (2013). Experimental and computational study of the energetics of hydantoin and 2-thiohydantoin. https://core.ac.uk/download/143403427.pdf (accessed on 4th October 2025).

16. Colussi, A. J., Enami, S., Hoffmann, M. R., Stewart, L. M., et al. (2010). Superacid chemistry on mildly acidic water. https://core.ac.uk/download/4887739.pdf (accessed on 4th October 2025).

17. Coradin, H., Guillemet-Fritsch, S., Navrotsky, A., Tailhades, P., et al. (2004). Thermochemistry of iron manganese oxide spinel's. https://core.ac.uk/download/12039038.pdf (accessed on 4th October 2025).

18. Devonshire, R., Loo, K. H., Stone, D. A., Tozer, R. C., et al. (2005). A dynamic conductance model of fluorescent lamp for electronic ballast design simulation. https://core.ac.uk/download/58541.pdf (accessed on 4th October 2025).

19. DiLabio, G. A., Koleini, M., & Torres, E. (2013). Extension of the B3LYP - dispersion-correcting potential approach to the accurate treatment of both inter- and intramolecular interactions. http://arxiv.org/abs/1307.1637 (accessed on 4th October 2025).

20. Drouet, C., Forray, F. L., & Navrotsky, A. (2005). Jarosite stability on Mars. https://core.ac.uk/download/12039042.pdf (accessed on 4th October 2025).

21. Drouet, C., Forray, F. L., & Navrotsky, A. (2005). Thermochemistry of yavapaiite $KFe(SO_4)_2$: Formation and decomposition. https://core.ac.uk/download/12039112.pdf (accessed on 4th October 2025).

22. Ebel, D. S. (2023). Condensation calculations in planetary science and cosmochemistry. http://arxiv.org/abs/2306.12645 (accessed on 4th October 2025).

23. Fabry, F., Fulcheri, L., Rehmet, C., Rohani, V., et al. (2013). Waste gasification by thermal plasma: A review.

24. Falcone, C. M., Zeleznik, F. J., Tandon, G. P., Fishtik, I., Prigogine, I., Samohýl, I., Nunziato, J. W., et al. (2010). A model for the degradation of polyimides due to oxidation. http://arxiv.org/abs/1011.5451 (accessed on 4th October 2025).

25. Ferrer Castillo, S., Kohen, A., Moliner, V., Wang, L., et al. (2013). QM/MM calculations suggest a novel intermediate following the proton abstraction catalyzed by thymidylate synthase. https://core.ac.uk/download/61429266.pdf (accessed on 4th October 2025).

26. Fisch, N. J., Gueroult, R., & Rax, J. M. (2017). Opportunities for plasma separation techniques in rare earth elements recycling. http://arxiv.org/abs/1710.04506 (accessed on 4th October 2025).

27. Fleming, R. M. T., Haraldsdottir, H., Milo, R., Noor, E., et al. (2013). Consistent estimation of Gibbs energy using component contributions. https://core.ac.uk/download/19335337.pdf (accessed on 4th October 2025).

28. Foley, R. T., Mitchell, G. D., & Roberts, G. N. (2025). Research into fundamental phenomena associated with spacecraft electrochemical devices, calorimetry of nickel-cadmium cells: Progress report, 1 Oct.–31 Dec. 1967. https://core.ac.uk/download/pdf/85242439.pdf (accessed on 4th October 2025).

29. Gilbert, L. M., & Scala, S. M. (2025). Theory of hypersonic laminar stagnation region heat transfer in dissociating gases. https://core.ac.uk/download/pdf/80680365.pdf (accessed on 4th October 2025).

30. Goddard, W. A., III, Xu, X., Zhang, Q., et al. (2009). Doubly hybrid density functional for accurate descriptions of nonbond interactions, thermochemistry, and thermochemical kinetics. https://core.ac.uk/download/4881733.pdf (accessed on 4th October 2025).

31. Goekoglu, S. (2025). Chemical vapor deposition modeling for high temperature materials. https://core.ac.uk/download/pdf/42814108.pdf (accessed on 4th October 2025).

32. Gregory, D. H., Gu, Q., & Mao, J. (2015). Revisiting the hydrogen storage behavior of the Na–O–H system. https://core.ac.uk/download/96883253.pdf (accessed on 4th October 2025).

33. Hamaekers, J., Israels, R., & Maaß, A. (2017). The octet rule in chemical space: Generating virtual molecules. http://arxiv.org/abs/1709.06746 (accessed on 4th October 2025).

34. Hong, Q. J., Navrotsky, A., Ushakov, S. V., van de Walle, A., et al. (2019). Carbides and nitrides of zirconium and hafnium. https://core.ac.uk/download/323067876.pdf (accessed on 4th October 2025).

35. Hu, R., Seager, S., & Yung, Y. L. (2015). Helium atmospheres on warm Neptune- and sub-Neptune-sized exoplanets and applications to GJ 436 b. https://core.ac.uk/download/33122345.pdf (accessed on 4th October 2025).

36. Le Méhauté, A., & Riot, P. (2017). From the arrow of time in Badiali's quantum approach to the dynamic meaning of Riemann's hypothesis. https://core.ac.uk/download/197460140.pdf (accessed on 4th October 2025).

37. Li, J. (2022). Oxygen evolution reaction in energy conversion and storage: Design strategies under and beyond the energy scaling relationship. Nano-Micro Letters, 14. https://doi.org/10.1007/s40820-022-00857-x (accessed on 4th October 2025).

38. Liao, X., Lu, R., Xia, L., Liu, Q., Wang, H., Zhao, K., Wang, Z., et al. (2021). Density functional theory for electrocatalysis. Energy & Environment Materials, 5, 157–185. https://doi.org/10.1002/eem2.12204 (accessed on 4th October 2025).

39. Marble, F. E. (1956). Flame theory and combustion technology. https://core.ac.uk/download/4887597.pdf (accessed on 4th October 2025).

40. Marshall, P., Glarborg, P., & Talbi, D. (2017). Importance of the hydrogen isocyanide isomer in modeling hydrogen cyanide oxidation in combustion. https://core.ac.uk/download/84003776.pdf (accessed on 4th October 2025).

41. Martin, J. M. L., & Parthiban, S. (2001). Assessment of W1 and W2 theories for the computation of electron affinities, ionization potentials, heats of formation, and proton affinities. https://core.ac.uk/download/25304517.pdf (accessed on 4th October 2025).

42. N/A. (1934). Special Libraries, December 1936. https://core.ac.uk/download/70425079.pdf (accessed on 4th October 2025).

43. Parrinello, M., Polino, D., Rizzi, V., Russo, A., et al. (2019). The onset of dehydrogenation in solid ammonia borane, an ab initio meta dynamics study. http://arxiv.org/abs/1809.06118 (accessed on 4th October 2025).

44. Penner, S. S. (1963). Combustion and Propulsion Research. https://core.ac.uk/download/4887476.pdf (accessed on 4th October 2025).

45. Ponniah, A. G. (1987). Determination of energy content of feeds. https://core.ac.uk/download/33015175.pdf (accessed on 4th October 2025).
46. Rissanen, M. (2012). Selected studies of carbon-centered free radical reactions using pulsed laser photolysis coupled to a photoionization mass spectrometer. https://core.ac.uk/download/14925572.pdf (accessed on 4th October 2025).
47. Rosdahl, J., & Teyssier, R. (2015). A scheme for radiation pressure and photon diffusion with the M1 closure in RAMSES-RT. https://core.ac.uk/download/388667390.pdf (accessed on 4th October 2025).
48. Taskinen, P., & Tesfaye, F. (2012). Phase equilibria and thermochemistry of selected sulfide systems in the pyrometallurgy of Ni and Cu. https://core.ac.uk/download/80702619.pdf (accessed on 4th October 2025).
49. Toure, O. (2014). Prédiction Des Propriétés D'équilibre Dans Les Milieux Biologiques Et Alimentaires Par Le Modèle COSMO-RS. https://core.ac.uk/download/46811380.pdf (accessed on 4th October 2025).
50. Woods, R. L. (2025). Publications of the Jet Propulsion Laboratory, July 1964 through June 1965. https://core.ac.uk/download/pdf/85253247.pdf (accessed on 4th October 2025).

CHAPTER 2

ENERGY, WORK, AND HEAT

Contents

Introduction 43
2.1. Types of Energy 48
2.2. The Concept of Work 53
2.3. The Relationship Between Energy and Work 57
2.4. Heat Transfer Mechanisms 61
2.5. The Laws of Thermodynamics 66
2.6. Practical Applications of Energy, Work, and Heat 71
Summary 76
References 78

INTRODUCTION

The subtle interaction between heat, work, and energy is the foundation of physical science as well as day-to-day living and thus a highly significant subject to learn. Energy is a key phenomenon that is seen in all fields of science, since it occurs in various forms such as kinetic, potential, thermal, and chemical energy. It is imperative to know how energy is transferred and transformed in physical systems, as these processes control both natural phenomena and technological innovation. As a way of understanding this connection, it is realized that work and heat are the two main ways in which energy is transferred between systems and the environment. This study will provide valuable information needed in the face of today's energy concerns, such as efficiency and sustainability, for zones of urban development where energy needs are always growing. The significance of controlling energy is also symbolized in the case of district heating systems, which is a model case of how mass consumer culture can propel energy needs.

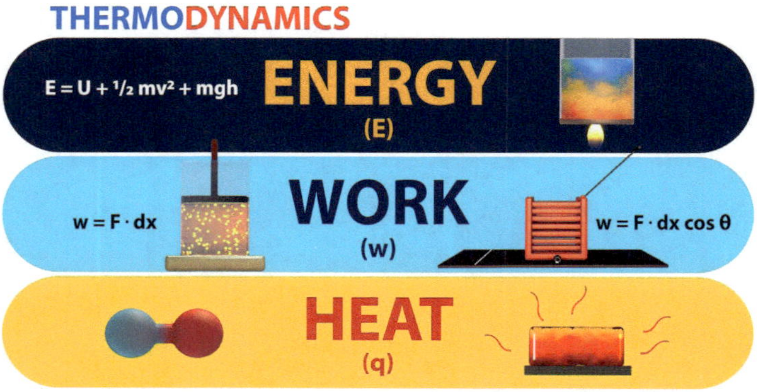

Figure 2.1. *Energy, work, and heat.*

Source: https://i.ytimg.com/vi/vlzYLk5ll14/maxresdefault.jpg.

The interactive interaction between domestic units and providers of energy shows the need to exchange information and data for the facilitation of sustainable action. As revealed by new studies, visualization and understanding of consumption patterns have the ability to advance informed decision-making on energy usage, thereby reducing overall ecological traces. The complexity of this association points to three key dilemmas facing sustainable design: the dynamics of personal behavior and system-wide energy processes, the multi-dimensionality of environmental impact as a behavioral driver, and ethical decision-making in the design process (Balksjö et al., 2013). In educational settings, emphasizing the notions of energy, work, and heat is important to fostering an energy-conscious citizenry capable of managing energy-related issues. Lectures designed exclusively for these concepts, like capstone projects for students in the science fields, are effective in bridging theory with practice. By incorporating hands-on experiments, demonstrations, and comprehensive assessments, teachers can enable students with a deeper understanding of how energy is conceived and applied in actual contexts. These teaching methods not only enhance understanding but stimulate critical thinking as students navigate the complexities of energy solutions. The evolution of curricular designs in this field also reflects the increasing awareness of the potential roles of energy educations in shaping future leaders to solve pressing global problems in sustainability (Arce et al., 2001) (Figure 2.1).

Definition of Energy

Energy is a fundamental concept that constitutes the backbone of physics and engineering studies and quantitative characterization of the capacity of systems to perform work. It exists in various forms, including kinetic, potential, thermal, and chemical energy, all of which can be converted from one form to another but not created or destroyed, based on the First Law of Thermodynamics. This conservation law is central to understanding energy transfer in physical systems. As recapitulated in thermodynamic accounts, energy must be accurately quantified to account for conversion and dissipation during processes, such as when mechanical energy is converted to heat. Knowledge of these interactions becomes an area of concentration in studying how systems work, particularly in non-equilibrium conditions, and establishes the need for constant tracking of energy within the context of energy, work, and heat. The term energy goes beyond mathematical definitions, encompassing the mechanisms of heat conduction, work done, and their corresponding thermodynamic principles. The Second Law of Thermodynamics introduces the concepts of entropy, linking energy conversions to a system's randomness. In particular, as it pertains to communication between driven and equilibrium systems, energy plays a significant role in identifying system response. For instance, in cases where quantum systems are interacting with thermal reservoirs, energy transport is required to facilitate thermodynamic equilibrium, a reflection of the delicate interplay between loss and energy storage. Conclusions derived from time-resolved energy transport analysis illustrate how energy transitions between reservoirs affect not only the efficiency of the system but also its conformity to thermodynamic laws, and thus understanding energy is essential for system optimization and sustainability (Arrachea et al., 2016). Moreover, energy's characterization is supplemented through its analysis in stochastic processes, whereby it exists in thermal-fluctuated systems and external forces. These systems exhibit complex interactions in which energy is passed on through mechanical work and heat, rendering classical notions of energy conservation challenging. Stochastic models offer a mathematically sound description of energy processes, accounting for the impact of random perturbations on energy distribution and transfer. The concept of non-Gaussian processes, e.g., in particle systems under stochastic forces, is more representative of the intricacies around energy definitions in presenting a picture of energy as a multi-dimensional quantity that remains strongly influenced by external conditions and internal states. This consolidation of the various aspects of energy brings out its importance in applications ranging from thermodynamics to statistical mechanics, highlighting its role both in theory and practice (Gudowska-Nowak et al., 2014).

Importance of Work in Physics

The physical idea of work is the foundation for investigating energy transfer in various systems. Work has been explained as the method of energy transfer that occurs when force is used to cause an object to move, with resulting displacement. This relationship serves as the foundation for investigating mechanical systems, whereby work gives rise to understanding energy conservation and conversion. For example, when a force is exerted on a piston in a car engine, the work done causes chemical energy to be transformed into mechanical energy that ultimately powers the automobile. This correlation between work and energy explains such basic principles of thermodynamics, particularly in heat transfer and engine efficiency, hence indicating the central position of work in practical applications, from car engineering to designing energy systems. Further, labor is critical in the field of advanced physics, as evident from its application in advanced technologies such as microelectronics and astrophysics. Thermal dynamics management plays a paramount role in microelectronics, particularly in the development of microevaporators and micro-condensers applied in high-performance devices. The challenge associated with the phase transitions of refrigerants, as evolved with study into annular flow, is significant to optimizing heat exchange processes in miniaturized cooling systems, essential to preventing overheating of computer chips and other components (4th Micro and Nano Flows Conference (MNF2014) et al., 2014). Such applications emphasize the reality that work is not abstract but a utilitarian tool which engineers employ to enhance the functionality and effectiveness of sophisticated technologies. In the universe, the laws of work manifest themselves in phenomena like the thermal evolution of neutron stars, where the interaction between spin, structure, and thermal dynamics at very high densities. The work involved in the cooling processes of these celestial objects is interrelated with their condition of rotation and internal phase transition, particularly when there is quark matter (Aguilera et al., 2009). The intricate relationship points out the manner in which effort, in all its categories, regulates not only earth systems but also the huge and intricate behavior of astral bodies. Identification of such relationships is of great significance to theoretical physics and to our comprehension of the universes evolution, and it demonstrates the intrinsic meaning of work as a unifying theme across different disciplines.

Overview of Heat as a Form of Energy Transfer

Heat, the core ingredient of thermodynamics, is actually a mode of energy transfer due to temperature differences between systems. This mode of energy transfer involves three dominant mechanisms: conduction, convection, and radiation. Conduction is the movement of heat through objects without the object actually moving, such as when a spoon gets hot when placed in boiling water. Convection is the flow of fluids in which the warmer, less dense regions move up and the cooler, denser ones move down. This occurs in atmospheric conditions and ocean currents. But heat transfer by electromagnetic waves is involved in radiation, whereby sunlight can warm the Earth's surface without there needing to be space as a medium. Knowledge of them is not just crucial to theoretical physics but also to applied applications across the whole of engineering and environmental science since they provide the foundation for energy efficiency and thermal control.

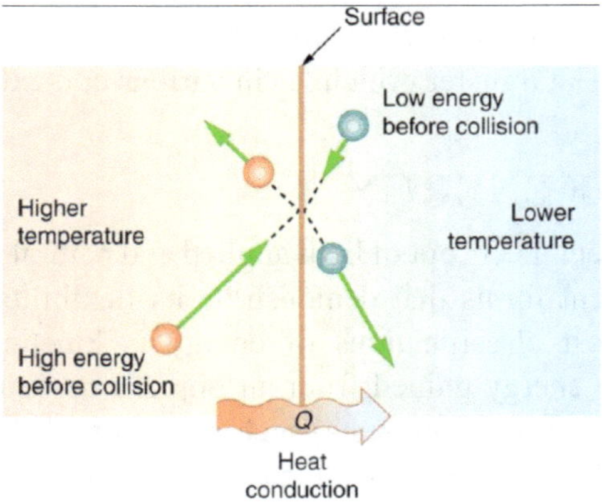

Figure 2.2. *Heat as a form of energy transfer.*

Source: https://storage.googleapis.com/static.prod.fiveable.me/search-images%2F%22Fundamental_principles_of_heat_transfer%3A_conduction_convection_radiation_temperature_difference_thermal_equilibrium_applications%22-figure-15-05-02a.jpg%3Frevision%3D1%26size%3Dbestfit%26width%3D358%26height%3D292.jpg.

The concept of heat as a means of energy transfer is related to the second law of thermodynamics, which stipulates that heat will spontaneously transfer from a hotter body to a colder one, a process that increases the overall entropy of an isolated system. This law establishes the spontaneous direction of thermal energy transfer, with emphasis that while heat can be converted into work, it cannot be reversed entirely without energy loss. In heat engines, for instance, conversion of heat to work is restricted by this law. Awareness of such limitations is vital in forming technologies to use thermal energy effectively. In addition, emerging applications in spin caloritronics illustrate control of heat through electron spin, showing emerging routes of energy transfer that have the potential to redefine conventional thermal control limits (Brataas et al., 2011). Educationally, comprehending the complexity of heat as a transfer mechanism for energy poses difficulty, primarily because energy and temperature are abstract entities difficult to envision. Emerging innovations in immersive technologies, such as augmented reality (AR), have been valuable tools in bridging this perception gap. By allowing students to visualize thermal phenomena and interactively experience temperature representations, these technologies make for better understanding and a deeper understanding of thermodynamics. A pilot study on the use of mixed reality in undergraduate labs was highly effective in improving student performance and comprehension of thermal conduction (Kapp et al., 2018). As energizing learning methods continue to be improved, they have the potential to empower learners with better ways of comprehending and implementing heat transfer principles in various contexts (Figure 2.2).

2.1. TYPES OF ENERGY

Energy is a fundamental concept of both applied and natural sciences that exists in numerous different forms that demonstrate its flexibility and significance. One of the foremost classifications of energy is kinetic energy, which is associated with the energy gained from an object's movement. This form of energy appears in natural phenomena, such as the rapid flow of rivers or the flight of birds, and is quantified by the equation $KE = \frac{1}{2} mv^2$, where m is mass and v is speed. The second form is associated with an object's position in a field, such as gravitational or elastic potential energy. Interchangeability between kinetic and potential energy is prevalent in mechanical systems, e.g., in pendulums and roller coasters, to demonstrate the transformation of energy from one type to another based on the law of conservation of energy. Other energies exist in particular scenarios, namely thermal, chemical, and electrical energies, besides kinetic and potential energy. Thermal energy, or the energy of the random motion of molecules and atoms, is at the heart of temperature and heat transfer, and plays a leading role in the laws of thermodynamics. Thermal energy is important to be grasped in all types of applications, ranging from engineering to environmental science, linking it to efficiency and energy storage. Moreover, chemical energy is locked away in the bonds of the chemical compounds and released or taken up in chemical reactions. This type of energy emphasizes the importance of chemical reactions in biological functions and industrial applications, unveiling the intricacy of energy conversions that occur in everyday life. The significance of energy extends beyond its kinds to the laws that govern its transfer and conversion. Thermodynamics laws, especially those that are defined in historical contexts such as Carnots efficiency and innovations witnessed in the 20th century, essentially create our understanding of energy systems. The second law of thermodynamics defines that energy conversions are not efficient, thereby pointing towards the fact that always there is waste heat lost in processes. This inefficiency is an important factor when designing energy systems, strongly impacting the design of engines and thermal devices, as demonstrated through research in finite-time thermodynamics. Finally, the identification of these kinds of energy, together with their laws, improves our understanding of energy interactions and their implications on technological innovation and sustainability (Apertet et al., 2015; Das et al., 2013).

2.1.1. Kinetic Energy and Its Applications

Kinetic energy, the energy possessed by an object due to its motion, is a crucial element of numerous physical phenomena and uses in many areas of science. In traditional mechanics, the kinetic energy (KE) of an object is quantitatively represented by the formula $KE = 1/2\ mv^2$, where m is mass and v is velocity. This is what draws attention to the fact that even small increases in speed cause significantly large increases in kinetic energy, affecting systems from basic mechanics to advanced fluid dynamics. These dynamics are fundamental in describing the motion of automobiles, aircraft aerodynamics, and even microscopic particle behavior. The significance of kinetic energy is also supported by its ability to forecast outcomes of collisions, which is crucial in safety engineering, sport science, and biomechanics. Applications of kinetic energy are demonstrated in recent innovations, namely in transport and power systems. For instance, vehicles convert fuels' chemical energy into kinetic energy to supply motion, highlighting the pragmatic coincidence of energy types. Secondly, technologies such as regenerative braking in electric and hybrid vehicles utilize kinetic energy by converting it into stored electrical energy. Such interaction not only enhances efficiency but emphasizes the contribution of kinetic energy in efficient practice. Thirdly, kinetic energy is also a concept applied in engineering design where information regarding particle movement can significantly impact design efficiency. Use of the lattice Boltzmann (LB) method in simulating kinetic phenomena is a glaring manifestation of this with its capability to precisely simulate fluid behavior in complex environments, such as fuel cells and energy storage devices (Chen et al., 2015). In the study of micro and nanofluids, kinetic energy emerges as a significant variable that influences thermal and hydrodynamic characteristics in various industrial and biological processes. The dynamic characteristics of charged micro and nanofluids are modeled by various competing mechanisms, such as electrostatic forces and viscous dissipation, which require sophisticated modeling for accurate representation. To this end, the thermal lattice Boltzmann model is a robust framework for studying micro/nano-flows, which enables thorough analysis and optimization of energy transportation and conversion devices (Barber et al., 2008). As scientists further explore and develop these techniques, the future of kinetic energy applications is full of potential for very significant improvements, particularly in maximizing energy efficiency, system performance, and assisting with problems in such areas as renewable energy to new materials science (Figure 2.3).

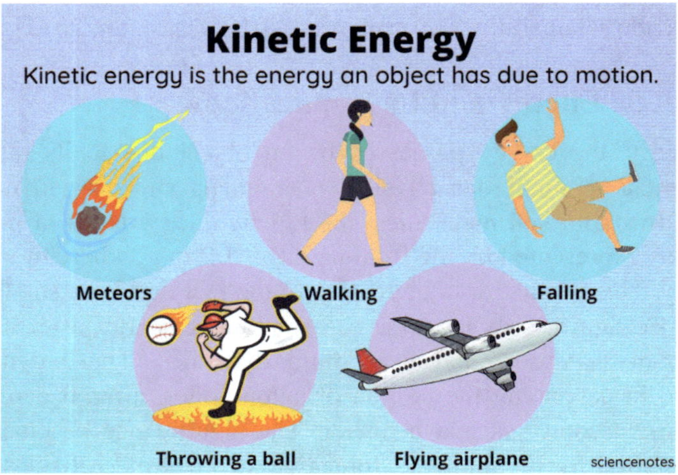

Figure 2.3. *Kinetic energy and its applications.*

Source: https://sciencenotes.org/wp-content/uploads/2021/03/Kinetic-Energy-Examples.jpg.

2.1.2. Potential Energy and Gravitational Effects

The concept of potential energy lies at the heart of understanding a broad array of physical systems, above all in the realm of gravitational interaction. Potential energy, or stored energy as a result of an object's position within a gravitational field, is of staggering significance when one contemplates phenomena of an astronomical nature. Within mega-scale objects, such as galaxy clusters, gravitational potential energy plays a significant role in determining the dynamics and thermodynamics of interstellar gas. As an illustration, intra-cluster medium (ICM) gravitational heating is a process of major significance whose net gravitational potential energy dissipated from the satellite galaxies heats surrounding gas. The mechanism not only suppresses cooling but also reflects energy contributions from other processes, highlighting the significance of gravitational effects in maintaining the thermal equilibrium in such large systems (Bower et al., 2008). Deeper investigation of gravitational effects introduces the distinction between various kinds of galaxy clusters. Research explains the distinction between cool core (CC) and non-cool core (NCC) clusters, each with varying energetics and thermal evolutionary histories. Gravitational heating and radiative cooling act in concert to regulate the entropy profiles of the clusters. Though gravitational heating adequately accounts for entropy distributions at greater radii, it fails to account for the behavior observed at smaller radii, particularly in cool core clusters.

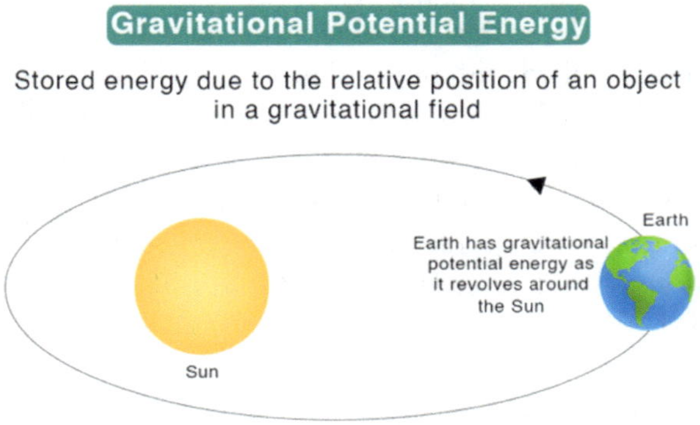

Figure 2.4. *Gravitational potential energy.*

Source: https://www.sciencefacts.net/wp-content/uploads/2022/08/Gravitational-Potential-Energy.jpg.

The latter is dependent on preheating of material prior to cluster collapse in a move to avert the formation of low-entropy regions. Such results point out the complexity of gravitational processes and how they are pivotal in the formation of diverse cluster environments (Babul et al., 2008). Ultimately, the understanding of gravitational influence and potential energy is beyond theoretical models and has real consequences for astronomical observations and models. Comprehension of how gravitational heating impacts the thermodynamic equilibrium of galaxy clusters can enable scientists to refine existing models as well as more accurately predict observed phenomena. The time-dependent and mass-dependent character of gravitational heating can account for trends in star formation rates, particularly in large galaxies. Moreover, exploring these gravity effects further contributes to understanding the distribution and movement of energy throughout the universe. This improved understanding not only impacts theoretical astrophysics but also favors a greater appreciation of the interdependent nature of energy, work, and heat within the universes larger tapestry (Babul et al., 2008; Bower et al., 2008) (Figure 2.4).

2.1.3. Other Forms of Energy (Thermal, Chemical, and Nuclear)

Thermal energy, a central aspect of the general energy debate, is a result of the kinetic energy of particles within an object. It occurs in various states, and they have an effect on temperature and changes in phase. Thermal energy is central to the process of heat transfer, where it moves from warmer bodies towards colder ones, and is a fundamental concept in thermodynamics. The interaction between thermal energy and other forms of energy demonstrates its significance; for example, in the case of neutron stars, their cooling and thermal evolution are closely linked with the evolution of their structure as well as their chemical composition. Advances in the understanding of the thermal accounts of these bodies suggest the manner in which latent heat at phase transitions can fundamentally alter temperature regimes, with superfluid states delaying cooling and resulting in temporary rises in temperatures prior to settling into an eventual pattern (Aguilera et al., 2009). Chemical energy, yet another vital source of stored energy, is significant for a broad variety of natural and industrial processes.

This energy is derived from interatomic bonds in molecules and released or absorbed during chemical reactions. These processes of energy transformation are fundamental within biological organisms, with glucose decomposition through cellular respiration rendering energy accessible for life. The combustion chemistry is also an example of energy transformation of chemical energy into heat energy, driving engines and providing heat for domestic and industrial purposes. The physical models that define combustion events also improve our understanding of fire spread and dynamics, showing how chemical reactions have the ability to cause energy state transitions and thermal distributions within a landscape at extremely high rates (Albini et al., 2007). Nuclear energy, the third to be discussed, comes from the energies which hold atomic nuclei together. This energy is stored in fission and fusion reactions, where there are very large releases of energy. Nuclear fission, for instance, is the origin of nuclear reactor power generation, where the nucleus of an atom is split to release vast amounts of thermal energy. Apart from that, the energy of fusion, the fuel of the sun, holds a glimmer of hope for a cleaner future but currently remains an issue for harnessing it for mass production on Earth. The link between nuclear energy and thermal energy suggests the stakes in harnessing different sources of energy and the fact that scientists are striving to harness nuclear reactions safely and efficiently for impending energy requirements. This entanglement of energies emphasizes the importance of new ways of controlling energy within the larger framework of work and heat (Aguilera et al., 2009).

2.2. THE CONCEPT OF WORK

Fundamentally, the definition of work can be defined as the basic aspect of energy transfer in thermodynamic systems. Work is done when a force is applied over a distance, leading to a transfer of energy from a system to another. This mechanism is essential in describing numerous physical processes, from basic mechanical systems to sophisticated biological functions. For example, when an engine runs, chemical energy is transformed into mechanical work, illustrating this interaction. The accurate measurement of work is necessary in thermodynamics studies because it serves to define systems' responses to forces. Furthermore, the work-energy relationship serves as the foundation for other scientific principles like the work-energy theorem that says work on an object equals the change in its kinetic energy. Thermodynamics' study of work reveals the significant link between transformations of energy and changes in state of materials. Work done in thermodynamics can be in various forms, such as pressure-volume work in gases, which is primarily related to change in internal energy. It can be shown in situations like isothermal expansion and adiabatic compression, where work done directly affects temperature and enthalpy of the system. Furthermore, work theory also extends its relevance to statistical mechanics, as the Gibbs-von Neumann entropy formalism brings macroscopic and intrinsic understanding of thermal systems together. As discussed in scholarly writings, recognition of the subtleties of work and how it engages with thermal states enables enhanced insight into energy's behavior in physical contexts (Maroney et al., 2007). Furthermore, the use of work is not limited to traditional physical sciences but also applies to economic and educational realms. Of special note is pedagogy in the teaching of scientific concepts related to energy and work; recent national standards have come under criticism for inadequate treatment of these subjects. Reports show that many states have failed to develop appropriate science standards, which can detract from students' understanding of fundamental scientific principles and inhibit future innovation within high-technology economies (Lynch et al., 2012). As such, rephrasing pedagogic practices to cultivate a good understanding of work and energy may confer much benefit not only in institutions of learning but also generally in the context of society development. Increased emphasis on clear and specific standards for instruction in the concepts is at the heart of promoting enhanced scientific literacy and technological advancement within the population.

2.2.1. Definition of Work in a Physical Context

Within the field of physics, the concept of work is fundamentally described in relation to its relationship with energy and movement. Work is performed whenever a force is exerted on an object in such a way that it travels through some distance. Mathematically, work (W) is defined as the product of force applied (F) and displacement of the object (d), with a dependency on the angle (θ) between the direction of force and displacement: $W = Fd\ cos(\theta)$. It has some important implications; in other words, work is not only a function of the applied force but also direction on which this force is being applied. Therefore, when a force causes no displacement, like in situations involving static equilibrium, work is not performed, demonstrating work depends on the motion facilitated by energy. Extending the definition of work even further closely follows laws of thermodynamics, particularly when writing about the work done within various processes thermodynamic.

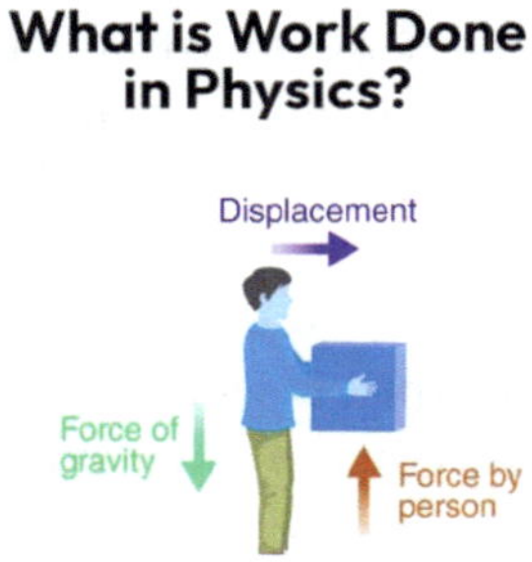

Figure 2.5. *Work done in physics.*

Source: https://blogcdn.aakash.ac.in/wordpress_media/2023/10/What-is-Work-Done-in-Physics.jpg.

In thermodynamics, work can be defined in terms of cycles, such as the Carnot cycle, which describes heat and work exchange in a theoretical engine. Efficiency in engines such as these depends heavily on the work per cycle, not just covering the transferred energy but also the inherent thermodynamic principles that direct the processes. Utilization of work in thermodynamic cycles is employed to describe the complex interaction between various energy types, emphasizing that work is a variable that changes with constraints imposed on a system, as indicated through examination both classically and quantum mechanically (Lucarini et al., 2009). In general terms, work analysis is carried over to other processes like mechanical systems and black hole mechanics. The analogy drawn between classical thermodynamics and black hole mechanics gives more insight into work in such systems. For instance, such black hole properties as surface gravity may be indicative of thermodynamic laws, and therefore work performed by a black hole can be understood in the same way as in classical thermodynamic systems. This perspective triggers the re-interpretation of work definitions and application outside traditional limits and threatens additional research on energy transfer in complex systems. Surprisingly, the investigation of work in these theories allows physicists to optimize systems, drawing analogies between mechanical work and energy production in thermodynamic systems, ultimately achieving a deeper understanding of energy's role in various physical spectrums (Curiel et al., 2014) (Figure 2.5).

2.2.2. The Work-Energy Principle

The work-energy principle is one of the cornerstones in physics that explains the link between work done on an object and the change in its energy as a result. According to this principle, the total work done on an object is equal to the change in its kinetic energy. Essentially, when work is exerted upon an object—be it by mechanical stress or by heat energy—its velocity will either grow or decrease, hence its kinetic condition changes. It is important to understand this concept in assessing many physical systems, especially those under the action of forces and movements. Within energy changes, this interdependence shows us how work into a system appears through energy conversions like alterations in velocity or direction following conservation and transfer principles in mechanical systems. Within thermodynamic systems, the work-energy principle serves to deepen our knowledge of energy conversions, especially where heat is added or taken away. This application not only meets the principles of the laws of thermodynamics but also pinpoints the way thermal energy and mechanical work relate in a system. For instance, in a closed cycle on a thermo-charged capacitor, one can see that energy is preserved while mechanical work and heat energy interact. These breakdowns, as the recent studies have shown, allow one to construct a harmonious image of how potential energy differences, such as the potential difference between capacitor terminals, are consistent with the work-energy principle (D'Abramo et al., 2015). The hypothesis thus goes beyond purely physical mechanical uses, understanding other physical phenomena by virtue of flexibility and underlying function of representing work and heat exchanges. Furthermore, when the work-energy principle is understood from a thermodynamic point of view, it provides deep insights into climate systems and broader environmental issues. In equating the ideas of work to energy fluxes through systems, scientists are able to construct models, such as an analogous Carnot engine, that display efficiency in energy conversions. These models apply the second law of thermodynamics to analyze the intricate relationship between work production and heat transport, illustrating the possibility for maximum utilization of energy to reduce climate effects (Lucarini et al., 2009). Eventually, the work-energy principle transcends traditional limits to become a governing paradigm for solving problems in contemporary energy efficiency and environmental sustainability, thereby establishing its usability in theoretical as well as practical physics (Figure 2.6).

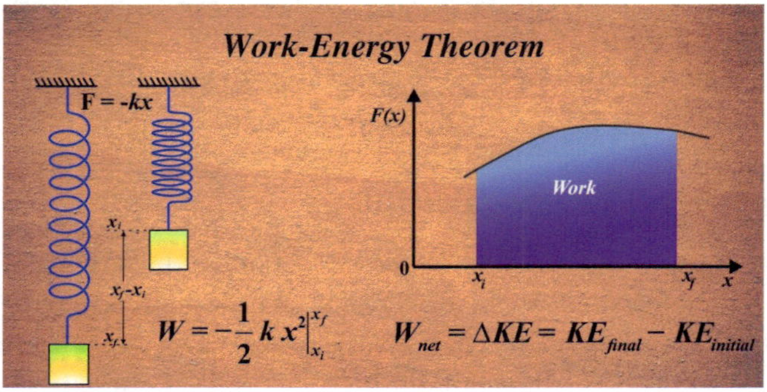

Figure 2.6. *Work-energy theorem.*

Source: https://unifyphysics.com/wp-content/uploads/2024/04/Work-Energy-Theorem-first.jpg.

2.2.3. Examples of Work in Everyday Life

Work is experienced in most physical events of daily life involving the transmission of energy, according to the principles of physics. For instance, when one raises a shopping bag from the floor to a shelf, work has been done as a result of energy being transmitted against gravity. This simple exercise demonstrates the relationship among force, distance, and energy, where the motion of lifting is a clear instance of positive work. As humans undergo various activities during their daily lives, like walking, climbing stairs, or utilizing machinery, they continually do work that assists them in keeping their individual health, productivity, and relation to their world. Every operation is a reminder that day-to-day processes are underlying the fundamental laws of energy and work that regulate physical movements in subtle and obvious ways. Furthermore, newer technologies in energy-conserving systems illustrate the application of work in achieving day-to-day functions without wastage of much energy. In modern buildings, for example, the application of energy-efficient lighting, heating, and air conditioning technologies optimize energy utilization while providing needed comfort levels to occupants. This strategy not only reduces overall energy consumption but also highlights the importance of analyzing work done by heterogeneous systems in a building. Systematic analysis of diverse energy end-use systems such as HVAC, lighting, and plug loads facilitates enhanced performance benchmarking and diagnostics that facilitate the explanation of the inherent energy consumption-work done relationship in built environments to aid in more sustainable lifestyle patterns (Fisher et al., 2011). Furthermore, measuring work performed in structures through key performance indicators (KPIs) has gained massive popularity as a way of enhancing energy efficiency to the fullest. Through the observation of the energy efficiency of different systems on a granular level, stakeholders are in a better position to implement improvements that eliminate waste and optimize output, ultimately benefitting both residents and nature. For instance, system-level KPI design provides insights on energy use, peak demand, and occupant comfort through various building typologies, facilitating data-based decision-making for energy management. The increasing integration of sensors and meters allows monitoring these KPIs, illustrating the shifting dynamics between energy, work, and heat generation in architectural scales (Hong et al., 2020). These advancements illustrate the way in which understanding work dynamics in daily life can lead to effective strategies for reducing energy consumption and enhancing operational efficiencies.

2.3. THE RELATIONSHIP BETWEEN ENERGY AND WORK

The energy-work relationship is central to physics as it encapsulates the manner in which energy is transferred and transformed within systems. Energy, which can be generally described as the ability to do work, may appear in a number of forms, including kinetic, potential, thermal, and chemical energy. If an external force is applied to an object, work is being done on the object, which leads to a transfer of energy. This equivalence is not a theoretical one but occurs in daily phenomena, for instance, in moving an object against gravity, where energy expended directly transforms into gravitational potential energy. Furthermore, energy conservation laws assert that the total energy of a closed system is fixed, such that all work is actualized within the systems energy balance, thereby filling in the connection between these two physics concepts. The laws of thermodynamics also illuminate the connection between energy and work, particularly against the backdrop of heat transfer and its implications. Work can also be defined as the energy transferred when force is applied over a distance, whereas heat is the energy movement with a temperature gradient. The interaction between work and heat energy comes to play in thermodynamic processes in so many different situations, as illustrated in heat engines that convert thermal energy to mechanical work. The connection must however be comprehended through investigating the statistical foundation of thermodynamics. Gibbs-von Neumann entropy, as elaborated in more recent literature, contributes to the knowledge of thermal states and how they can be translated into work through Hamiltonian dynamics, supporting the contention that work and energy are intrinsically related in both macroscopic and microscopic contexts (Maroney et al., 2007). In addition, historical works cite the significance of creating mathematical formulations for energy and work. For instance, George A. Linhart's' article sets a precedent for the determination of formulas that express heat capacities in terms of variations in temperature using a stringent method based on the principles of thermodynamics and theory of probability (Adkins et al., 2010). His article is an example of how theoretical advancement in the understanding of thermodynamics can find practice in material science and engineering with emphasis placed on the practical application of the energy-work relationship. Through the synthesis of these theoretical and historical perspectives, a thorough appreciation for how energy and work are related can be achieved, with an emphasis on their essential roles in physical systems and the overall principles of energy conservation and conversion.

2.3.1. Conservation of Energy Principle

The conservation of energy principle is a cornerstone of physical science, which asserts that energy cannot be created or destroyed but only converted from one type to another. This principle provides a model of understanding for numerous various energy interactions, particularly work and heat. For instance, in mechanical systems, kinetic and potential energy exchange as something travels within a gravitational field. When the object accelerates, potential energy is lost but kinetic energy is gained, showing the phenomenon of energy transformation. The vigilant study of energy conservation has served to bring significant advances in various fields, including engineering and thermodynamics, ultimately leading to a better grasp of natural processes and technology. Furthermore, the law of conservation of energy is closely related to the second law of thermodynamics, a principle that describes the direction of energy change. Heat energy in thermodynamic changes transfers from more to less energetic bodies in the process of work interactions. Especially, the concept of local equilibrium becomes very vital in such processes and has been dealt with rigorously in various different contexts. In non-equilibrium systems with delay responses, objections have been raised over traditional assumptions made on the basis of local equilibrium, and hence, entropy production here has been questioned. As noted in non-equilibrium thermodynamics studies, generalization to the second law is possible to acknowledge the character of real-world systems, where circumstances at any moment may not always persist (Glavatskiy et al., 2015). Furthermore, energy conservation is not only confined to classical mechanics but also extends to fields such as spintronics, whose energy exchange dynamics between sub-systems can lead to exceptional outcomes and applications. Here, the reciprocal interaction between ferromagnetic and electric systems is an archetype of the conservation rule by spin transfer processes. Theoretical approaches applied, particularly stochastic models and thermokinetic calculations, demonstrate that even unconventional parameters like effective temperature can occur in energy transactions at interfaces. Knowledge of these interactions in conservation illuminates not only our comprehension of fundamental physics but also lays the foundation for emerging technologies, highlighting the principle's relevance in a number of scientific and engineering disciplines (Aharoni et al., 2006) (Figure 2.7).

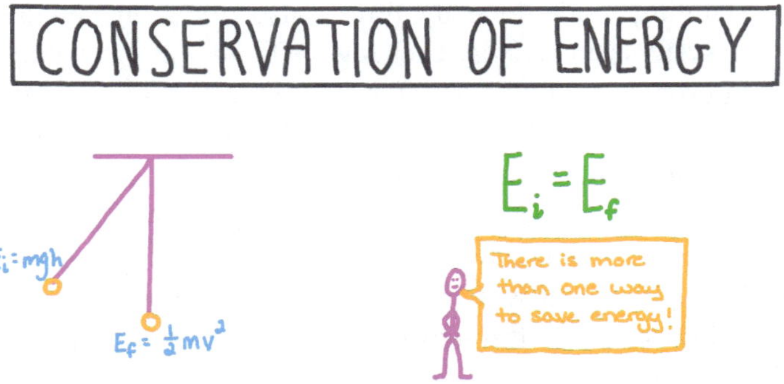

Figure 2.7. *Principle of conservation of energy.*

Source: https://s3.amazonaws.com/bucketeer-6a6b5dd7-82e9-48dd-b3be-ec23fe6cc180/notes/images/000/000/994/original/thumbnail_l.jpeg?1623429157.

2.3.2. Work Done by Forces

Forces' work done is an important element in grasping the ideas of energy change and transfer. Work in classical mechanics is the applied force to an object multiplied by the resulting displacement along the direction of applied force. Observing systems that show rotational motion, the application of Newtons second law is required. The relationship between force and work is inextricably linked to rotational dynamics through the angular impulse-momentum equation, which allows for a formal exploration of the manner in which work is accountable for energy change in systems. Furthermore, where non-conservative forces come into the equation, frictional forces being a case in point, the analysis must extend into the domain of thermodynamics and the significance of the interaction between mechanical work and heat introduced to a system (Fiolhais et al., 2013). When mechanical systems are moving, forces will automatically impact their energy and motion states and therefore perform work. This performed work can lead to a variety of phenomena, such as altering kinetic energy or generating heat. In the context of an energy conservation principle, the work-energy theorem demonstrates this relationship by indicating that the work done on a body is equal to the difference in its kinetic energy. Furthermore, in nonequilibrium systems, this interaction is more complicated, as external work can induce entropy changes beyond the capability of simple mechanical interactions. In these instances, the inclusion of non-mechanical work, in the form of heat application or the action of dissipation forces, can render the simple use of conservation principles untenable, necessitating a more refined approach to exploring energy transformations (Evans et al., 2003). Understanding work done by forces also gives understanding about thermodynamic systems and energy exchange in them. For instance, when there is work done in processes of heat transfer, it is important to apply the first law of thermodynamics. The total energy of an isolated system is constant as per this law, which includes mechanical work and thermal energy. The less obvious interaction that rules such energies highlights the prominence of entropy and its conversion, which, as an easy, definable process in states of equilibrium, manifests itself in a less clear way in nonequilibrium states. Particularly where mechanical work is returned through thermal processes, the implications for the generation and transport of entropy must be thoroughly investigated to determine the full energy topology of the system (Evans et al., 2003; Fiolhais et al., 2013) (Figure 2.8).

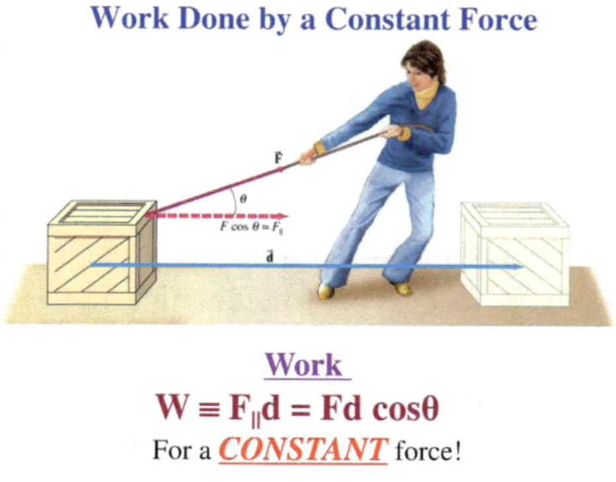

Figure 2.8. *Work done by a constant force.*

Source: https://image2.slideserve.com/4339679/work-done-by-a-constant-force-l.jpg.

2.3.3. Energy Transfer Through Work

Transfer of energy through work is one of the fundamental concepts in thermodynamics that explains how systems distribute energy in an effort to do work or generate heat. Work refers to the process of energy transfer that occurs when a force causes an object to move along the direction of the applied force. In thermodynamic systems, this transfer can be in numerous forms, such as mechanical work performed by engines or electrical work performed by circuits. Efficiency and effectiveness of such transfers of energy are very essential since they make or break the performance and output of the working system. Understanding these dynamics allows scientists and engineers to design and refine systems for practical application, whether for industrial machinery or renewable energy hardware. The intricate dance between energy transfer through work and system conditions is not well understood. Since a system is in non-equilibrium, work done upon the system introduces peculiar challenges in energy management. For instance, fluctuations in energy transfer across such states indicate a necessity to reexamine classical thermodynamic laws. Notably, fluctuation theorems illuminate work done in closed and open systems under conditions where external forces can move the system away from equilibrium. Principles observed here illustrate how work and heat are distributed out of energy, revealing a higher-level interaction that defies the traditional separation of thermodynamic processes and needs a rethink of underlying theoretical frameworks, confirming the significance of work in these dynamic systems (Esposito et al., 2009). Removing the ambiguities arising from nonequilibrium work and heat, scientists are keen on parameter definition in thermodynamic descriptions. According to the generalized Onsager-Machlup theory, without clear demarcation of work contributions to energy, one can encounter issues that hide application of the first and second laws of thermodynamics. The lack of distinctive definition of work and heat in these states may result in a large number of interpretations of thermodynamic laws, which complicates our ability to describe energy transfer processes. Taking the motion of a Brownian particle in an electric field as an example, the fineness of energy dynamics demonstrates that distinctive relations are established only when physical parameters are considered explicitly. Hence, careful study of energy transfer through work is essential not just for theory but also for practice across many domains (Gomez-Marin et al., 2007).

2.4. HEAT TRANSFER MECHANISMS

Heat transfer is a fundamental concept in thermodynamics that describes the transfer of heat energy from one body or system to another. The process of this transfer occurs through three dominant mechanisms: conduction, convection, and radiation. Conduction refers to the manner in which the heat is transferred through direct contact among materials based on the microscopic interaction and collision of particles. The process is in gases less effective due to lower particle density, while in solids like metals, heat conduction is quicker due to closely packed atoms that support more effective energy transfer. Transfer of heat by conduction is governed by Fourier's law, which delineates a linear relationship between heat flux and temperature gradient. It is a significant concept in thermal insulation engineering and materials science, in which heat flow has to be regulated for efficiency and safety. Convection is fluid bulk flow, which also facilitates heat transfer through advection and diffusion. It is subdivided again into natural and forced convection.

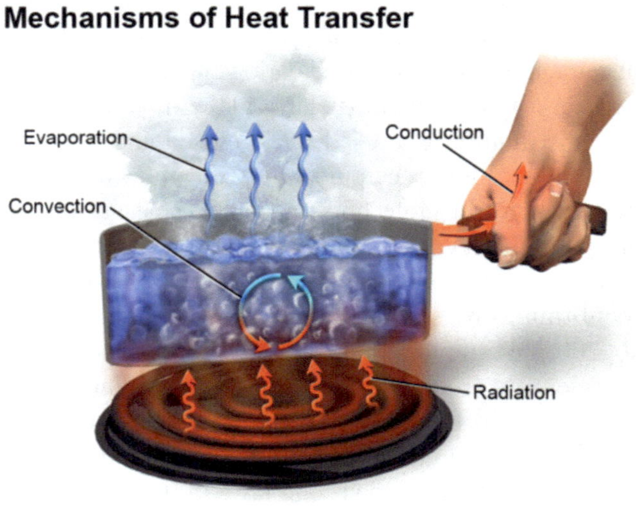

Figure 2.9. *Heat transfer mechanisms.*

Source: https://thermtest.com/wp-content/uploads/heat-transfer-e1733152711997.png.

Natural convection results from the flow induced by buoyancy, with the warmer, less dense fluid rising and cooler fluid falling, creating a circulation. In forced convection, fluid motion is created using forces externally in the form of fans or pumps, increasing the rate of heat transfer. The efficiency of convective heat transfer could be defined by factors such as fluid viscosity and surface area contact. Notably, studies on SAR pattern-based channel geometry have highlighted the ability of chaotic flow patterns to significantly enhance heat transfer rates. This study reveals the subtle interplay between thermal energy transport and fluid flow, explaining heat exchanger and other thermal device design optimization (Anxionnaz-Minvielle et al., 2016). The third process of heat transfer is radiation, which is the emission of electromagnetic waves, like infrared radiation. Unlike conduction and convection, radiation does not depend on a medium and has the ability to travel through a vacuum, which makes it a consideration in space technology or thermal insulation technology. The heat transfer by radiation follows the Stefan-Boltzmann law, an important fact that the energy radiated varies as the fourth power of the absolute temperature of the emitting surface. Such a process is of utmost significance in the global energy balance, influencing climate regimes and design in thermal systems in engineering. Further, more recent studies of kinetic plasma turbulence have revealed richness in processes of energy dissipation and conversion at scale, emphasizing the significance of determining how the radiative heat transfer sits in the larger context of energy dynamics of thermal surroundings (Chen et al., 2017) (Figure 2.9).

2.4.1. Conduction and Its Significance

Conduction is an elementary process in energy transfer that facilitates the thermal energy to move from one substance to another. Conduction is the method by which the heat energy migrates from one particle to another by direct contact without the bulk material moving. This brief contact is highly significant in solids, where particles are closely packed, thus able to transfer energy effectively. Material properties such as thermal conductivity, which indicates the ability of a substance to conduct heat, also determine efficiency in conduction. As seen in most numerical analyses, temperature-dependent viscosity and thermal conductivity are inherent parameters in enhancing heat transfer, particularly for spot-heated systems, where their behavior varies prominently (2nd Micro and Nano Flows Conference (MNF2009) et al., 2009). The implications of conduction extend beyond the realm of pure thermal transfer; it is critical in most real-world and industrial applications. For instance, in microfluidic devices, the understanding of conduction can be applied to design improved heat exchangers and cooling elements. The coupling between convection and conduction can play a vital role in system efficiency in thermal management, particularly in low Reynolds number flows where temperature has to be controlled very precisely. Experiments have established that interactions between velocity fields and temperature structures become increasingly sophisticated with varying thermal conditions, indicating the importance of conduction within complex fluid dynamics systems (Anderson et al., 2006). This sophistication in the treatment of thermal surroundings justifies the importance of conduction for engineering applications aimed at energy efficiency. Conduction is also important in astrophysical phenomena, such as accretion near black holes. The ability of thermal conduction to facilitate lateral energy transport in accretion disks corroborates the dynamics necessary for the formation of outflows, significant for research into the evolution of galaxies. The use of such a phenomenon of conduction in astrophysics goes further to define its significance over traditional levels of heat transfer. While studies persist in exploring the intricate behaviors of thermal energy within various settings—industrial systems to cosmic structures—conductive heat transfer remains a top area of study, being the foundation behind many of the forces that function both terrestrial and extraterrestrial energy processes (Anderson et al., 2006).

2.4.2. Convection and Fluid Dynamics

Convection and fluid dynamics need to be explained when studying energy transfer, particularly in the scenarios of heat and work. Convection is the process of heat transfer by means of the motion of fluids, which can be either liquid or gas. Essentially, when a fluid gets heated, its density decreases so that it expands and rises, while the heavier, cooler fluid descends to take its place. This kind of process creates a convective current, and this optimizes the flow of thermal energy in the fluid. In addition, the effectiveness of convection depends on a variety of factors such as fluid properties of viscosity, temperature, and the transfer medium of heat. Therefore, models can be formulated to capture the complexity of such interactions, which in turn helps in the effective design of heat systems and energy-saving technologies based on such principles, thereby impacting industries ranging from HVAC to renewable energy systems. The study of computational methods applied, such as the Finite Element Method (FEM), allows for a greater understanding of fluid heat transfer. The numerical solution can deal with complex geometries and boundary conditions that often occur in real-world applications. For instance, renewable energy application, solar collectors have become more prominent due to rising energy demands and climatic concerns regarding the utilization of fossil fuels. As quoted by Allegrini et al. (2016), such systems rely on a profound understanding of thermodynamic theory, which may be efficiently modeled using FEM. Designers are able to optimize designs so as to maximize heat absorption and minimize losses by simulating the thermal performance of such prototypes as evacuated tubes. Such innovations are paramount in achieving the utmost energy efficiency and meeting global sustainability initiatives committed to promoting cleaner sources of energy. Further, the use of new technology in applying fluid dynamics has facilitated extensive modeling of various phenomena, including wildfire and its spread. Computational techniques have been employed in experiments to investigate fluid dynamics in the context of combustion and heat transfer, which are essential in understanding fire behavior across landscapes, as observed by Albini et al. (2007). The understanding of heat transfer dynamics is crucial in the prevention and management of wildland fires, wherein convection plays a significant role in the spread of flames. It illustrates the broader applications of fluid dynamics for both energy transport and environmental protection, revealing the complementarity of these concepts. As ongoing research and modeling efforts progress, the applications of convection and fluid dynamics will continue to reshape our approach to energy management and environmental conservation (Figure 2.10).

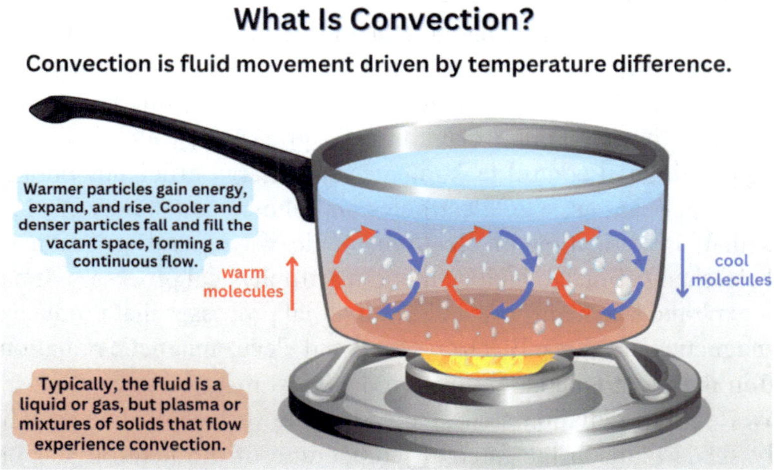

Figure 2.10. *Convection.*

Source: https://sciencenotes.org/wp-content/uploads/2024/11/What-Is-Convection-1024x683.png.

2.4.3. Radiation and Electromagnetic Waves

Understanding the intricacies of radiation and electromagnetic waves is crucial to grasping their role in energy transfer and thermal dynamics. Radiation produces a broad spectrum of electromagnetic waves ranging from radio waves to gamma rays, each with specific properties and applications. Among these, thermal radiation is a fundamental concept in that objects emit energy proportional to their temperature. This energy transfer occurs via electromagnetic waves, allowing for the process of heat exchange within any given system, either in everyday life like when touching the sun or in advanced technology like infrared thermography. The quantifiable nature of this radiation is what allows scientists to develop models that can predict energy distributions and patterns of radiation within different settings, allowing for a greater comprehension of phenomena such as blackbody radiation and its uses in astrophysical applications. The interaction between electromagnetic radiation and materials is a complex phenomenon that is a cornerstone in material science applications through to quantum mechanics. When electromagnetic waves interact with a medium, the process can lead to absorption, reflection, or transmission and ultimately decide the mechanisms of energy transfer involved. This relation is particularly relevant in the analysis of near-field effects, where the proximity of a probe particle, for example in scanning optical near-field microscopy, actually alters the nearby electromagnetic conditions significantly. In such situations, the calculated EM-LDOS becomes central, providing insight into material characterization and facilitating improved understanding of thermal conduction. As described in recent research, this proximity effect provides a near investigation of heat transfer processes, which is important in the design of thermal control technologies and energy efficiency optimization for various applications, e.g., electronics and photonics (Babuty et al., 2013). Besides that, investigation of electromagnetic waves and radiation extends beyond classical notions to include modern astrophysical processes. It has been recently explored that neutron stars have internal processes that produce certain electromagnetic emissions. Electron sound and electromagnetic radiation being coupled in these extreme conditions produce fast magnetosonic waves, which are converted into radiation as they travel to the surface of the star. This link not only sheds light on the internal temperature of the neutron star but also suggests a dominant mechanism of heat transfer through vortex radiation rather than normal thermal conduction. These results bring a new insight into neutron stars energetic emissions, making us better understand superdense matter and enabling improved models for stellar behavior and evolution considerations of energy loss as well as heat dissipation (Amundsen et al., 2002).

2.5. THE LAWS OF THERMODYNAMICS

The Laws of Thermodynamics are basic principles that address energy exchanges, work, and heat in physical systems. The first law, which is also referred to as the Law of Energy Conservation, asserts that energy cannot be created or destroyed, but can only transform from one form to another. This law lays the ground for understanding how energy is transferred in systems and indicates the connection between heat and work. By asserting that all types of energy can be transformed into all other types, the first law supports the postulate that the sum of energy in an isolated system remains constant with time, in close accord with empirical findings in classical and contemporary physics. It forms a foundation for the investigation of numerous physical and chemical phenomena, largely acting as an intermediary between theoretical principles and practical applications in many fields, such as engineering and environmental science. The second law of thermodynamics introduces the concept of entropy, a measure that describes a system's measure of disorder or randomness. According to it, in any energy transfer or transformation, the isolated system's total entropy will never be lesser, i.e., natural processes move in the direction of a state of greater disorder. This legislation has profound implications, particularly in determining the efficiency of energy conversions, as it illustrates that energy is always dissipated as heat and cannot be fully converted into work. The second legislation thus points out the limits of irreversibility in processes to shape the idea of efficiency in engines and technological systems. By combining the theory of thermodynamics with concepts of entropy and disorder, scientists can design better systems while still realizing the natural limitations dictated by the rules of nature. The third law of thermodynamics also states that if a system is brought close to absolute zero, the entropy of an ideal crystal approaches a minimum constant value. This legislation references the systems' behavior at the quantum scale and carries significant implications for low-temperature and cryogenics investigations of material properties. When cooled, systems experience reduced molecular motion, causing peculiar behavior that cannot be explained classically by thermodynamic principles. Particularly, the third law is a foundation pillar in statistical mechanics, where it provides insight into the microstates of a system. The exploration of such laws in conjunction with emerging theories in statistical analysis of thermodynamic systems, such as the model postulated by Das et al. (2013) and Gay-Balmaz et al. (2018), warrants the ongoing use of thermodynamic principles to broad aspects of scientific inquiry, especially in analyzing intricate systems and their behaviors under various conditions.

2.5.1. First Law: Conservation of Energy

The First Law of Thermodynamics, or the Conservation of Energy Law, dictates that energy cannot be created or destroyed but can only be transformed from one form to another. This law underpins the underlying physics of energy systems such that total energy in a closed system remains constant. For instance, in mechanical systems, potential energy is converted to kinetic energy and vice versa. This change is important in most applications, such as roller coasters or pendulums, where the total mechanical energy is preserved during the movement, provided that there is no energy loss by friction or air drag. The simplification of energy conversion processes by this law enables scientists and engineers to design systems maximizing energy usage and minimizing energy waste, hence advancing technology in renewable energy and mechanical engineering. In higher-level energy interactions, such as those for quantum systems, the First Law's applications find themselves in the realms of heat and work. As shown in quantum coherent electron systems, energy transactions are intimately associated with the heat fluxes that pass through different components. The treatment of such systems reveals that energy conversion must take various factors into account, including interaction with numerous thermal reservoirs at finite temperatures. Doing this, it is revealed that the First Law of Thermodynamics is not a stationary concept but a dynamic theory which holds true at each instant of time (Arrachea et al., 2016). This echoes the importance of understanding the nuances of energy conservation in contemporary quantum mechanics, where classical macroscopic principles assume new meanings due to quantum interactions and behaviors. Aside from classical and quantum mechanics, the First Law of Thermodynamics is the basis for conceptualizing the interplay among work, heat, and entropy. Heat transfer and work in a system can be quantified while retaining the idea of conservation of energy.

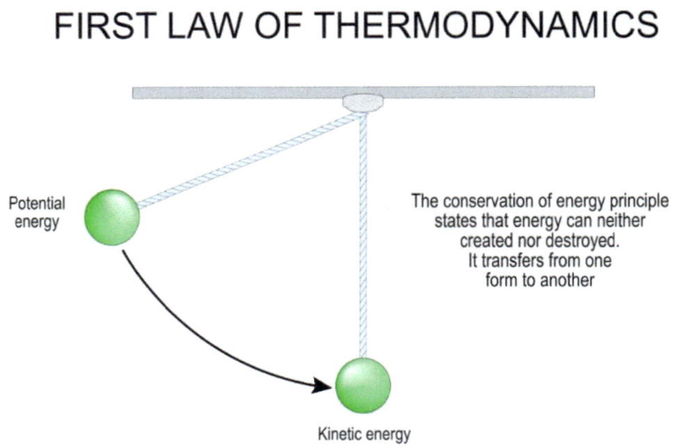

Figure 2.11. *First law: Conservation of energy.*

Source: https://images.shiksha.com/mediadata/images/articles/1747299507phpPTJAdn.jpeg.

In bipartite quantum systems, heat and work interactions become intricate, with some constraints being placed by the local Hamiltonians and their commutative relations. Formalism derived in recent work demonstrates how work is connected to energy variations that do not affect local entropy, thereby being in agreement with principles of thermodynamics (Hossein-Nejad et al., 2015). By reconciliation of such factors, scientists add rich meaning to the recalcitrance of the problem of complexity of energy interaction in non-equilibrium situations, reflecting the ongoing utility and vitality of the First Law in explaining current physical phenomena (Figure 2.11).

2.5.2. Second Law: Entropy and Energy Quality

The Second Law of Thermodynamics brings into play vital concepts of energy quality and entropy that deeply influence understanding energy conversion and transfer processes. Entropy is traditionally defined as a measure of disorder in a system; however, its significance is far more than chaos. As a factor for thermal energy capacity, entropy accounts for constraints on energy available for work. When energy is converted from one form to another, particularly from more ordered forms to more disordered forms, the quality of such energy diminishes. The Second Law thus indicates this diminution of quality in the usefully available side of energy. Thus, entropy not only forms a part of pillars of principles of thermodynamics but also forms a guide framework to evaluate the efficiency of energy systems and thus confirms the necessity to consider entropy while framing concepts of sustainability and resource management (Barnes et al., 2000). The dual dependency of free energy on entropy is important in understanding the working capabilities of thermodynamic systems. Free energy, as it is defined in thermodynamic environments, refers to the amount of available energy deployed for work after suitable adjustment has been made for the energy dissipated to entropy. The distinction between common free energy and free energy of concentration further explains this relationship, which is that not all energy is available to perform work. In biological systems, for instance, free energy facilitates metabolic reactions in that it is transferred and stored in biochemical pathways. This connection between entropy and free energy demonstrates how the Second Law prescribes the boundaries on which energy conversions are permissible, thereby prescribing the workability and efficiency of energy systems under real conditions (Barnes et al., 2000). Beyond thermodynamic fundamentals, enhancing interactions between thermoelectric systems and their surroundings also sheds light on practical uses of the Second Law. The efficiency of thermoelectric generators, for instance, strictly relies on thermal and electrical impedance matching, and this is similarly dependent on energy quality. Through the study of this interaction from the perspective of linear nonequilibrium thermodynamics, researchers reveal the complexity behind energy transfer processes. These findings stress the significance of careful control of thermal resistance and heat transfer in attaining higher efficiency in thermoelectric devices. They further highlight the overriding role of entropy in the system, which compels engineers and designers to consider entropy's impact while seeking optimal performance in energy conversion technology (Figure 2.12).

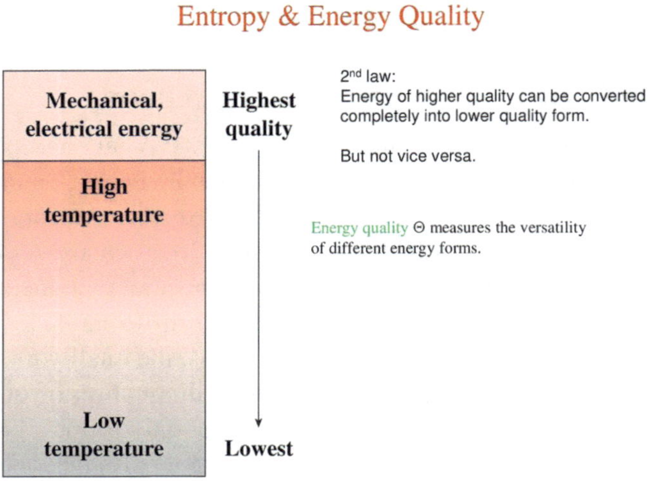

Figure 2.12. *Entropy and energy quality.*

Source: https://slideplayer.com/14613517/90/images/slide_1.jpg.

2.5.3. Third Law: Absolute Zero and Its Implications

The third law of thermodynamics that the entropy of a system will reach a minimum value when the temperature of a system comes closer to absolute zero has significant implications on numerous fields of science and engineering. This law asserts that it is impossible to cool a system to absolute zero by any finite series of processes, thereby imposing a theoretical limit on the efficiency of a heat engine. Consequently, systems at temperatures close to this temperature have peculiar properties, such as less molecular motion and increased order, that can have significant influences on thermal properties. The fact that absolute zero is a point at which entropy is at minimum highlights its importance to explain the behavior of matter in extreme conditions, giving core knowledge to fields like low-temperature physics and material science (Klimenko et al., 2012). The third law of thermodynamics finds its applications in practical problems in many engineering problems, particularly in thermal system problems and the development of efficient heat engines. The reality that absolute zero is impossible to attain modifies how engineers approach designing energy systems because it informs them what parameters need to be minimized to minimize energy loss and maximize work output. In the case of heat engines, for instance, temperatures play a crucial role in calculating the efficiency, as explained in the equation $\eta=1-T\_l/T\_h$ that looks very much like concepts in statistical mechanics. These laws not only influence the design of modern engines but also the production of materials that can withstand highly elevated temperatures, thus promoting technological innovation and energy efficiency (Arnaud et al., 2008). Other research into the third law indicates its influence on the nature of negative temperatures and systems with unusual thermodynamic characteristics. Negative temperatures are found in highly controlled systems and exhibit counterintuitive behavior according to some research. These systems, and principles relevant to ideas of negative heat capacity, contradict classical ideas of energy distribution and thermodynamics. The existence of these states provides useful paradigms for the explanation of complex thermal interaction and underlines the applicability of entropy outside conventional definitions. Through the examination of such properties under the third law, not only does one have greater understanding of thermal dynamics, but new uses for quantum mechanics and information theory are also found, thereby bridging gaps between remote scientific disciplines (Klimenko et al., 2012).

2.6. PRACTICAL APPLICATIONS OF ENERGY, WORK, AND HEAT

The experimental applications of work, energy, and heat are very much applicable in most industrial situations, especially when dealing with the operation of thermal systems. For instance, complex heat exchangers are required in achieving maximum energy transfers and losses. Microchannel heat exchangers provide one of the finest illustrations of innovative thinking in the field of thermal control since novel designs facilitate cost-effective fabrication without reducing effectiveness. Brandemuehl et al. (2012) point out that such designs not only allow for various configurations—counter-flow, cross-flow, or parallel-flow—but also hold promise in other materials, including polymers and metals. Such flexibility in design and material provides engineers tools to enhance energy efficiency in systems requiring precise heat control, ultimately creating enormous energy savings on a huge scale. In food processing, the utilitarian feature of energy analysis and modeling plays a key role in guaranteeing improved operating efficiency. Industrial fry operations, for example, are prone to being plagued by inefficiencies that lead to wasteful use of energy. The use of energy balance models as a corrective feature operates by enabling manufacturers to identify energy flows and losses. An overview of past research and experimental findings, as documented in Jouhara et al. (2010), calls for the efficacy of such analyses in optimizing frying processes without compromising product quality. By evaluating and enhancing operating parameter interactions, manufacturers can effectively minimize energy wastage, thereby facilitating sustainability while maintaining throughput. Not only is this environmentally friendly but also lowers operating costs, thereby making the food manufacturing sector more competitive. Moreover, implementation of energy-saving technologies has the potential to create revolutionary impacts in domestic settings. Presence of smart home systems revolutionized the dynamics between energy, labor, and heat in domestic settings. Smart home systems implement sensors and smart meters to optimize energy consumption by adjusting heating, cooling, and other energy demands based on real-time data. With the automation of energy management, homeowners are able to minimize losses considerably while optimizing efficiency. Moreover, as more and more citizens learn about their carbon footprint, the adoption of such technologies becomes in line with greater sustainability goals. Per a variety of studies, such as those with findings similar to those presented in Jouhara et al. (2010), the application of energy efficient designs and smart technology has the potential to bring about collective change toward living in a more sustainable way, highlighting the critical role energy principles play within everyday life.

2.6.1. Energy Efficiency in Machines

In today's situation of energy management, the efficiency of equipment has become a top priority for minimizing the usage of energy and reducing environmental impact. Energy efficiency in equipment is the ability to perform useful work using input energy with minimal loss. Efficiency is predetermined based on factors including design, material, and operating conditions. For the case of electrical machines, developments in thermal analysis have been key contributors to the enhancement of energy efficiency. Techniques such as lumped-parameter thermal networks, finite-element analysis, and computational fluid dynamics enable designers to reduce machine design while subsequently improving thermal performance and reducing energy losses during operation. Knowing these thermal dynamics is crucial for both the efficiency improvements and the lifespan, part of an enhanced energy future as has been noted in research on the development of thermal analysis methods (Boglietti et al., 2009). The addition of exergetic analysis to machine efficiency assessment also adds another layer of understanding to energy consumption in many systems. Through the focus on energy quality and the identification of irreversibilities in machines, the additional awareness of energy efficiency can be achieved. For systems like nuclear reactors with high energy, the identification of exergy destruction is critical in leading improvement of performance. As indicated by the related research studies, analysis of the nuclear steam supply system revealed that accurate fission temperature calculations greatly enhance the estimation of exergetic efficiency. This approach encourages designers and engineers to look for inefficiencies and strive to optimize operation performance, ensuring that the machines not only provide energy requirements but do so using maximum efficiency and minimal waste (Ferroni et al., 2018). Also, the consequences of energy efficiency are not solely based on mere mechanical performance but also extend to economic and environmental implications. Efficiency at high levels in machines translates into significant savings and reduced greenhouse emissions, and the argument for spending on the newest technology and best practice is all the more compelling. As industries become more concerned about sustainability, demand for energy-efficient equipment can only increase, promoting innovations that bring together technological advancement and protection of the environment. By establishing an environment in which machines work in an efficient manner, industries can contribute to a more sustainable future and enhance competitiveness in a more dynamic marketplace. Thus, energy efficiency in machinery is not only a technical necessity but also a necessity for the global climate change mitigation strategy as a whole (Figure 2.13).

Figure 2.13. *Energy efficiency in machines.*

Source: https://img.oemmagazine.org/files/base/pmmi/all/image/2022/03/Fluke_EnergyMonitoring.6228c0fc1ca9d.png?auto=format%2Ccompress&fit=max&q=70&w=1200.

2.6.2. Renewable Energy Sources and Their Impact

Transition to clean energy is vital in climate change and the sustainability of the future. As the world's cities embrace cleaner energy technologies, the seismic implications for urban energy demands are reeling. For instance, Boston's transition to clean energy will purportedly transform its fuel and electricity needs. By 2050, although demand for electricity is expected to rise, the reliance on petroleum fuels and natural gas is expected to drop significantly. This transition requires city planners to integrate practices that not only restrict the utilization of fossil fuel but also guarantee that the transition facilitates economic development and social equity. Carbon neutrality can be realized through the use of advanced technologies and energy efficiency in buildings—approaches that support the general objectives of environmental sustainability as well as improved urban quality of life (Cleveland et al., 2019). In addition to local interventions, the global context of renewable energy emphasizes the importance of cross-country technology diffusion. Patent records in most countries indicate that recipient countries' capability is a factor in climate-friendly technology adoption. This indicates that preparedness in the economic and technological spheres significantly decides the successful entry of renewable energy ventures. To everyone's shock, foreign direct investment barriers may sometimes facilitate technology transfer. Such dynamics indicate that there needs to be a multidimensional policy, such as economic policies and investment promotions, in order to stimulate the mass diffusion of renewable energy innovations. By strategic alliances and reduction of trade barriers, countries are able to better leverage technology breakthroughs for their energy sectors, thereby accelerating the transition to the low-carbon economy (Dechezlepretre et al., 2009). The impacts of renewable energy extend beyond energy use; they also create economic activity and generate jobs in emerging industries. The construction and maintenance of renewable energy facilities require expert staff, thus driving employment in the clean energy economy. In addition, as individuals invest in technologies such as wind turbines and solar panels, local economies are able to save money on their utility bills and become more energy independent. Such localized energy generation is not only environmentally sustainable but also socially equitable because it brings the access to low-cost energy within reach of marginalized groups. Finally, energy conversion to renewable energy sources guarantees a net improvement in community resilience and sustainability, demonstrating the interdependence of heat, work, and energy in modern environments.

2.6.3. Heat Engines and Thermodynamic Cycles

The field of thermodynamics underpins the science of heat engines, which are required to transform thermal energy into mechanical work. Thermodynamic cycles, such as the Rankine and Otto cycles, are at the core of this process that describes how energy is transformed and utilized by various engines. Heat engines operate through the transfer of heat from a hot reservoir, doing work, and transferring some energy to a cold reservoir. The performance of such engines is dependent on this cycle process, illustrating the energy input vs. conversion vs. output play. Also, with technology advancement, special systems like the Organic Rankine Cycle (ORC) are now available with innovative solutions for low-to-medium-grade thermal-energy conversion. ORC technology becomes increasingly applied to practical applications, converting geothermal and other heat into useful work, eventually making energy more efficient across a broad spectrum of industries (Markides et al., 2015). A closer examination of heat engines reveals more broad varieties of thermodynamic cycles that are each optimized for different types of energy sources and working conditions. For instance, although the Rankine cycle is the foundation of conventional steam engines, modern advances have included other cycles such as the Otto and Diesel cycles used primarily in internal combustion engines.

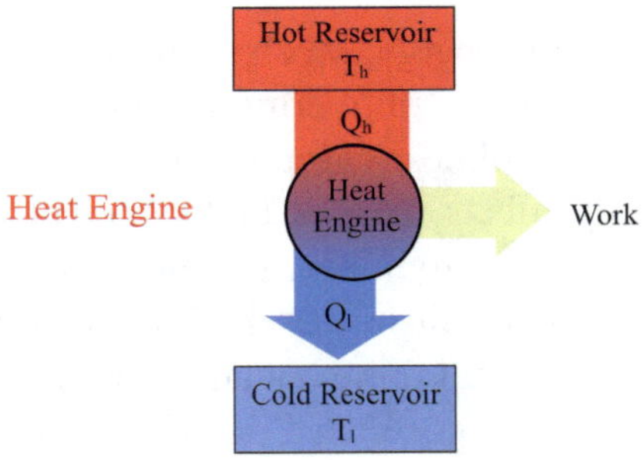

Figure 2.14. *Heat engines and thermodynamic cycles.*

Source: https://external-content.duckduckgo.com/iu/?u=https%3A%2F%2Fwww.eigenplus.com%2Fwp-content%2Fuploads%2F2022%2F12%2Fcarnot_cycle_carnot_engine.jpg&f=1&nofb=1&ipt=940d2df8128feab37bde7e98648b864bbd98d4a3e213b01087f4a38a47e6f82f.

All these cycles handle particular parameters such as pressure, volume, and temperature in order to maximize energy from fuel. Furthermore, advances in nano heat engines are a clear example of further advancement in the field where quantum resources are an important basis. For instance, recent research reveals that the performance of single-ion engines going through quantum Otto cycles closely matches the Curzon-Ahlborn efficiency, the connection between classical thermodynamic theory and quantum mechanics (Deffner et al., 2018). This convergence between disciplines points towards a bright new future for the development of heat engines. The implications for future practical applications of better heat engine technology are immense, in particular concerning the sustainability of energy and environmental considerations. Through optimizing thermodynamic cycles, engineers can optimize the performance of heat engines to gain higher efficiencies and thereby reduce fuel consumption as well as associated emissions. Technologies like the Organic Rankine Cycle that can recover low-temperature waste heat for useful energy are the pinnacle of this movement towards efficient energy systems. In addition, the flexibility of ORC technology for renewable energy sources such as solar energy facilitates its use in seeking alternatives for sustainable energy. Further development aims not only to meet traditional efficiency limitations but to evolve to respond to shifting energy needs through the integration of renewable energies. Along with emerging quantum technologies, the future of heat engines is one of greater functionality and sustainability in energy production (Markides et al., 2015) (Figure 2.14).

SUMMARY

The interrelationship among energy, work, and heat lies at the heart of an understanding of the principles of thermodynamics and their applications across various scientific disciplines. Through an examination of such things as entropic forces and gravity effects, it can be understood just how complex energy conversion is. As an instance, arguments pertaining to the entropic origin of gravity introduce a paradigm that dismisses traditional thought by proposing that gravity may not be a fundamental force but an emergent property derived from thermodynamic principles. The postulated theories in current research indicate the limitations of correlating gravitational attraction with entropic forces and proclaiming that gravity exists ultimately because of the geometric nature of spacetime, thus maintaining the inherent correlation between energy processes and gravitational phenomena (Shan et al., 2010). Furthermore, the investigation of heat transfer mechanisms, particularly in mixed convection, emphasizes the intricacies of fluid behavior under diverse temperature regimes. Experiments by numerical methods, such as those which take into account thermocapillary forces and buoyancy-induced forces, exhibit richness of flow structures that dictate the efficiency of heat transfer. Development of stationary longitudinal convective rolls, for example, illustrates the complex interaction of a number of variables, such as Reynolds and Rayleigh numbers, which dictate flow stability and energy transport in a system. These findings are vital to the creation of efficient thermal systems and understanding natural processes, showing the methods by which thermodynamic principles can be applied to both engineering devices and natural processes (Bammou et al., 2012). Lastly, the integration of energy, work, and heat not only increases theoretical knowledge but also enhances practical applications in all areas, from engineering to environmental sciences. The knowledge acquired from both the ancient thermodynamic models and modern interpretations, such as the emergent theory of gravity and advanced computational studies of heat transfer, underlines the need for an integrated view of energy systems. With advancing research, the debate surrounding these ideas is poised to yield innovative solutions to existing problems, from energy sustainability to climate change adaptability. Through the interconnection of the subject presented in this study, society can better understand the role of energy and how it affects our world as well as our prospective technology.

The connection between energy, work, and heat is a basic principle that serves the foundation of numerous scientific and engineering disciplines. Energy can be considered as the capability to do work, and its transfer is a physical concept that serves the basis for the operation of physical systems. Work is performed in mechanics when force is applied over a distance, causing a direct change in the energy state of an object. Heat is energy transferred due to temperature differences between systems, as defined by thermodynamics. Understanding these interactions allows scientists and engineers to optimize processes and systems, ranging from mechanical motors to biological processes. For example, knowing that the energy conversion efficiency determines how much work is produced can lead to developments in energy conservation and system design in life in general and hence the importance of mastering these concepts in real life. The law of energy conservation is important in studying work and heat. The law says that energy cannot be created or destroyed but can be converted. This basic principle has profound implications for closed and open systems. For instance, in a closed system, the conservation of energy informs us regarding the processes involved whenever work is performed or heat is conveyed, hence enabling one to examine system dynamics to the fullest extent. Additionally, the magnitude of the potential of this principle is in areas such as mechanical engineering and environmental science, where efficiency in the use of energy matters. As scientists continue to work on creating energy transformations, the necessity to adhere to this principle becomes increasingly apparent, particularly in the context of renewable resources and sustainable forms of energy production and usage. New technology continues to push and recontextualize paradigms of energy, work, and heat, with potential for revolutionary breakthroughs. As an illustration, space-based power stations have the potential to revolutionize long-distance delivery of energy. The technologies are founded on the understanding and utilization of energy transmission principles in effective setups. Research has alluded to energy beaming from space as having potential economic benefits, while still insisting on the necessity of advanced systems such as lasers and receivers for efficiency. As outlined in the studies, complexities involved with implementing such technologies within existing paradigms highlight the need to learn basic principles of energy transfer and propulsion methods (Garrett et al., 2025). The ongoing investigation across these fields reflects the dynamic nature of energy research, constantly updating our understanding of work, heat, and how they contribute to next-generation technological innovation.

REFERENCES

1. Adkins, C. J., Ben-Naim, A., Bersuker, I. B., Brissaud, A., Bustingorry, S., Caves, C. M., de Vega, H. J., et al. (2010). Many faces of entropy or Bayesian statistical mechanics. http://arxiv.org/abs/1007.1773 (accessed on 4th October 2025).
2. Aguilera, D. N., Baldo, M., Baym, G., Blaschke, D., Drago, A., Fernández, R., Freire, P. C. C., et al. (2009). Signature of deconfinement with spin down compression in cooling hybrid stars. http://arxiv.org/abs/0801.0358 (accessed on 4th October 2025).
3. Aharoni, A., Awschalom, D. D., Baxter, R. J., Bazaliy, Y. B., Bertotti, G., Blatt, F. J., Brown, W. F., et al. (2006). Spin-transfer in an open ferromagnetic layer: From negative damping to effective temperature. http://arxiv.org/abs/cond-mat/0610264 (accessed on 4th October 2025).
4. Albini, F. A., Anderson, H. E., Sullivan, A. L., et al. (2007). A review of wildland fire spread modelling, 1990–present. 1: Physical and quasi-physical models. http://arxiv.org/abs/0706.3074 (accessed on 4th October 2025).
5. Allegrini, E., Micangeli, A., Michelangeli, E., Ricci, R., et al. (2016). Heat transfer simulation of evacuated tube collectors (ETC): An application to a prototype. https://core.ac.uk/download/154947586.pdf (accessed on 4th October 2025).
6. Amundsen, L., Svidzinsky, A. A., Becker, W., Blandford, R. D., Golden, A., Gottfried, K., et al. (2002). Radiation of neutron stars produced by superfluid core. https://core.ac.uk/download/25365302.pdf (accessed on 4th October 2025).
7. Anderson, M., Cho, J., Falcke, H., Honma, F., Menou, K., McKinney, J. C., et al. (2006). Hot accretion with conduction: Spontaneous thermal outflows. https://core.ac.uk/download/44108022.pdf (accessed on 4th October 2025).
8. Anxionnaz-Minvielle, Z., Cabassud, M., Couturier, R., Gourdon, C., et al. (2016). Implementation of ‘chaotic’ advection for viscous fluids in heat exchanger/reactors. https://core.ac.uk/download/83534070.pdf (accessed on 4th October 2025).
9. Apertet, Y., Goupil, C., Lecoeur, P., Ouerdane, H., et al. (2015). Continuity and boundary conditions in thermodynamics: From Carnot’s efficiency to efficiencies at maximum power. http://arxiv.org/abs/1411.4230 (accessed on 4th October 2025).
10. Arce, J., & Betancourt, R. (2001). The design, implementation, and assessment of a new capstone course aimed at science education majors.

https://core.ac.uk/download/215487647.pdf (accessed on 4th October 2025).

11. Arnaud, J., Chusseau, L., & Philippe, F. (2008). A simple model for Carnot heat engines. http://arxiv.org/abs/0812.2757 (accessed on 4th October 2025).
12. Arrachea, L., Ludovico, M. F., Moskalets, M., Sanchez, D., et al. (2016). Dynamics of energy transport and entropy production in AC-driven quantum electron systems. https://core.ac.uk/download/132611543.pdf (accessed on 4th October 2025).
13. Babul, A., Binney, J., Bondi, H., Bower, R. G., Edge, A. C., Fabian, A. C., McCarthy, I. G., et al. (2008). Towards a holistic view of the heating and cooling of the intracluster medium. http://arxiv.org/abs/0706.2768 (accessed on 4th October 2025).
14. Babuty, A., Ben-Abdallah, P., Chapuis, P. O., De Wilde, Y., et al. (2013). Strong tip-sample coupling in thermal radiation scanning tunneling microscopy. https://core.ac.uk/download/47095893.pdf (accessed on 4th October 2025).
15. Balksjö, T., de Jong, A., & Katzeff, C. (2013). Challenges in energy awareness: A Swedish case of heating consumption in households. https://core.ac.uk/download/11435470.pdf (accessed on 4th October 2025).
16. Bammou, L., Benhamou, B., Blancher, S., Guer, Y., et al. (2012). A numerical study of the longitudinal thermoconvective rolls in a mixed convection flow in a horizontal channel with a free surface. http://arxiv.org/abs/1206.1398 (accessed on 4th October 2025).
17. Barber, R. W., Emerson, D., Gu, X. J., Zhang, Y., et al. (2008). A thermal lattice Boltzmann model for micro/nano-flows. https://core.ac.uk/download/9019260.pdf (accessed on 4th October 2025).
18. Barnes, B., Benedicks, M., Bent, H. A., Born, M., Carnot, S., Clausius, R., et al. (2000). Inquiries into the nature of free energy and entropy in respect to biochemical thermodynamics. https://core.ac.uk/download/25283819.pdf (accessed on 4th October 2025).
19. Bejan, A., Ioffe, A., Andresen, B., Van Wees, B. J., Van den Broeck, C., et al. (2013). A linear nonequilibrium thermodynamics approach to optimization of thermoelectric devices. http://arxiv.org/abs/1308.6735 (accessed on 4th October 2025).
20. Boglietti, A., Cavagnino, A., Mejuto, C., Mueller, M., et al. (2009). Evolution and modern approaches for thermal analysis of electrical machines. https://core.ac.uk/download/pdf/11409050.pdf (accessed on 4th October 2025).

21. Bower, R. G., Cattaneo, A., Cooper, M. C., Dekel, A., Ostriker, J. P., Khochfar, S., Madau, P., et al. (2008). Adding environmental gas physics to the semi-analytic method for galaxy formation: Gravitational heating. http://arxiv.org/abs/0704.2418 (accessed on 4th October 2025).
22. Brandemuehl, M. J., Denkenberger, D. C., Pearce, J. M., Zhai, Y., et al. (2012). Expanded microchannel heat exchanger: Design, fabrication and preliminary experimental test. https://core.ac.uk/download/151508534.pdf (accessed on 4th October 2025).
23. Brataas, A., Einstein, A., Slachter, A., Sugihara, A., Kovalev, A. A., et al. (2011). Spin caloritronic. https://core.ac.uk/download/pdf/148239576.pdf (accessed on 4th October 2025).
24. Chen, Q., He, Y. L., Kang, Q. J., Li, X., et al. (2015). Lattice Boltzmann methods for multiphase flow and phase-change heat transfer. https://core.ac.uk/download/79517637.pdf (accessed on 4th October 2025).
25. Chen, S., Daughton, W., Haggerty, C. C., Matthaeus, W. H., et al. (2017). Energy transfer, pressure tensor and heating of kinetic plasma. http://arxiv.org/abs/1705.02054 (accessed on 4th October 2025).
26. Cioncolini, A., & Thome, J. R. (2014). Unified modeling suite for two-phase flow, convective boiling and condensation in macro- and micro-channels. In 4th Micro and Nano Flows Conference (MNF2014). https://core.ac.uk/download/29139824.pdf (accessed on 4th October 2025).
27. Cleveland, C. J., Fox-Penner, P., Pollack, A., Walsh, R., et al. (2019). Carbon Free Boston: Energy technical report. https://open.bu.edu/bitstream/2144/39216/2/CFB_Energy_Technical_Report_190514.pdf (accessed on 4th October 2025).
28. Curiel, E. (2014). Classical black holes are hot. https://core.ac.uk/download/33752611.pdf (accessed on 4th October 2025).
29. D'Abramo, G. (2015). "Closed-loop" analysis of a thermo-charged capacitor. http://arxiv.org/abs/1407.6627 (accessed on 4th October 2025).
30. Das, J., Seok, S. C., Hodge, S. E., & Vieland, V. J. (2013). Measurement of statistical evidence on an absolute scale following thermodynamic principles. https://core.ac.uk/download/18998984.pdf (accessed on 4th October 2025).
31. Dechezleprêtre, A., Glachant, M., & Menière, Y. (2009). What drives the international transfer of climate change mitigation technologies? Empirical evidence from patent data. https://core.ac.uk/download/220255.pdf (accessed on 4th October 2025).

32. Deffner, S. (2018). Efficiency of harmonic quantum Otto engines at maximal power. http://arxiv.org/abs/1810.09371 (accessed on 4th October 2025).

33. Esposito, M., Harbola, U., & Mukamel, S. (2009). Nonequilibrium fluctuations, fluctuation theorems, and counting statistics in quantum systems. https://core.ac.uk/download/18438534.pdf (accessed on 4th October 2025).

34. Evans, D. J., Ruelle, D., Cohen, E. G. D., Bonetto, F., et al. (2003). Note on nonequilibrium stationary states and entropy. http://arxiv.org/abs/cond-mat/0312306 (accessed on 4th October 2025).

35. Ferroni, L., & Natale, A. (2018). Exergy analysis of a PWR nuclear steam supply system–II part: A case study. https://core.ac.uk/download/188824902.pdf (accessed on 4th October 2025).

36. Fiolhais, M., & Güémez, J. (2013). Thermodynamics in rotating systems–analysis of selected examples. http://arxiv.org/abs/1404.1786 (accessed on 4th October 2025).

37. Fisher, D., Hand, J., Liesen, R., Wilson, T., et al. (2011). Screening of energy efficient technologies for industrial buildings' retrofit. https://core.ac.uk/download/9033399.pdf (accessed on 4th October 2025).

38. Gao, S. (2010). Why gravity is not an entropic force. https://core.ac.uk/download/11922773.pdf (accessed on 4th October 2025).

39. Garrett, L. B., & Holloway, P. F. (2025). Comparative analyses of space-to-space central power stations. https://core.ac.uk/download/pdf/42858285.pdf (accessed on 4th October 2025).

40. Gay-Balmaz, F., & Yoshimura, H. (2018). From Lagrangian mechanics to nonequilibrium thermodynamics: A variational perspective. http://arxiv.org/abs/1904.03738 (accessed on 4th October 2025).

41. Glavatskiy, K. S. (2015). Local equilibrium and the second law of thermodynamics for irreversible systems with thermodynamic inertia. https://core.ac.uk/download/43382568.pdf (accessed on 4th October 2025).

42. Gomez-Marin, A., Beck, C., Jarzynski, C., Andrieux, D., Jou, D., et al. (2007). Nonequilibrium steady state thermodynamics and fluctuations for stochastic systems. http://arxiv.org/abs/0708.2940 (accessed on 4th October 2025).

43. Gudowska-Nowak, E., Kuśmierz, Ł., & Rubi, J. M. (2014). Heat and work distributions for mixed Gauss–Cauchy process. http://arxiv.org/abs/1307.6392 (accessed on 4th October 2025).

44. Hong, T., Lee, S. H., Li, H., Sofos, S., et al. (2020). System-level key performance indicators for building performance evaluation. https://core.ac.uk/download/288432586.pdf (accessed on 4th October 2025).

45. Hossein-Nejad, H., O'Reilly, E. J., & Olaya-Castro, A. (2015). Work, heat and entropy production in bipartite quantum systems. https://core.ac.uk/download/79497964.pdf (accessed on 4th October 2025).

46. Jouhara, H., Karayiannis, T. G., Tassou, S. A., Wu, Y., et al. (2010). Analysis of energy use in crisp frying processes. https://core.ac.uk/download/2733850.pdf (accessed on 4th October 2025).

47. Kapp, S., Klein, P., Knierim, P., Kuhn, J., et al. (2018). Physics holo.lab learning experience: Using smartglasses for augmented reality lab work to foster the concepts of heat conduction. http://arxiv.org/abs/1711.05087 (accessed on 4th October 2025).

48. Klimenko, A. Y. (2012). Teaching the third law of thermodynamics. http://arxiv.org/abs/1208.4189 (accessed on 4th October 2025).

49. Liu, J. T., & Peng, X. F. (2009). Local heating effects on flow and heat transfer in microchannels. In 2nd Micro and Nano Flows Conference (MNF2009). https://core.ac.uk/download/9633452.pdf (accessed on 4th October 2025).

50. Lucarini, V. (2009). Thermodynamic efficiency and entropy production in the climate system. https://core.ac.uk/download/1443035.pdf (accessed on 4th October 2025).

51. Lynch, J., Lerner, L. S., Schwartz, M., Gross, P. R., Schwartz, R., & Goodenough, U. (2012). The State of State Science Standards 2012. https://core.ac.uk/download/71359935.pdf (accessed on 4th October 2025).

52. Markides, C. N. (2015). Low-concentration solar-power systems based on organic Rankine cycles for distributed-scale applications: Overview and further developments. https://core.ac.uk/download/77007564.pdf (accessed on 4th October 2025).

53. Maroney, O. J. E. (2007). The physical basis of the Gibbs–von Neumann entropy. https://core.ac.uk/download/44164588.pdf (accessed on 4th October 2025).

54. N/A. (2007). The Development of Student Conceptions of The Earth–Sun Relationship in An Inquiry-Based Curriculum. https://core.ac.uk/download/pdf/5207049.pdf (accessed on 4th October 2025).

CHAPTER 3

ENTHALPY CHANGES IN CHEMICAL REACTIONS

Contents

Introduction 83
3.1. The Concept of Enthalpy 88
3.2. Types of Enthalpy Changes 92
3.3. Measuring Enthalpy Changes 97
3.4. Factors Affecting Enthalpy Changes 101
3.5. Applications of Enthalpy Changes 106
3.6. Case Studies of Enthalpy Changes 111
Summary 115
References 118

INTRODUCTION

The concept of enthalpy is at the core of the explanation of chemical reactions because it embraces the total heat content of a system, both internal energy and pressure-volume work. Enthalpy changes during the progress of reactions provide essential details on the thermodynamic feasibility and amount of energy consumed or released as reactants are transformed to products. Such transformation occurs in various processes such as combustion, phase transformations, and sophisticated catalysis. For example, in reactions catalyzed by zeolites, studies have shown that structural understanding of zeolites inherently affects reaction kinetics, so parameters like Kads-H+ and k(int) must be examined with great care to clarify their functions towards overall enthalpy changes (Bell et al., 2018). Not only is this significant from a theoretical point of view but also of practical value as it allows chemists to streamline for favorable outcomes within industrial processes. In the study of enthalpy change, catalysts become a necessity, particularly as they accelerate reactions without long-term modification. Metal-organic frameworks (MOFs) are perfect catalysts as they are defined by their adjustable structures and active metal centers, which optimize molecular interactions. Investigation into the electronic structure of MOFs reveals the way protonation and metal cation substitution alter their catalytic function (Aziz et al., 2015).

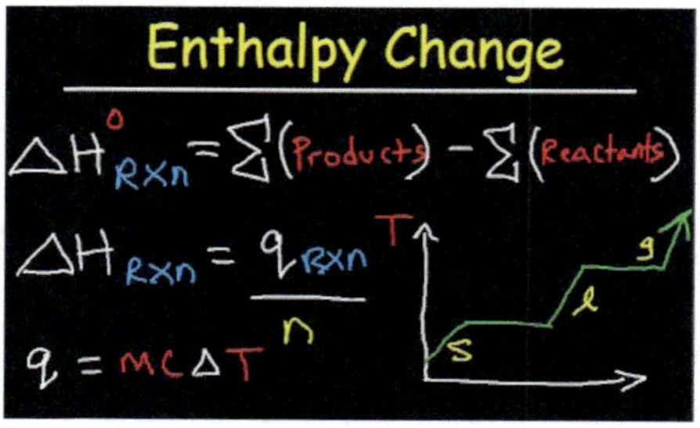

Figure 3.1. *Enthalpy change of reaction.*

Source: https://i.ytimg.com/vi/Xe_N7WYl5BI/hqdefault.jpg.

It is understanding these interactions that is so important for applying solar energy to fuel synthesis and related reactions, where precise enthalpic measurement is what enables functionality. This discovery of the energetic landscape of reactions made possible through MOFs profoundly advances the understanding of how enthalpy changes can be controlled to enable beneficial chemical transformation. The investigation of enthalpy changes during chemical reactions is a sophisticated endeavor that teeters on the precipice of both theoretical and practical facets of chemistry. The exploration of the relationship between reaction mechanisms and enthalpy changes enables scientists to predict the likelihood of spontaneous reactions and quantify the energy demand of endothermic and exothermic reactions. For instance, the energy processes of alkane cracking reactions influence efficiency in fuel production but also distill the broader ideals of sustainability and maximized resource utilization. By exploring how zeolite and MOF catalysts influence these energy transformations, we gain extremely pertinent information to enable more efficient synthetic pathways (Aziz et al., 2015; Bell et al., 2018). This research ultimately comes to energy technology advancements and highlights the need for constant research into enthalpy changes in an effort to foster innovation in chemical sciences (Figure 3.1).

Definition of Enthalpy

Enthalpy, a fundamental concept of thermodynamics, is the measure of the total amount of heat in a system at constant pressure and can be mathematically represented as $H = U + PV$, with U representing the internal energy, P being the pressure, and V being the volume of the system. This definition emphasizes the notion that enthalpy not only describes the internal energy of a substance but also accounts for the energy associated with variations in its pressure and volume. In chemical reactions, measurement of enthalpy change allows chemists to understand the thermal behavior of reactions, whether they are exothermic, giving out heat, or endothermic, absorbing heat. This information is crucial when predicting reaction behavior under various conditions, which applies in fields such as biochemistry and engineering, in which reactions can be controlled for a particular outcome under specific conditions. Entropy's significance is also evident when applied to predict the spontaneity and viability of chemical reactions. According to the laws of thermodynamics, a reaction will be spontaneous when it accompanies a decrease in enthalpy, leading to a more stable system. One common type of reactions is reactant conversion to products, which is typically accompanied by the release of energy due to product formation through the development of more stable bonds than their counterpart in the reactants, with a negative change in enthalpy. Study of enthalpy change through calorimetry gives precise values, enabling scientists to associate heat data with processes of reaction. Such thermodynamic principles directly correlate with current research in combustion, in which enthalpy has a direct bearing on the efficiency and viability of energy conversion processes, as observed in the evaluation of combustion theory (Marble et al., 1956). In more complex circumstances, enthalpy interactions also offer useful indications of molecular behavior. For instance, in organometallic complex research, the enthalpy of bond dissociation reactions gives insight into the stability and reactivity of chemical species. The calorimetric measurements of C–Au bonds highlight how reactions can be measured by examining heat evolved or absorbed during bond making and breaking. These very specific investigations not only contribute to research but also find application in practice, such as in catalyst engineering and the production of new materials. By understanding these changes in enthalpy, scientists can optimize conditions favoring desired reactions, illustrating the dual function of enthalpy as a theoretical and practical tool within chemistry (González et al., 2018).

Importance of Enthalpy in Chemistry

The use of enthalpy in chemistry has to do directly with the energy changes in chemical reactions. Enthalpy, which refers to the total heat of a system, is a significant parameter in quantifying the energy dynamics of reaction. By Hess's law, i.e., stating that the net enthalpy change in a reaction is a sum of single steps enthalpy changes, chemists can make predictions about the thermochemistry of reactions with precision. This even applies to predicting spontaneous processes, where decreasing enthalpy values indicate exothermic reactions releasing energy into the surroundings. Knowledge of these energy transfers is essential to many an application, from the optimization of industrial synthesis to increasing the efficiency of energy in chemical reactions.

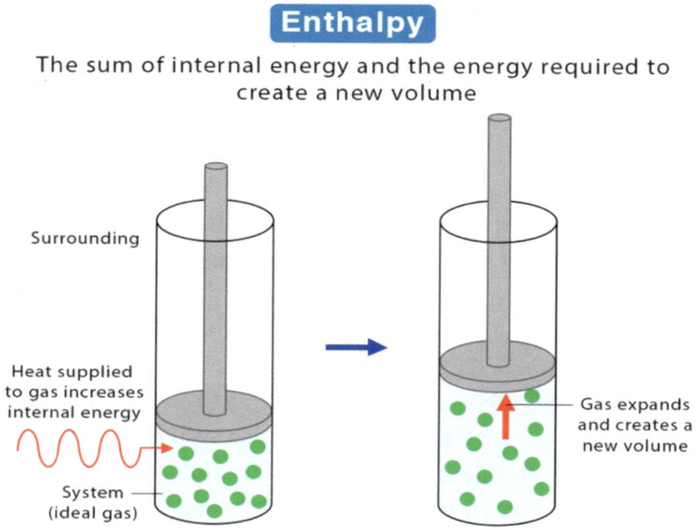

Figure 3.2. *Enthalpy in chemistry.*

Source: https://www.chemistrylearner.com/wp-content/uploads/2022/04/Enthalpy.jpg.

In addition, the concept of enthalpy has a direct correlation with entropy and dictates the overall spontaneity of reactions according to the second law of thermodynamics. The relationship of enthalpy and entropy supports the significance of enthalpy in the assessment of free energy changes that propel reaction processes. As Barnes et al. (2000) further elaborates, free energy serves as an indicator to gauge the feasibility of chemical reactions, highlighting the significance of enthalpy in biological processes and designed reactions too. The relationship becomes particularly evident in biochemical reactions, where enzymes catalyze reactions by lowering the activation energy as well as altering enthalpic contributions, eventually determining the direction and rate of metabolic processes. Another aspect of the importance of enthalpy is its use in creating supramolecular organization and phase transformation. Measurement of enthalpy changes during these processes is crucial to advancing knowledge in areas such as materials science and biochemistry. As explained in existing research, the study of enthalpy-entropy compensation sheds light on cooperative processes within supramolecular aggregates, especially with reference to water interaction at hydrophobic interfaces and conformational change in biological molecules. The referred micro-phase transitions outlined in literature highlight the way in which enthalpy is involved in poly-faceted interaction in chemical systems, which validates its significance in the understanding of complex reactions and material behavior (Nordén et al., 2012). Thus, a deeper insight into enthalpy not only augments real-life applications in chemistry but also enhances our theoretical understanding of molecular behavior (Figure 3.2).

Overview of Enthalpy Changes in Chemical Reactions

Understanding enthalpy changes in chemical reactions is crucial for describing reaction mechanisms and energy profiles. Enthalpy, defined as the heat content of a system under constant pressure, is a determining factor in whether a reaction is thermodynamically favorable or not. Where bonds are broken and made in a chemical reaction and also where heat is released or consumed, there is a change in enthalpy of the system. The enthalpy between the initial and final states of the reaction can be expressed as an endothermic process, where heat is added, or an exothermic process, where heat is released. This alteration provides chemists with data regarding reaction spontaneity and stability, enabling them to predict reaction conditions and potential products. Such data are the basis of numerous applications, ranging from industrial chemistry to environmental science, as it allows the reaction to be optimized at various conditions. Quantitatively, the enthalpy change can be expressed through the equation $\Delta H = H\_products - H\_reactants$, where ΔH is the enthalpy change. This equation is indicative of the necessity of precise thermodynamic measurements, which can aid in the design of thermochemical cycles. For instance, accurate enthalpy values can be used to assess reaction pathways and intermediates such that our understanding of dynamic processes in chemistry is significantly enhanced. Advanced techniques such as calorimetry enable chemists to determine directly the transferred heat while undergoing a chemical reaction, providing an experimental avenue in which to determine ΔH. Advances in computational chemistry also enable simulations and predictions of enthalpy change, making it more plausible to predict complex reaction mechanisms. As the field advances, these approaches call attention to the significance of accurate thermodynamic data in constructing meaningful predictive models for reaction performance, which leads to breakthroughs in materials science and sustainable chemistry. Incorporating enthalpy changes into chemical reactivity theories not only improves our theoretical description but also directs actual practice in environmental chemistry and biochemistry. Modeling of systems, for instance, the HPx multicomponent reactive transport model used in geochemical engineering, highlights enthalpy change integration with transport processes. Coupling of transport properties with geochemical state variables, for instance, solute uptake and transport processes, demonstrates the complexity involved in enthalpy changes in real applications (Jacques et al., 2018). Moreover, the enthalpy values derived from advanced experimental techniques allow the refinement of thermochemical networks, emphasizing their pivotal role in the examination of reaction energetics (Bodi et al., 2012). Overall, research into enthalpy changes of chemical reactions not only serves as a theoretical framework but also offers valuable insights concerning applied problem-solving in scientific pursuits.

3.1. THE CONCEPT OF ENTHALPY

The concept of enthalpy is relevant to the study of chemical reaction energy change, basically outlining the interaction between heat, pressure, and work. Enthalpy, H, is a measure of the internal energy of a system and the product of the pressure and volume ($H = U + PV$). This concept is essential as it may establish the transfer of energy without necessarily defining the nature of the process—isothermal or adiabatic. Changes in enthalpy are then pivotal in defining the absorption or release of energy as reactions occur, leading to complete knowledge of endothermic and exothermic processes. This understanding aids in describing the chemical conversion processes and is a foundation for thermodynamic calculations in numerous fields of study, including chemistry, physics, and engineering. Another instance of the significance of the change in enthalpy is in combustion reactions, where there is massive interaction between work and heat. Combustion, as much as it is a complex process, is simply a process of conversion of chemical energy into heat energy, and hence enthalpy is a significant variable to study. The application of enthalpy here not only helps in measuring energy efficiency but also helps in understanding the stability of reaction products. As researchers venture into the theory of combustion, it is clear that a reasonable correlation between combustion theories and practicality is a necessity. This information is constantly evolving, with theorists and researchers striving to relate theoretical models to improvements in the technology of combustion chamber operation and design, demonstrating enthalpy change knowledge as invaluable to further progress in this engineering field (Marble et al., 1956). Entropy changes are also directly related to the concept of exergy, which quantifies usable energy content of a system relative to its environment. While exergy considers all forms of energy, chemical exergy purposely includes enthalpic contributions that add a system's character to its workability. The interdependence of exergy and enthalpy compels the establishment of a standard chemical reference state, even if some scholars doubt convention necessity while approximating exergy losses in steady-flow processes. They contend that the use of a less complex expression for particular exergy accurately reflects the underlying efficiency of material streams in chemical reactions so that system performance may be more easily evaluated. This exchange provides greater insight into enthalpic behavior, thus enriching the conceptual basis for enthalpy changes in chemical reactions and their application in various scientific fields (Siemons et al., 1986).

3.1.1. Explanation of Enthalpy as a Thermodynamic Quantity

Enthalpy, one of the fundamental thermodynamic parameters, is an important indicator of the heat content of a system. It is defined as the sum of the system's internal energy and the product of its volume and pressure, expressed mathematically as $H = U + PV$. This definition highlights that enthalpy both encompasses energy to create a system at constant pressure and temperature as well as work to create space for it. In the case of chemical reactions, enthalpy change (ΔH) refers to the heat supplied or released when reactants are transformed into products under conditions of constant pressure. Understanding enthalpy is critical since it forms the basis for the differentiation between exothermic and endothermic processes, thus explaining how energy is transferred in chemical change. Enthalpy is more significant than mere examination of heat exchange; it has a close relationship with the general principles of thermodynamics, especially with free energy and entropy. Through the observation of how enthalpy handles these concepts, researchers are in a better position to grasp the nature of energy transformation in chemical reactions. For example, the equilibrium between free energy and enthalpy points out that all heat changes do not have the potential to generate useful work, as stipulated by the Second Law of thermodynamics. The concept that free energy aids in thermodynamic work underscores the significance of understanding enthalpy in connection with entropy, which is a measure of disorder within a system. Elaboration on the existence of such a connection can show the intricate mechanism of conservation of energy and how it applies in realistic conditions in biological systems and other real systems (Barnes et al., 2000). On a more applied level, enthalpy is at the heart of meteorological science. Specific thermal enthalpy, in particular within moist air masses, becomes a foundation measurement upon which to study atmospheric phenomena. It entails the use of latent heat considerations within phenomena such as condensation and evaporation. To calculate some such specific thermal enthalpy one needs really to integrate content of specific heat in accounting for many latent heats. The intricate interplay between wet thermal enthalpy and other parameters like wet static energy highlights the delicacy of thermodynamic systems, especially in atmospheric dynamics. Examining these interplays underscores inconsistency within formulations of thermodynamics and draws attention to the necessity for sophisticated understanding of enthalpy when evaluating energy budgets in a complex system. This multifaceted approach depicts the significance of enthalpy as not merely a concept but as a significant tool in chemical and environmental science (Ambaum et al., 2014) (Figure 3.3).

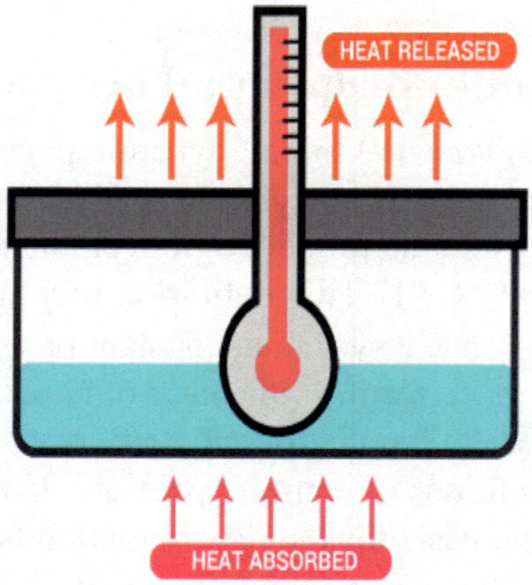

Figure 3.3. *Explanation of enthalpy as a thermodynamic quantity.*

Source: https://cdn1.byjus.com/wp-content/uploads/2018/12/enthalpy.png.

3.1.2. Relationship Between Enthalpy and Internal Energy

It is crucial to understand the relationship between enthalpy and internal energy to comprehend the ideas of principles of controlling chemical reactions. Internal energy (U) is the total amount of energy present within a system, including kinetic and potential energies of particles at the microscopic level. Enthalpy (H), however, is the addition of internal energy and the product of pressure and volume ($H = U + PV$). Under constant-pressure conditions, which occur in the majority of chemical reactions, changes in enthalpy (ΔH) are proportional to heat exchange with the surroundings. Therefore, through a greater understanding of enthalpy, chemists can predict the manner in which energy transfer occurs during reactions, providing information regarding the exothermic or endothermic nature of a reaction. The relationship of internal energy and enthalpy not only serves as a foundation for thermodynamics but also interrelates with valuable applications in other sciences as well. In life sciences, for instance, a syllabus that incorporates concepts of chemical energy in combination with those of physical science can potentially enhance understanding in students concerning energy change in biological reactions. This combined method deepens a multi-disciplinary perspective of energy, which is key to understanding how chemical reactions drive physiological processes. By integrating concepts from physics, biology, and chemistry, learners are able to comprehend the dynamic nature of enthalpy and internal energy, ultimately making the study of energy more relevant to actual issues in chemical reactions (Dreyfus et al., 2013). Besides, chemical reactions are followed by changes in enthalpy and internal energy, depending on some structural characteristics of the catalysts, i.e., acidic zeolites. These types of catalysts have the ability to significantly affect the reaction paths and energies involved in the hydrocarbon processing. The correlation between changes in internal energy and enthalpy in such situations can be understood if the kinetics of the reactions under various conditions is studied. For example, alkane cracking mechanism involves a consideration of not just the energy outcome of the reaction, but also the manner in which the zeolite framework alters these energy factors, including equilibrium constants and reaction rates. These studies highlight the importance of enthalpic and internal energy considerations, as they have critical roles in optimizing catalytic efficiency as well as enhancing production of useful lower molecular weight product from high hydrocarbons (Bell et al., 2018) (Figure 3.4).

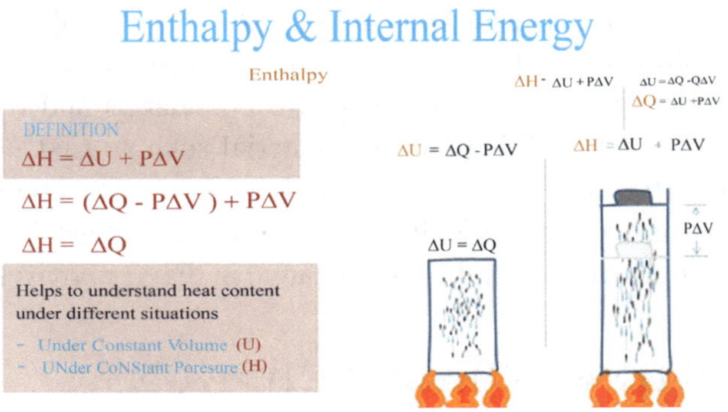

Figure 3.4. *Relationship between enthalpy and internal energy.*

Source: https://i0.wp.com/boarding-school.in/wp-content/uploads/2024/01/Class-11-internal-energy-and-enthalpy.png?fit=450%2C253&ssl=1.

3.1.3. Units of Measurement for Enthalpy

In thermodynamics, the measurement of enthalpy is important in determining changes in energy in chemical reactions. The primary unit of measurement of enthalpy is the joule (J), which is the transferred energy within the system. In practical applications, kilojoules (kJ) are employed, predominantly for dealing with reactions involving significant energy changes. This scaling is vital in computations involving reactions that release or gain heat, thus making ease of communication and comparison of such values easy. The accuracy of units for energy makes it easier for chemists to relate reaction spontaneity and equilibrium conditions with enthalpy change and gain useful insight into the thermodynamic behavior of reactants and products. Measurement of enthalpy, particularly in relation to moisture and phase changes, also needs special modifications. For instance, the individual enthalpy of wet air is not simply a one-to-one correspondence of the traditional binary systems of gases. Rather, it encompasses additional variables such as specific heat contents and phase-change latent heats like evaporation and condensation. This demands the addition of specific heat content as a function of absolute temperature, along with water vapor interaction-specific variables. Such precise formulations ensure that behavior of moist air can be modeled properly, particularly in meteorology, where tiny variations in enthalpy produce giant differences in weather and climate response. Further, computation and comparison of enthalpy across different materials or phases may produce large differences if not approached with great caution. For instance, for electrochemical systems, it was observed that enthalpy changes during lithiation are heavily based on electronic and configurational entropy that may vary for each phase and material details. Accordingly, the measurement unit must also reflect these variations to avoid over-simplification. Moreover, combining measurements with theoretical predictions enhances understanding of phase boundaries and transitions like the type occurring during lithiation in a material such as $LixCoO_2$. This dual approach demonstrates the demand for subtle units of measurement and their applications in different scientific disciplines and applications, leading to an in-depth understanding of chemical energy change in diverse environments.

3.2. TYPES OF ENTHALPY CHANGES

Familiarity with the types of enthalpy changes is fundamental to the appreciation of thermodynamic descriptions of chemical reactions. Enthalpy, or the overall heat content of a system, can change with different processes, and these differences are divided into many types. Some of the most significant are the enthalpy of formation, combustion, and reaction. The enthalpy change when one mole of the compound is formed from elements in their standard state is referred to as the enthalpy of formation. The enthalpy of combustion, on the other hand, is the heat that is released when a substance is fully burned in the presence of oxygen. Both are the major reference points in calculating the enthalpy change for most chemical reactions and provide insight into the energy change during these processes. Yet another critical type of change in enthalpy is the enthalpy of solvation, or the heat change on dissolving a solute in a solvent. This is particularly important for influencing the interactions between the solvent and the solute molecules to have a profound impact on the net thermodynamic stability of a solution. The thermal capacity of solutions of solutes, for instance, nickel and cobalt salts, can be significantly different depending on the presence of ionic interactions and solute concentrations, as revealed in high pressure acid leaching process-based research. These phenomena highlight the need for accurate thermodynamic data, e.g., metal sulfate solution densities and heat capacities, in trying to move forward our understanding of solvation enthalpies and their implications for chemical engineering and industry applications, as depicted in Chen et al. (2003). Also, complexation reactions possess their very own unique enthalpy changes that need to be considered. These reactions are generally characterized by coordination complex formation in which transition metal ions are coordinated by ligands with significant enthalpy changes. The heat of reaction in these processes can be characterized by methods like titration calorimetry that provides significant information regarding stability of the complexes. In studies of molecular interaction with molecules like Norfloxacin and DNA, enthalpy and entropy changes related to their binding interactions reflect complex information in terms of molecular affinity and stabilization. Information on the enthalpy changes when complex forms demonstrates the core relevance of thermodynamics in biochemical processes, as in Davies et al. (2006).

3.2.1. Enthalpy of Formation

Enthalpy of formation, a central principle of thermodynamics, measures the change in enthalpy upon the formation of one mole of a compound from its elements in their standard states. This value gives vital information regarding the stability and reactivity of compounds and, therefore, is crucial in anticipating the effects of chemical reactions. For example, a negative enthalpy of formation means that formation of a compound is exothermic, with the implication that the formed compound is more stable than the elements separately. However, a positive sign means an endothermic process, and it implies that energy has to be provided for the compound to form. These thermodynamic accounts not only allow the explanation of synthesis routes but also allow the estimation of energy requirements in industrial processes, emphasizing the fundamental importance of enthalpy in chemical reactions. The inclusion of enthalpy of formation in reaction thermodynamics allows a basis for investigation of energy changes in chemical systems. It makes it possible for chemists to calculate the overall enthalpy change in a reaction with Hesss Law, which says that overall enthalpy change is pathway-independent. Through standard enthalpy of formation for products and reactants, one can develop a uniform method of deciding whether a reaction is endothermic or exothermic. For instance, enthalpy change contrasts derived from individual formation enthalpies can inform about the thermodynamic favorability of a reaction and how this can inform decision-making within synthetic chemistry. Other studies into specific material properties, e.g., stability of perovskite structures, demonstrate how enthalpy and entropy considerations might meet, as illustrated in recent work pointing out inconsistencies in calorimetric data that can influence predictions of stability (Ciccioli et al., 2018).

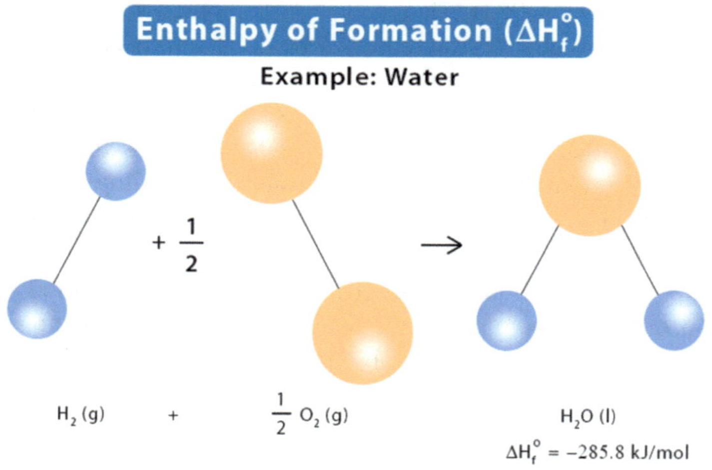

Figure 3.5. *Enthalpy of formation.*

Source: https://www.chemistrylearner.com/wp-content/uploads/2022/10/Enthalpy-of-Formation.jpg.

Moreover, formation enthalpy has implications beyond just chemical synthesis; it also plays a crucial role in kinetics and reaction mechanisms. Only by being aware of the enthalpic changes associated with various pathways can scientists realize the kinetic parameters able to dominate a reaction under specific conditions. The study of degree of rate control, as abstracted by more recent work, gives a nuanced treatment of thermodynamic and kinetic control in complex chemical networks (Andreasen et al., 2011). Such frameworks enable chemists to predict the product distribution and selectivity in reactions and, in turn, inform experimental design and optimization. With the growing need for efficient and sustainable chemical processes, a good understanding of enthalpy of formation and allied concepts will remain an essential prerequisite to formulate theoretical as well as experimental aspects of chemical thermodynamics (Figure 3.5).

3.2.2. Enthalpy of Combustion

The enthalpy of combustion is a significant thermodynamic property used to describe the energy released when a material, typically a fuel, undergoes complete combustion in the presence of oxygen to form water and carbon dioxide. It is a significant figure for application in many fields like engineering, chemistry, and the environment. The enthalpy of combustion can forecast the energy available from a specific fuel, deciding on the material selection for energy production process. Additionally, the enthalpy of combustion guides the design of combustion systems by enabling true calculations of fuel efficiency and emissions, thereby enabling engineers to create cleaner technology. The combustion enthalpy is typically found by a calorimetric technique where heat given off in the reaction is quantitated under controlled conditions. The laboratory technique not only provides insight into the energetic characteristics of fuels but also allows extrapolation of data for related chemicals according to Hesss law. As an example, hydrocarbon combustion enthalpies can be predicted by correlating with molecular structure and hence predictions made of new or ill-characterized compounds. Computational models, as previously defined in the literature, have now begun to predict enthalpic values based on simulations that account for complex interactions at the molecular level, closing the gap between theoretical comprehension and practical engineering applications (Badhuri et al., 2004). Furthermore, the value of enthalpy of combustion extends beyond the mere energy payback; it does have significant relevance in environmental science. Fuel with high energy will produce high CO_2 emissions upon combustion, which has made research on alternatives that can deliver similar energies but lower environmental footprints imperative. While earlier models of combustion focus mainly on the recovery of energy, subsequent studies give greater emphasis to the coordination of combustion theory with technological advancements toward the design of better combustion chambers with a view to minimizing atmospheric pollution (Marble et al., 1956). Continuous advancements in experimental procedures and computational modeling are required to further refine combustion processes so that they serve energy needs while being environmentally friendly in our changing world (Figure 3.6).

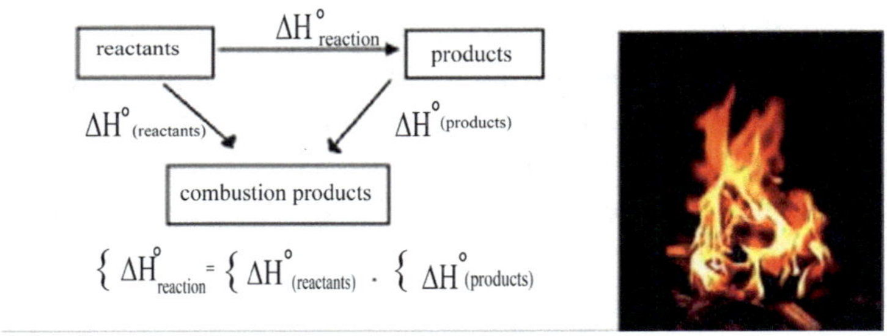

Figure 3.6. *Standard enthalpy of combustion.*

Source: https://encrypted-tbn0.gstatic.com/images?q=tbn:ANd9GcTHNDM-y_HbBIsXxPYB_K8tYaqB2mjjTvsOIg&s.

3.2.3. Enthalpy of Reaction

Reaction enthalpy is a key concept of thermochemical study because it gives the heat evolved or consumed in a chemical process when conducted at constant pressure. The change in enthalpy is denoted as ΔH and aids in the description of energy trends in reactions, influencing reaction rates and equilibria. The separation of reactions into exothermic and endothermic depends upon the sign of ΔH; an exothermic reaction, defined by the liberation of heat, possesses a negative ΔH, whereas an endothermic reaction traps heat, resulting in a positive ΔH. The separation enables chemists to predict the viability of reactions as well as manipulate conditions to allow their optimization, hence connecting thermodynamic principles to practical applications in a variety of contexts such as industrial synthesis and environmental chemistry. The role of catalysts in reactions indicates the complexities associated with enthalpy changes. Catalysts like zeolites are capable of significantly impacting the kinetics of reactions without altering the net change in enthalpy. Acidic zeolites, for instance, have been extremely helpful in cracking hydrocarbons to enhance the efficiency of fuel production. Analysis of the reaction enthalpy of these catalytic processes illustrates the interrelatedness of structural properties and catalytic activity, underlining the need for precise measurements of rate coefficients and equilibrium constants. Theories obtained from complex computational models, such as configurational-bias Monte Carlo simulations, also offer greater insight into how changes in zeolite structure influence enthalpy-related parameters, ultimately resulting in the design of more effective catalysts (Bell et al., 2018). Among various reaction processes, self-propagating combustion synthesis (SHS) serves as a great example of how changes in enthalpy can impact compound formation, such as for TiC nanosized aggregates. The multifaceted interrelation between reactants, substrate temperature, and heat generation in SHS illustrates the dynamic function of enthalpy in chemical reactions. Since heat arises from the reactive interactions, it influences not only the formation of the desired product but temperature gradients in the reaction medium as well. Computational models used to simulate such reactions are insightful as they plot temperature gradients and assess how these influence reaction kinetics and enthalpy changes in real-time, guiding innovation in fields like materials science (Badhuri et al., 2004).

3.3. MEASURING ENTHALPY CHANGES

Measurement of enthalpy change during chemical reactions forms the core of thermodynamics in chemical reactions. Among the most usual methods of measurement of these changes is calorimetry, providing quantitative data on heat transfer during reaction. Differential Scanning Calorimetry (DSC) is particularly useful as it measures the heat flow of phase transitions and chemical reactions under temperature control. This technique allows scientists to make precise measurements that make it possible to ascertain enthalpy changes as reactants are transformed to products. Through the careful examination of thermal profiles, scientists can make valid conclusions regarding reaction mechanisms and kinetic parameters, laying the groundwork for new innovations in theoretical and applied advancements in chemistry and materials science. In addition to calorimetric methods, a number of other methods for measuring enthalpy changes are applicable to certain types of reactions and materials. Dynamic methods, for instance, will more likely be applied to assess exothermic reactions, particularly in polymerization where enthalpy changes can significantly influence final material properties. The curing of thermosetting resins is a typical case, since the exothermic reaction causes chemical shrinkage that influences bond performance in applications such as adhesive anchors. The inclusion of such techniques as digital image correlation enables measurement of shrinkage behavior in situ, enhancing understanding of enthalpic change in real time. By employing different methodologies, scientists can obtain a general understanding of enthalpy change in various contexts to enhance material design and chemical synthesis. Furthermore, information about the correlation between enthalpy change and chemical reactivity and stability is necessary to enhance knowledge on various areas of study like environmental chemistry and materials science. Experiments on compounds like humic acids, which exhibit complex interaction with metal ions, illustrate the importance of quantifying enthalpy change to be in a position to predict stability when subjected to heat stress. Through techniques like differential scanning calorimetry and thermogravimetric analysis, researchers can examine the oxidative stability of compounds like these and their potential reactivity. These results may lead to new applications in soil chemistry, wastewater treatment, and organic synthesis, eventually bringing fundamental thermodynamic principles to bear on practical problems of environmental significance. By investigating the many methods for measuring enthalpy change, scientists contribute significantly to knowledge of chemical reactions and properties of materials at large (Figure 3.7).

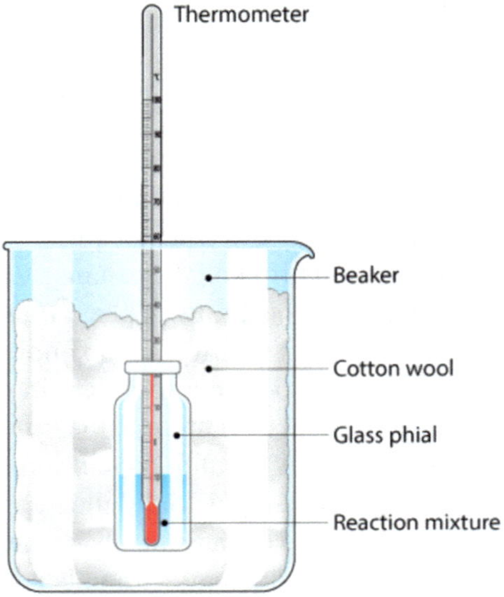

Figure 3.7. *Measure enthalpy changes.*

Source: https://d2fc7ojl9ttfsq.cloudfront.net/web/LC_Assets/frg/images/MEC_EU_04.jpg.

3.3.1. Calorimetry Techniques

Calorimetric techniques play an essential role in enthalpy change measurement in chemical reaction, and they are invaluable tools for measuring the quantification of energy transfer. The most sensitive of them is differential scanning calorimetry (DSC), for it measures the passage of heat associated with the temperature difference in a sample. DSC accurately measures thermodynamic data that is of key relevance for the discovery of reaction kinetics and mechanism. The method involves the subject of a reference and a sample to regulated temperature variations to allow the measurement of thermal events such as phase transformations or chemical reactions. Investigation of the data obtained from DSC has the ability to shed significant insight into endothermic or exothermic reactions and ultimately advance our knowledge of the underlying principles governing chemical reactions, e.g., humic acid-metallic ion interaction, as indicated by recent studies (Nováčková et al., 2009). Another significant technique in calorimetry is isothermal titration calorimetry (ITC), in which one is able to directly witness heat change in the process of interaction between molecules dissolved in solution. By detecting the heat evolved or absorbed in the process of titrating a reactant into a solution with a desired species, ITC allows for the precise determination of thermodynamic parameters such as binding affinities and enthalpy changes. This approach is particularly useful in biochemical interaction studies, where delicate bonding and structural change can be of extreme importance in the determination of reaction pathways. The capability of ITC to provide a complete thermodynamic profile for complex systems positions it as a cornerstone in the evaluation of chemical reaction enthalpy change, also demonstrating its value in advancing knowledge in basic and applied chemistry situations (González et al., 2018). Also, the application of traditional calorimetry methods, such as bomb calorimetry, is still crucial, especially in reaction systems involving gaseous reactants or combustible materials with high energy. Bomb calorimetry consists of burning a sample in an insulated vessel with water around it, thereby allowing precise measurement of heat released by the reaction. This technique not only provides precise data regarding enthalpy changes but also allows researchers to conduct comparative analyses of various fuels or reactive materials under controlled laboratory conditions. The data obtained through bomb calorimetry proves extremely helpful in fields like material science and energy research, where heat dynamics knowledge is crucial for designing more efficient processes and materials. Collectively, these calorimetric techniques significantly enhance our understanding of chemical reaction enthalpy change, underpinning many scientific advances across a wide range of disciplines.

3.3.2. Hess's Law and Its Applications

Hesss Law is an important concept in thermochemistry that enables one to compute reaction enthalpy change via an anchor of energy conservation. This postulate states that the net enthalpy change during a chemical process is path-independent, providing a solid foundation for the calculation of reaction enthalpies from ones that are known. Using Hesss Law, chemists can break down complex reactions into a series of simpler steps that have measurable enthalpy changes. This approach not only renders calculations simple but also ensures precision and reliability in determining enthalpy changes. Furthermore, Hesss Law proves useful where measurement is hard directly, allowing scientists to complete the research gaps in experimental data through known reactions, thus making further research possible in the field of studying enthalpy changes in chemical reactions. One of the major applications of Hesss Law is in synthetic chemistry, and it is employed here to optimize reaction pathways and increase yield outcomes. By calculating the changes in enthalpy during intermediate steps, chemists can make informed decisions regarding reagent choice and reaction conditions. This is a crucial facility in designing eco-friendly processes, particularly as the industry is put under increased pressure to minimize waste and energy consumption. The application of Hesss Law makes it possible to evaluate energy inputs and outputs through a synthetic route, thus guiding scientists towards better methodologies. Specifically, through coupling this law with computer tools, such as Density Functional Theory (DFT), scientists are able to predict efficiencies and selectivities in complex reactions, thus showing the immense scope of Hesss Law beyond standard textbook applications (Hall et al., 2012). Furthermore, Hesss Law is applicable in other interdisciplinary fields, for instance, environmental science, where reaction enthalpy plays a pivotal role in energy transformation and storage technologies. The fact that today's renewable energy challenges in reality further emphasize the importance of proper calculations of energy storage systems. For instance, battery and hydrogen energy storage technologies hybridization stands to gain a great deal from Hesss Law, as this aids in determining energy balances associated with these systems. Through the examination of enthalpy changes, efficiency and sustainability potential of such technologies can be identified while addressing environmental concerns and meeting energy requirements. Hence, the use of Hesss Law in this instance defines the interconnectedness of chemical reaction principles and the pressing need for novel energy solutions (Guo et al., 2024). By its extensive uses, Hesss Law continues to be an essential instrument in advocating for scientific research and addressing pressing issues (Figure 3.8).

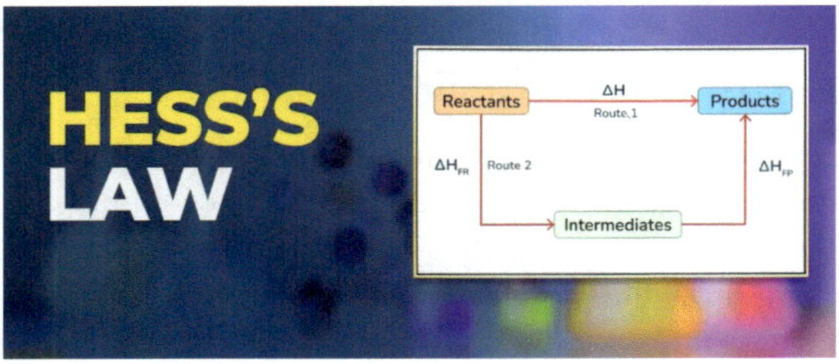

Figure 3.8. *Hess's law and its applications measuring enthalpy changes.*

Source: https://media.geeksforgeeks.org/wp-content/uploads/20240219190120/Hess's-law.webp.

3.3.3. Standard Enthalpy Changes

Standard enthalpy changes are key concepts in the thermodynamics of chemical reactions and are useful indicators of the energy changes occurring during the formation or decomposition of a compound. Enthalpy, *H*, is the heat content of a system and its change is measured under constant pressure conditions, which are termed as standard conditions. This change, given as $\Delta H°$, is an important factor in approximating the energetics of a reaction. For instance, exothermic processes in which heat is evolved will be described by negative changes in enthalpy, and endothermic reactions that absorb heat possess positive values. Such energy transformations are responsible for determining the directionality and spontaneity of chemical processes and subsequently defining the enthalpy-chemical stability relationship based on both reaction thermodynamics and reaction kinetics. Deeper, the calculation of standard enthalpy changes relies heavily on experimental data, including calorimetric determinations of heat flow in a reaction. Reliable enthalpy values provide insight into the reactivity and stability of compounds, as reflected in studies employing zeolites as catalysts. Bell et al. (2018) report that the structure of zeolites with which hydrocarbons interact emphasizes the need to understand enthalpic differences in order to maximize a catalytic process. The adsorption equilibrium constant, Kads-H+, and the intrinsic reaction rate coefficient, k(int), are both demonstrated to be significantly affected by enthalpic considerations. This correlation highlights the significance of enthalpy in applications, including petroleum refining, where catalytic efficiency can be maximized through precise adjustment of standard enthalpy changes. Moreover, thermodynamics is involved in assessing stability and decomposition behavior of various materials, including perovskite frameworks. An example from the research of $CH^3NH_3PbX_3$ perovskites reflects the intricate connection between enthalpic and entropic stability contributions (Ciccioli et al., 2018). The assessment of both kinetic and thermodynamic control in establishing decomposition mechanisms mirrors the nature involved in enthalpy changes. These intrinsic properties not only affect the performance of the materials but also set a course for future studies by establishing a need for more experimental data in thermodynamics. This sustained research on enthalpic behavior ultimately expands our understanding of chemical interaction and challenges innovation in materials science and reaction engineering.

3.4. FACTORS AFFECTING ENTHALPY CHANGES

Enthalpy change affecting factors are multifaceted, involving factors such as temperature, pressure, and chemical constitution. Temperature plays a crucial role in influencing the kinetic energy of reactants and therefore the reaction rate as well as the energy exchange magnitude that takes place. As temperature rises, enthalpic contributions to the overall energy balance also rise, having a tendency to produce more reaction advancement and energy release in exothermic reactions. However, endothelial reactions may require high temperatures in order to provide sufficient energy to overcome activation barriers for the reactants. Therefore, the best temperature conditions must be ensured for correct determination of enthalpy changes in various chemical systems, supporting the role of controlled experimental conditions to yield reproducible results and information on the energetics of reactions. Pressure is another critical factor determining enthalpy changes, particularly for gaseous reactants and products. Pressure changes are recognized to induce volume changes within a system, impacting internal energy and subsequently the enthalpy of a system. Thermodynamics stipulates that rising pressure will typically follow the reduction in volume, triggering an adjustment of energy levels between reactants. This is particularly observed in gas reactions, where the change in pressure may impact equilibrium positions and directly influence enthalpy values. It is important to consider the pressure factor in handling reactions based on the enthalpy concept since deviations from standard atmospheric pressure will lead to different enthalpic results and interpretations. The chemical composition of reactants also plays an important role in enthalpy changes in reactions. The diversity of elements and the characteristics of chemical bonds formed between them determine the energy for bond formation and breaking. For instance, transition metal reactions, such as those in metal-organic frameworks, exhibit unique enthalpic characteristics due to complex interactions between the metal centers and organic ligands. Experiments have observed the significant impact of metal coordination on catalytic reactions, with variations in enthalpy changing substantially based on specific metals. The intricate relationship between molecular structure, bond interactions, and enthalpy brings into perspective the significance of profound understanding of chemical composition in evaluating energy changes in chemical reactions, as demonstrated by current research on photocatalysis on metal-organic frameworks (Aziz et al., 2015).

3.4.1. Temperature and Pressure Effects

Effects of pressure and temperature on enthalpy changes during chemical reactions are profound, influencing reaction kinetics, position of equilibrium, and thermodynamic stability. With increasing temperature, it will try to make reactant molecules possess increased kinetic energies and a greater probability of effective collisions that could overcome activation energy barriers. Increased energy can lead to reaction equilibria shifts according to Le Chateliers principle such that higher temperature favors endothermic reactions. Besides, specific systems may exhibit drastic volume change with temperature variations, and such is liable to alter pressure conditions in a closed system, making the thermodynamic surroundings even more complicated. For example, within iron in magnesium silicate perovskite, temperature sensitivity plays a key role in regulating iron oxidation states and thus both structural and physical properties in geochemical processes (Morgan et al., 2015). Other than temperature, pressure influences should not be downplayed when examining enthalpic change within chemical reactions. As pressure increases, the volume of gaseous reactants diminishes, stabilizing certain product formations that may be unfavorably favored at lower pressure intensities.

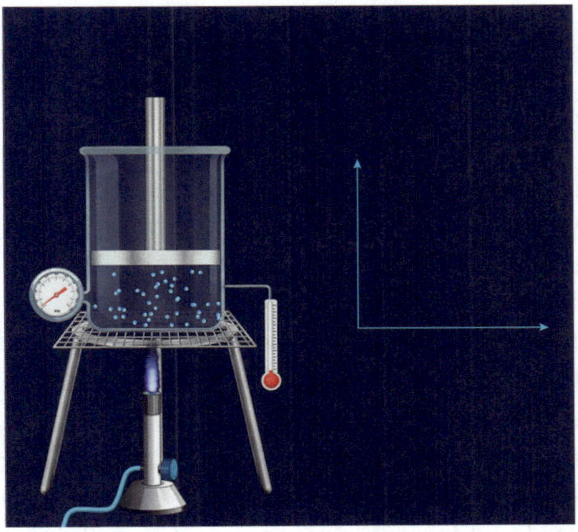

Figure 3.9. *Temperature and pressure effects.*

Source: https://d29zwu34e1ouns.cloudfront.net/inline-images/aakash_8749774.png.

Due to the elevated compressibility of gases over liquids or solids, reactions of gas-phase reactants are particularly sensitive to pressure changes. Such behavior is often observed in systems in which structural reorganization is suppressed at increased pressures, as is normally the case in geological repositories or catalytic processes. For instance, the structure of zeolites in hydrocarbon cracking processes is greatly influenced by the pressure applied, thus greatly influencing the catalytic activity. Such pressure-induced effects can potentially influence rate coefficients and equilibrium constants, which serve as a foundation for assessing reaction outcome in natural and industrial settings (Bell et al., 2018). In addition, the temperature-pressure interaction is more than a simple binary effect; it is an intricate thermodynamic relationship that controls the energy landscape of chemical reactions. When high-pressure and high-temperature reactions are in question, such as those of the Earth's mantle, increasing configurational entropy can play a large role in affecting species stability and formation. For example, under extreme conditions, the presence of differing iron oxidation states alters material bulk modulus and conductivities for minerals such as magnesium silicate perovskite. Thus, the additive effect of pressure and temperature not only affects our theoretical understanding of enthalpy changes but also has experimental use in materials science and geology, guiding the development of catalysts and the recovery of valuable resources from the Earth's crust. An understanding of these interdependencies is central to the refinement of theoretical models as well as experimental techniques in thermochemistry (Bell et al., 2018; Morgan et al., 2015) (Figure 3.9).

3.4.2. Concentration of Reactants

Concentration of reactants in chemical reactions is an important determinant of the enthalpy change of the reaction. Concentration influences the rate at which reactant molecules collide and hence react and therefore the kinetic course of the reaction. Higher concentrations favor more frequent and shorter collision between reactant molecules, therefore a rapid advancement of the reaction. This dynamic demonstrates the role of concentration in the activation energy barrier, wherein an increased number of reactant particles can result in more efficient collisions that are capable of crossing this barrier. Therefore, comprehension of this dynamic becomes critical in order to predict the manner in which a reaction will proceed and the enthalpy change that will occur. In addition, the concentration of the reactants can profoundly influence the thermodynamic stability of the products resulting from a chemical reaction. In systems of high-reactant-concentration, equilibrium will be moved in the direction of increased product formation and can be contributed to a rise in enthalpy change. With the aid of strict models and thermodynamic mechanisms, scientists can quantify these concentration shifts and their relation to enthalpic shifts. For instance, it has been shown that concentration fluctuations of reactants may alter the thermodynamic favorability of certain pathways, revealing why reactions prefer to be more exothermic under varying conditions (Morgan et al., 2015). The connection illustrates the nuanced interaction of concentration and enthalpy, showcasing them as integral parts of reaction mechanism knowledge. Apart from that, configurational entropy must also be considered when making judgments on reactant concentration and the effect of enthalpy changes. Because concentration does affect the arrangement and spatial organization of molecules in a system, it therefore affects the reaction's entropy as well. For example, temperature and pressure conditions may influence the molecular arrangement of reactants, thus their resultant potential energy landscape and free energy behavior as a whole (Barnes et al., 2000). This relationship underlines the reality that enthalpy changes are not only a function of bond energies but also correlate with reactant spatial arrangement and energy levels. Therefore, enhanced understanding of reactant concentration and how it correlates to thermodynamic concepts can provide greater insight into reaction enthalpies and how they contribute to chemical and biological processes.

3.4.3. Physical State of Reactants and Products

An understanding of the physical state of reactants and products throughout chemical reactions is important when considering enthalpy changes. The contrast between solids, liquids, and gases significantly influences reaction dynamics and thermodynamic properties. For instance, reactions with gaseous reactants have higher rates of reaction than those confined to solid-state interactions, mainly due to the greater mobility and greater concentration of reactant molecules in the gaseous phase. This is a significant kinetic characteristic as it corresponds exactly to the enthalpy changes shown by a reaction. The physical state of the reactants and products further influences their interaction energies, which determines ultimately the direction and magnitude of enthalpy change associated with the reaction. Thus, a careful probe of the physical state yields crucial information about the thermodynamic behavior of a chemical system and basically defines our conception of reaction energetics. The thermodynamic quantities and constants of critical relevance to being physical emerge from multiple studies and affect theoretical and practical areas. For example, the hydroperoxy radical reaction with acetone illustrates how states, i.e., gaseous and condensed phases, modify equilibrium character and energy profiles. According to findings from studies, such a response is by means of rapid equilibrium of a hydrogen-bonded complex that may be significant in atmospheric chemistry but whose existence is influenced by the phases of the reacting species. The net enthalpy change of the reaction, obtained at Δ_(*r*)H°_(245) = −*35.4* ± *2.0* kJ mol^(−*1*), illustrates how the physical phase may influence the energy return of a particular chemical interaction (Atta et al., 2011). This underscores the significance of considering state differences in the broader context of thermodynamic assessment and its implication in environmental and industrial processes. Moreover, the free energy and entropy thermodynamic principles are directly related to physical states of products and reactants. Change of state is accompanied by changes in these states of energy, which influence reaction spontaneity and accessible work energy. Classically, free energy change ΔG becomes crucial in evaluating reaction feasibility and extent, with entropy measurements contributing to evaluating system disorder at these various states. Specifically, entropy change in phase transitions, for instance, solid to liquid or liquid to gas, affects thermal energy distributions, as represented in most thermodynamic formalisms (Barnes et al., 2000). Consequently, the examination of physical states is more than mere classification as it constitutes an integral component in thermodynamic calculations that account for the complexity of enthalpy change in chemical reactions (Figure 3.10).

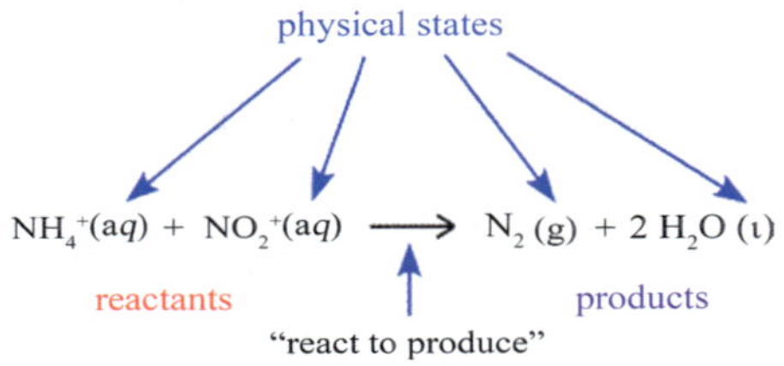

Figure 3.10. *Physical state of reactants and products.*

Source: https://lh3.googleusercontent.com/proxy/zNHpvoOQLKY5SPTsJR1jZqM-mvAVknqUiUlWtI6UtoQ_yV3htG0lCNAgyeYE1UAlQuHwDBftOPsCloNQz_RFX6RgV--Pqblvx9tMGARjLgkedcPkmnBFS7Rg_W_WgxQtb56lS4oolvutbzU7T.

3.5. APPLICATIONS OF ENTHALPY CHANGES

Applications of enthalpy change extend beyond laboratory reactions to actual industrial processes, particularly catalysis. In hydrocarbon fuel refining, acidic zeolites are excellent catalysts according to the structural properties of such materials, which influence catalytic activity through the enthalpy change process. Understanding the enthalpic processes involved in such a reaction is critical in optimizing processes such as dehydrogenation and cracking of alkanes. In particular, when the adsorption equilibrium constant and intrinsic rate coefficients are examined, it is possible to delineate accurately the enthalpy changes for these reactions. Current studies point out that changes in enthalpy may play a decisive role in determining the rate coefficients, disproving earlier presumptions and showcasing the significance of quantitatively precise enthalpic analysis in catalysis (Bell et al., 2018). This serves to emphasize the importance of enthalpy in enhancing efficiency in fuel production and directs future industrial chemistry developments. Apart from catalysis, enthalpy changes are fundamentally important functions in the atmospheric sciences, particularly when the thermal dynamics of moist air are considered. Thermal enthalpy for a wet-air parcel is described in terms of specific content of heat, integrating over latent heats associated with phase transformations—i.e., condensation, sublimation, and evaporation. This approach allows for a greater understanding of thermodynamic phenomena in the atmosphere, particularly regarding weather events. By delineating thermal enthalpies as not being equal to moist static energy (MSE), researchers cite the weaknesses of traditional models in the accurate calculation of atmospheric budgets. Enthalpy variations play most critical roles during mixing of dry-air and humid air masses, where very accurate calculations have to be made to predict weather phenomena and assess energy balances in cloud systems (Ambaum et al., 2014). Thus, application of enthalpy in atmospheric science carries significant ramifications for meteorology and climate studies. Furthermore, applications of enthalpy variations go beyond particular chemical systems as they shed light on broader environment and energy problems. For instance, for sustainable energy, enthalpy changes in various chemical reactions are knowledge that is essential in optimizing the processes of biomass gasification and biofuel production. It is through measuring enthalpy changes accurately that scientists can increase energy conversion processes' efficiency, thereby leading to greenhouse gas emission reduction and enhanced energy sustainability. This focus on enthalpy also encourages innovation in alternative energy technologies. With the increasing energy needs of the world, application of enthalpy-based techniques stimulates more efficient and cleaner forms of energy systems. Therefore, the study of changes in enthalpy not only improves our knowledge of chemical reactions but is also of vital importance in addressing the pressing environmental concerns (Ambaum et al., 2014; Bell et al., 2018).

3.5.1. Role in Predicting Reaction Spontaneity

Spontaneity of prediction of reaction lies primarily on the basis of thermodynamics, where changes in enthalpy play a central role. Enthalpy is denoted by ΔH and is the measure of heat content of a system under constant pressure. A reaction is spontaneous when there is a negative change in free energy (ΔG) in total. According to the Gibbs free energy equation, $\Delta G = \Delta H - T\Delta S$, where ΔS is entropy change and T is the temperature in Kelvin. Thus, spontaneity can't be determined based on the change in enthalpy alone, one must also include entropy in the calculation. For instance, if the reaction is exothermic ($\Delta H < 0$) and results in an increase in disorder ($\Delta S > 0$), then the reaction would generally be spontaneous. This interaction between enthalpy and entropy becomes critical in explaining why some reactions are spontaneous under natural conditions and others are not. Further, the thermodynamic properties of a variety of compounds play a critical role in determining their ability to predict spontaneity. The interaction of molecules, as found when in different solvents, can exhibit different probabilities of spontaneous reaction based on their enthalpic and entropic contributions. For instance, zwitterion ionic liquids (ZIs), which are highly polar and possess high hydrogen-bonding capacity, can influence reaction mechanisms and spontaneity by virtue of their typical thermodynamic behavior (Ferro et al., 2019). The enthalpic alteration of dissolving or engaging with them suggests how their molecular structure defines reaction potential. Therefore, spontaneity in response becomes a refined evaluation that factors in not only enthalpy but also the thermodynamic properties that are derived from molecular interactions to identify the tendency of a reaction to proceed spontaneously. Kinetic factors must also be factored into an investigation of reaction spontaneity, particularly in reactions that may have favorable thermodynamic characteristics but must occur under specific conditions. Investigation of a novel biorefinery route to the sustainable production of bio-aviation fuel provided intriguing findings on spontaneity in a catalytic environment (Mary-Jane et al., 2025). The reaction process involved an optimized reaction system to give a negative Gibbs free energy, which is a measure of spontaneity. However, the enthalpy was endothermic, indicating that although the reaction had the potential to occur spontaneously under suitable conditions, it was not thermodynamically favorable under standard conditions without application of heat. This indicates that although enthalpy changes provide intrinsic information regarding spontaneity prediction, the intricacy of reaction conditions and mechanisms is equally important in determining the feasibility of chemical transformation in practical applications (Figure 3.11).

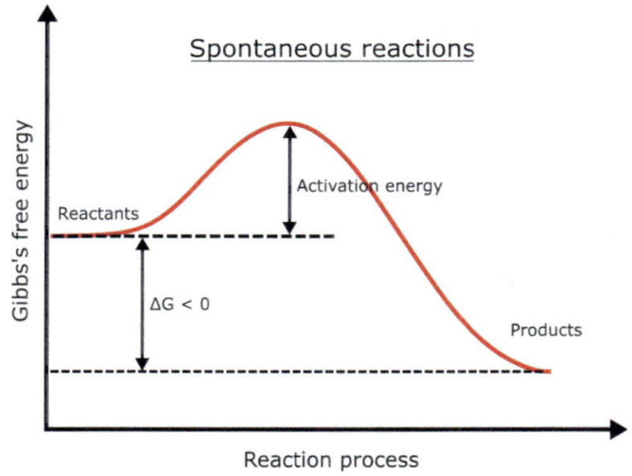

Figure 3.11. *Role in predicting reaction spontaneity.*

Source: https://psiberg.com/wp-content/uploads/2021/10/Spontaneous-reactions.svg.

3.5.2. Importance in Industrial Processes

The significance of understanding enthalpy changes in industrial operations cannot be overestimated, particularly in connection with energy efficiency and conservation of resources. During the chemical manufacturing process, understanding how to accurately calculate enthalpy changes in reactions allows engineers to optimize conditions that minimize energy consumption while maximizing output. For instance, processes such as the Haber process for ammonia production are largely based on the thermodynamic principles, where the values of certain enthalpies can dictate reaction routes. Measuring these changes in enthalpy enables industries to implement strategies that not only maximize production rates but also reduce costs of operations and environmental impacts. This fine adjustment of thermodynamic variables later leads to cleaner industrial processes, facilitating harmony between economic viability and environmental sustainability. Industrial catalysts also have a very significant impact on enthalpy change during chemical reactions, thereby the efficiency and selectivity of production processes. The introduction of catalysts has a tendency to adjust the energy landscape of reactions so that activation energy paths can be established. This is a description of what Campbell's expansion of the scope of rate control addresses, that being how important the kinetic and thermodynamic factors are in the design of catalytic processes (Andreasen et al., 2011). As industries find out how to manage these variables, they are able to construct catalysts that promote selectivity to the product of interest while restricting the formation of by-products. In the process, firms are not just optimizing their operations for yield but also positioning themselves competitively based on product purity and process innovation. Besides catalysis, enthalpy changes play a crucial role in determining the stability of materials and products produced in industrial environments. The thermodynamic characteristics of compounds largely determine how they behave under conditions of production and storage. For instance, metal ions' reaction with humic materials illustrates the importance of taking into account enthalpy so as to understand stability and reactivity of material (Nováčková et al., 2009). Such thermodynamic interactions are important for industries that rely on raw material stability in producing quality final products. By studying the enthalpy changes in such interactions in detail, manufacturers can improve quality control and extend shelf life. The role of enthalpy in industrial process thus transcends mere reaction kinetics to a broader aspect of process efficiency and product integrity.

3.5.3. Use in Biochemical Reactions

In enzymatic reactions, the use of enthalpy changes is of highest significance in the calculation of biochemical pathways that regulate cellular functions. Enthalpy as a measure of the heat content of a system is not only a stipulation with regard to predicting the spontaneity of reaction but also plays a major role in stabilizing the enzyme-substrate complex. The stabilization of the enzyme-substrate complex is essential since it affects the kinetics of biochemical processes. For example, precise enthalpic change measurements via heat of reaction enable an understanding of how enzymes lower activation energy barriers and therefore accelerate metabolic processes. Advances in microfluidic calorimetry allow researchers to precisely measure such enthalpic changes, furthering reaction dynamics knowledge in cellular environments. These microfluidic calorimeters can measure small samples, which can allow for high-throughput measurements critical for the exploration of a wide range of biochemical reactions, including the urea hydrolysis reaction heat and the mixing enthalpy of various solvents (Axelrod et al., 2009). Furthermore, biochemical reactions are intimately connected with the concepts of free energy and entropy, which complement each other to regulate reaction spontaneity and equilibrium. The second law of thermodynamics introduces the idea that systems would always seek to increase entropy, thus tending towards the spreading out of energy in available space. Biologically coupled reactions tend to have intermediates which enable the transformation of energy from one form to the other, and this emphasizes the importance of understanding free energy change through such processes. As described under thermodynamic principles, free energy needs to be evaluated in order to understand how energy is conserved and transferred within biological organisms.

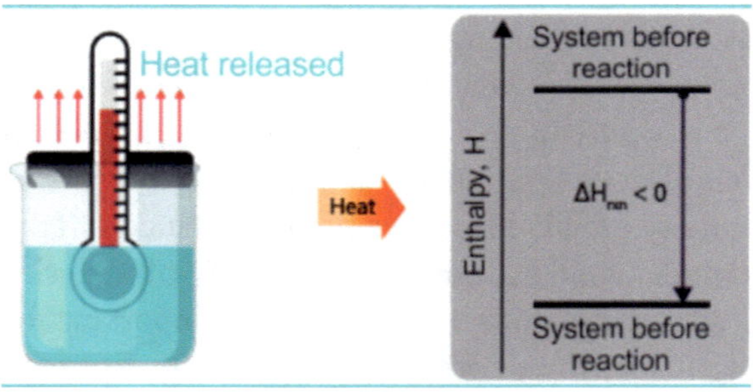

Figure 3.12. *Enthalpy changes in use in biochemical reactions.*

Source: https://cdn.testbook.com/1711056041082-enthalpy%20change.png/1711056042.png.

This is of vital importance when considering metabolic pathways that are either endergonic or exergonic, as these reactions determine how organisms can use energy for growth and repair. The knowledge of interaction between free energy and enthalpy changes allows researchers to predict the feasibility of biochemical reactions under various conditions (Barnes et al., 2000). The applicability of these concepts to biochemical reactions has implications in practical applications in drug design and bioengineering. Researchers can adjust reaction conditions by manipulating enthalpy and free energy to achieve maximal efficacy of therapeutic drugs or maximal metabolic pathways in engineered organisms. Interpreting the subtleties in enthalpy changes indicates what molecular interactions control the rates and product yields of reactions. Applying microfluidic devices to measure enthalpic change at previously unattained resolutions allows multidisciplinary teams to explore complex biochemical pathways more effectively. This not only avails the possibility of metabolic engineering breakthroughs but also emphasizes the ongoing demand for precision in the quantification of thermodynamic properties, thus energizing breakthroughs in biochemical reactions at an atomic level (Axelrod et al., 2009) (Figure 3.12).

3.6. CASE STUDIES OF ENTHALPY CHANGES

Enthalpy change knowledge is fundamental to the study of chemical reactions because it dictates the spontaneity and direction of reactions. Enthalpy change case studies present adequate understanding of not just theoretical frameworks but also practical applications in areas like catalysis and astrochemistry. One significant example is catalytic cracking of hydrocarbons, where acidic zeolites play a significant role to play. Research has shown that the zeolite framework can influence the catalytic action of Brønsted protons, which are essential for the cracking reaction. Through rate coefficient analysis of such reactions at various temperatures, researchers can better define the correlation between adsorption kinetics and variations in enthalpy, showcasing the delicate interplay between the energy dynamics and molecular structure (Bell et al., 2018). This research not only facilitates understanding of catalytic systems but also offers potential to tailor reaction conditions for enhanced efficiency. A second strong case study concerns astrochemical reactions in the extreme environments of dense molecular clouds. Such environments present unique challenges for understanding enthalpy changes, particularly due to the low temperatures that reign in such settings. One of the key works is concerned with a specific reaction which has been termed endoergic, i.e., one that uses energy and thus is repressed under typical dark cloud circumstances. This finding is of massive importance from the perspective of understanding chemical equilibria in space, as it can drastically alter gas-phase abundance predictions for a number of interstellar species. The revised models propose that altering our understanding of enthalpy changes will have the ability to dramatically impact the perceived presence of various molecules within the interstellar medium, which in turn suggests energetic considerations must be used in modeling astrochemical systems to accurately model them (Bhaskar et al., 2017). In total, the exploration of enthalpy changes through case studies illustrates the rich complexity found in chemical reactions across fields. These investigations not only illustrate the theoretical basis for chemical energetics but also emphasize the practical implications which arise from a basic understanding of enthalpy fluctuations. Whether in the confines of catalysis, for example, zeolite-catalyzed hydrocarbon conversion, or the outer limits of interstellar chemistry, the analysis of enthalpy changes provides decisive information on reaction mechanism and product formation. By putting the importance of enthalpy in the spotlight, these case studies all contribute to further discussion of chemical reactions and their implications, laying the groundwork for future basic and applied research.

3.6.1. Example of a Combustion Reaction

Combustion reactions are simple examples of chemical reactions that release energy as light and heat and emphasize the concept of changes in enthalpy. A classic instance is combustion of hydrocarbons, such as methane (CH_4), in the presence of oxygen (O_2). The balanced chemical equation for the process is $CH_4 + 2O_2 \rightarrow CO_2 + 2H_2O$. During the process, reactant strong bonds are broken, while new bonds form in the products. Energy difference between the states leads to the creation of an exothermic reaction with the change in enthalpy representing heat released. Understanding heat of combustion is relevant from energy production to the internal operation of combustion engines, where proper calculation governs efficiency and environmental acceptability. Closer scrutiny of combustion reactions reveals their complexity, particularly with respect to reaction conditions and products obtained. Combustion that is complete, as demonstrated by the methane example, produces carbon dioxide and water, which are typically harmless products. Nevertheless, there are cases of incomplete combustion under conditions with limited oxygen levels, leading to the formation of carbon monoxide (CO) and leftover hydrocarbons. All these nuances in combustion processes indicate the necessity for profundity in studying chemical reactions and changes in their enthalpy. For instance, the combustion and heat transfer phenomenon associated with small rocket engines utilizing gaseous hydrogen and oxygen illustrate how heat flux and combustion efficiency vary due to varying mixture ratios and thereby altering the overall combustion process behavior (Reed et al., 2025). Considering enthalpy change during the combustion reaction is essential not only in theoretical problems but even in practical applications in industry. In technological utilization, for instance, in rocket engine combustion chamber design, the linkage between combustion science and technological practice is most significant. The studies present the challenge of the inability to establish a strong linkage between combustion science and its engineering counterpart, particularly in laminar flame front dynamics (Marble et al., 1956). As scientists continue to probe more deeply the intricate workings of combustion reactions, an improved understanding of the role of changes in enthalpy on performance and emissions can help advance the development of cleaner, more efficient energy systems. Advances in this field work to underscore the utility of combining thermodynamic laws with imaginative engineering to meet the challenge of energy in the contemporary period (Figure 3.13).

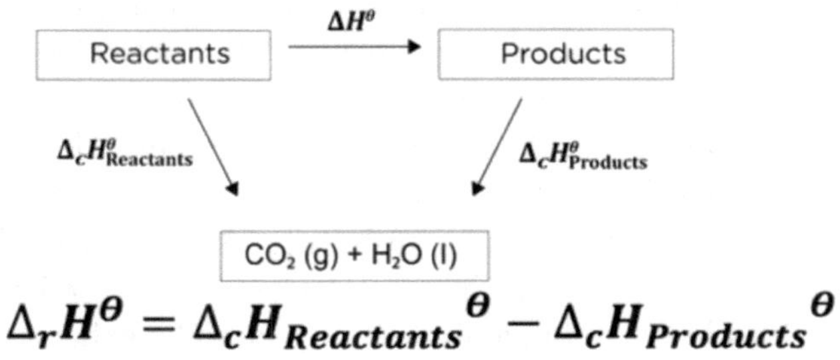

Figure 3.13. *Example of a combustion reaction.*

Source: https://dr22kca9mbyqg.cloudfront.net/a-levelnotes/images/uploads/9/6/0/2/96027112/published/screen-shot-2019-09-09-at-18-38-25.png?1568050732.

3.6.2. Example of a Formation Reaction

Formation reaction study is a significant feature in the identification of enthalpy change in chemical reactions. Formation reaction, a reaction that produces one mole of a compound from its elements in standard state, is fundamental in providing information on thermodynamic basics. The production of water from hydrogen and oxygen gas is a case: $2H_2(g) + O_2(g) \rightarrow 2H_2O(l)$. The enthalpy change in this reaction, the standard enthalpy of formation, plays a crucial role in determining the stability and energy relations of the resulting products. This enthalpy change not only signifies the energy that is released or taken up during the course of the reaction but also serves as a crucial reference point for comparison with other chemical reactions in the field of thermochemistry. The role of polyunsaturated fatty acids (PUFAs) in biological systems also indicates the significance of formation reactions for enthalpy changes. PUFAs, including docosahexaenoic acid (DHA), as emphasized in recent studies, exhibit unique structural features pertinent to their formation and application in biological membranes. PUFAs are specifically significant as antioxidants, particularly in the human brain, emphasizing their thermodynamic relevance. Phospholipid synthesis involving PUFAs involves intricate conformational changes on the basis of molecular interaction and energetics. Such information demonstrates how conformational demands affect molecular enthalpy, impacting structural stability as well as biochemical function of cell membranes, thereby linking formation reactions to basic biochemical processes (Csizmadia et al., 2007). Research into the enthalpy of formation of iron's manganese oxide spinels also demonstrates the intricacies of formation reactions within inorganic chemistry. For instance, the enthalpy of formation of spinels derived using oxide melt solution calorimetry varies with varying manganese composition. Negative enthalpies of formation values between 0 and 0.67 are characteristic of a thermodynamically stable reaction, but trends towards slightly positive values predict stability changes at higher compositions. These results point towards cation distribution and oxidation-reduction kinetics as key factors in the synthesis of such compounds, significantly affecting enthalpy. Thereby, the research emphasizes the necessity of understanding the enthalpy change of formation reactions in providing a hint on the material's behavior and stability and is indispensable in the quest for developing academic and industrial utilization (Coradin et al., 2004).

3.6.3. Example of a Phase Change Reaction

Phase change reactions provide the best demonstration of the intricate relationship between matter and energy, particularly in the course of chemical reactions. One of the most typical examples is the solid-liquid phase transition of titanium under a self-propagating combustion synthesis process. Upon heating titanium particles, they may undergo a solid-liquid phase transition where the enthalpy of fusion becomes the main consideration to focus on. This alteration not only identifies the energy required to break intermolecular bonds in the solid phase but also indicates that the system is heat-absorbing up to the point of melting. Such phase transitions, thus, constitute perfect indicators of enthalpic changes, demonstrating how energy phenomena dictate material properties and behavior for chemical synthesis processes. In the SHS case, the interaction of solid titanium with carbon illustrates how phase transition facilitates chemical reactions. If a critical temperature is reached, liquid titanium can react with particles of carbon, which, when in contact, yield titanium carbide (TiC). The effectiveness of this reaction relies on thermal conditions since the localized heat from the transformation not only facilitates the phase transition but also promotes the reaction. The fine interplay of the thermal energy and the material state enables fine planning necessary for productive reaction conditions in industrial applications. Such interactions bring into focus the manner in which the energetics of phase transition can lead to the constructive formation of compounds of essential significance in various technological applications. Moreover, such phase change reactions can be better understood with frameworks that compare kinetic and thermodynamic factors. Here, Campbells rate of control informs us about what different influences do to reaction speed and selectivity, such as those in the production of TiC. For instance, the kinetic rate control level can dictate how quickly the titanium particles transfer into the liquid state and engage in subsequent reactions. The inclusion of frameworks that study both kinetic and thermodynamic events allows for a full understanding of enthalpy change in phase transitions. This is particularly important in complexity reaction networks with multiple pathways, illustrating that the energy landscape surrounding phase changes not only provides practical insights into synthesis but also fundamental insight into underlying chemical principles, reinforcing the delicate balance between material evolution and energy conversion in chemical reactions (Andreasen et al., 2011; Badhuri et al., 2004) (Figure 3.14).

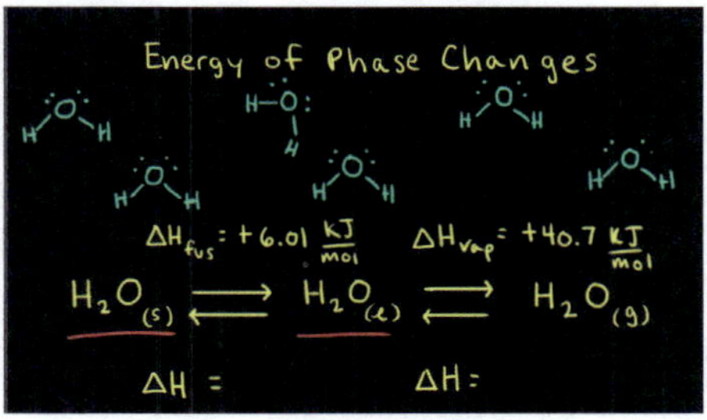

Figure 3.14. *Example of a phase change reaction.*

Source: https://cdn.kastatic.org/ka_thumbnails_cache/79213b66-1851-41f8-b13e-cacebad3962c_1280_720_base.png.

SUMMARY

The investigation of chemical reaction enthalpy changes is an important demonstration of the intimate coupling of reaction mechanism and thermodynamics. With careful examination, it has come to be appreciated that the thermodynamic variables, particularly enthalpy, not only dictate the viability of chemical processes but also the rate of such processes. The impact of enthalpic disparity is most important in areas such as catalysis, where adjustments in enthalpy can be a matter of making or breaking reactivity and selectivity. As seen from current studies on acidic zeolites, identification of the relationship between structure and enthalpy change can be utilized to develop better catalytic operations, and consequently more efficient chemicals and fuel production. This intersection of principles of thermodynamics and their engineering use is an indication of the necessity for further research seeking to optimize reaction pathways in green chemical engineering. Furthermore, enthalpy change studies offer evidence in relation to stability and degradation processes in materials. For instance, the investigation of perovskite structures, where it has been established that enthalpic and entropic effects significantly dictate the stability of the compounds. The subtleties incorporated into materials like $CH_3NH_3PbX_3$ show that enthalpy needs to be thought about in combination with others; thermodynamic stability is strongly linked with kinetic considerations, thereby impacting long-term material performance in applications such as photovoltaics. In addition, discrepancies between calorimetric measurements of such perovskites show that stronger experimental approaches are required on an immediate timescale—highlighting the way in which changes in enthalpy are key to chemical understanding and technological advancement. These findings necessitate an integrated approach involving both thermodynamics and kinetics in future work. Lastly, the difference in enthalpy changes via chemical reactions validates the need for comprehensive and in-depth research to examine their implication on reaction dynamics. As discussed, the phenomenon of alkane cracking in zeolites and perovskite degradation reactions describes a broader theme concerning how enthalpic factors affect not only reaction efficiency but material stability too. It is integration of theoretical considerations and empirical data that is required to construct predictive models capable of informing industrial practice. The necessity for precise thermodynamic and kinetic studies emphasizes the role of enthalpy as the cornerstone of chemical thermodynamics. With an improved understanding of this, chemists can approach more sustainable processes and material discoveries, leading to opportunities for future breakthroughs in chemistry and materials science.

Enthalpy change in chemical reactions is significant in predicting what occurs in energy during reactions. The concept of enthalpy, denoted as H, is the heat present in a system and is a significant parameter in thermodynamic evaluation. In a chemical reaction, the change in enthalpy (ΔH) quantifies energy expended or released and reflects the spontaneity and stability of the reaction. Surprisingly, endothermic processes that absorb heat possess a positive ΔH, while exothermic processes release heat and possess a negative ΔH. These variations are inherently related to the molecular interactions and bond formations that occur throughout the reaction process. Hence, understanding these energy transitions is not only useful for theoretical prediction but also gains additional applicability for practical uses in a wide array of scientific fields, particularly chemical engineering and materials science. Inclusion of enthalpy changes transcends theoretical models and influences many practical applications. In fields such as combustion science, an understanding of enthalpy is essential to the optimization of reaction conditions and effective design of energy systems. Combustion processes, for instance, being often at the cutting edge of energy conversion technology, require proper estimates of enthalpic change to optimize fuel economy and reduce environmental impact. This has been made possible by advanced computational models that have allowed scientists to model and predict the behavior of complex systems. As identified in recent advancements, the aerothermochemistry equations inform the development of combustion chamber technologies, with an even deeper connection between theoretical knowledge and engineering applications (Marble et al., 1956). This shows that a comprehensive understanding of enthalpy changes is needed to drive innovation in energy production and safety. Greater technological advancements in simulation technologies have immensely fostered greater understanding of enthalpy changes in chemical processes. As explained, the construction of numerical means for chemical vapor deposition (CVD) allows room for comprehensive examination of reaction dynamics, and more so in complex geometries. Using computational fluid dynamics codes like FLUENT, researchers can simulate main phenomena, including temperature oscillations and chemical species distributions, which are crucial to optimizing reaction atmospheres in real applications (Childs et al., 2025). These advances not only facilitate a better appreciation of the link between enthalpy change and reaction mechanism but also facilitate more efficient designs in the industrial setting. The higher accuracy in the simulation of enthalpy-associated phenomena as the technology continues to evolve offers promise to produce additional innovation, eventually resulting in more efficient and sustainable operations in chemical engineering and related disciplines.

REFERENCES

1. Ambaum, M. H. P., Arakawa, A., Bannon, P. R., Bechtold, P., Betts, A. K., et al. (2014). On the computation of moist-air specific thermal enthalpy. http://arxiv.org/abs/1401.3125 (accessed on 4th October 2025).
2. Andreasen, A., & Stegelmann, C. (2011). Finding the key transition states and intermediates controlling net reaction rates and selectivity. https://core.ac.uk/download/pdf/290114.pdf (accessed on 4th October 2025).
3. Axelrod, B. W., Johannessen, M., Roukes, M. L., Murphy, K., Recht, D., Thorsen, T., Fon, W., et al. (2009). High-sensitivity microfluidic calorimeters for biological and chemical applications. https://core.ac.uk/download/4882823.pdf (accessed on 4th October 2025).
4. Aziz, A., Calero, S., Cruz Hernández, N., Grau-Crespo, R., et al. (2015). Electronic structure of porphyrin-based metal–organic frameworks and their suitability for solar fuel production photocatalysis. https://core.ac.uk/download/185608498.pdf (accessed on 4th October 2025).
5. Badhuri, R., Bardenhagen, S., Burgess, D., Carslaw, H. S., de Gennes, P. G., Dunmead, S., Frenkel, D., et al. (2004). Particle-in-cell simulation of combustion synthesis of TiC nanoparticles. http://arxiv.org/abs/cond-mat/0402005 (accessed on 4th October 2025).
6. Barnes, B., Benedicks, M., Bent, H. A., Born, M., Carnot, S., Clausius, R., et al. (2000). Inquiries into the nature of free energy and entropy in respect to biochemical thermodynamics. https://core.ac.uk/download/25283819.pdf (accessed on 4th October 2025).
7. Bell, A. T., Head-Gordon, M., Janda, A., Lin, L. C., et al. (2018). Understanding Brønsted-acid catalyzed monomolecular reactions of alkanes in zeolite pores by combining insights from experiment and theory. https://core.ac.uk/download/158345582.pdf (accessed on 4th October 2025).
8. Bhaskar, R. G., Savin, D. W., Urbain, X., Vissapragada, S., et al. (2017). On the energetics of the $HCO^+ + C \rightarrow CH^+ + CO$ reaction and some astrochemical implications. https://core.ac.uk/download/161457412.pdf (accessed on 4th October 2025).
9. Bodi, A., Baboul, A. G., Baer, T., Berry, R. J., et al. (2012). A halomethane thermochemical network from iPEPICO experiments and quantum chemical calculations. https://core.ac.uk/download/pdf/9427899.pdf (accessed on 4th October 2025).
10. Casares González, J. Á., Espinet Rubio, P., & Pérez Iglesias, M. (2018). Comparing proteolysis and trans metalation reactions: Microcalorimetric studies of C–Au(I) bonds in [AuRL] complexes. https://core.ac.uk/download/225143515.pdf (accessed on 4th October 2025).

11. Chen, T. (2003). Physicochemical properties of nickel and cobalt sulphate solutions of hydrometallurgical relevance. https://core.ac.uk/download/11231289.pdf (accessed on 4th October 2025).

12. Childs, E. P., & Jasinski, T. J. (2025). Numerical modeling tools for chemical vapor deposition. https://core.ac.uk/download/pdf/42809940.pdf (accessed on 4th October 2025).

13. Ciccioli, A., & Latini, A. (2018). Thermodynamics and the intrinsic stability of lead halide perovskites $CH_3NH_3PbX_3$. https://core.ac.uk/download/159268116.pdf (accessed on 4th October 2025).

14. Coradin, H., Guillemet-Fritsch, S., Navrotsky, A., Tailhades, P., et al. (2004). Thermochemistry of iron manganese oxide spinels. https://core.ac.uk/download/12039038.pdf (accessed on 4th October 2025).

15. Csizmadia, G. I., Penke, B., & Viskolcz, B. (2007). A gyökfogó dokozahexaénsav mint agyvédő = Docosahexaenoic acid (DHA) as a free radical scavenger brain protector. https://core.ac.uk/download/11853925.pdf (accessed on 4th October 2025).

16. Davies, D. B., Evstigneev, M. P., & Rybakova, K. A. (2006). Complexation of norfloxacin with DNA in the presence of caffeine. https://core.ac.uk/download/7205.pdf (accessed on 4th October 2025).

17. Davis-Van Atta, C., Grieman, F. J., Noell, A. C., Okumura, M., et al. (2011). Determination of equilibrium constants for the reaction between acetone and HO_2 using infrared kinetic spectroscopy. https://core.ac.uk/download/4891946.pdf (accessed on 4th October 2025).

18. Dreyfus, B. W., Geller, B. D., Gouvea, J., Redish, E. F., et al. (2013). Chemical energy in an introductory physics course for the life sciences. http://arxiv.org/abs/1308.3667 (accessed on 4th October 2025).

19. Ferro, V. R., Gonzalez-Miquel, M., Moreno, D., Palomar, J., et al. (2019). Molecular and thermodynamic properties of zwitterions versus ionic liquids: A comprehensive computational analysis to develop advanced separation processes. https://repositorio.uam.es/bitstream/handle/10486/685268/exploring_moreno_CPC_2018ps.pdf?sequence=1&isAllowed=y (accessed on 4th October 2025).

20. Fultz, B., Graetz, J., Reynier, Y., Rez, P., et al. (2004). Entropy of Li intercalation in Li_xCoO_2. https://core.ac.uk/download/4871545.pdf (accessed on 4th October 2025).

21. Guo, T., Jiang, L., Li, B., Ren, Y., et al. (2024). An overview of application-oriented multifunctional large-scale stationary battery and hydrogen hybrid energy storage system. https://core.ac.uk/download/596422343.pdf (accessed on 4th October 2025).

22. Hall, M. L. (2012). Density Functional Theory: Development and Applications. https://core.ac.uk/download/161439672.pdf (accessed on 4th October 2025).
23. Jacques, D., Mallants, D., Šimůnek, J., van Genuchten, M. T., et al. (2018). The HPx software for multicomponent reactive transport during variably-saturated flow: Recent developments and applications. https://core.ac.uk/download/286361774.pdf (accessed on 4th October 2025).
24. Lichtenegger, H. C., Singer, G., Sinn, G., Veigel, S., et al. (2019). Evaluation of *in-situ* shrinkage and expansion properties of polymer composite materials for adhesive anchor systems by a novel approach based on digital image correlation. https://core.ac.uk/download/287939523.pdf (accessed on 4th October 2025).
25. Marble, F. E. (1956). Flame Theory and Combustion Technology. https://core.ac.uk/download/4887597.pdf (accessed on 4th October 2025).
26. Morgan, D., Shim, S. H., & Xu, S. (2015). Origin of Fe^{3+} in Fe-containing, Al-free mantle silicate perovskite. http://arxiv.org/abs/1607.02777 (accessed on 4th October 2025).
27. Nordén, B., Starikov, E. B., & Einstein, A. (2012). Entropy–enthalpy compensation may be a useful interpretation tool for complex systems like protein-DNA complexes: An appeal to experimentalists. http://arxiv.org/abs/1202.5734 (accessed on 4th October 2025).
28. Nováčková, K. (2009). Stability of salts and complexes of humic acids. https://core.ac.uk/download/30278457.pdf (accessed on 4th October 2025).
29. Reed, B. D. (2025). Small hydrogen/oxygen rocket flow field behavior from heat flux measurements. https://core.ac.uk/download/pdf/42806223.pdf (accessed on 4th October 2025).
30. Siemons, R. V. (1986). Interpretation of the exergy equation for steady-flow processes. https://core.ac.uk/download/pdf/11473275.pdf (accessed on 4th October 2025).

CHAPTER 4

CALORIMETRY AND EXPERIMENTAL METHODS

Contents

Introduction to Calorimetry....................121
4.1. Basic Principles of Calorimetry....................126
4.2. Types of Calorimeters....................131
4.3. Experimental Methods in Calorimetry....................136
4.4. Factors Affecting Calorimetric Measurements....................141
4.5. Case Studies in Calorimetry....................146
4.6. Advances in Calorimetric Techniques....................151
Summary....................155
References....................158

INTRODUCTION TO CALORIMETRY

Calorimetry is a fundamental experimental method applied in chemistry and materials science for the measurement of heat transfer associated with physical and chemical reactions. The technique enables one to calculate energy change during reactions, phase transitions, and material thermal stability quantitatively. Through various types of calorimeters—e.g., bomb calorimeters for combustion processes or differential scanning calorimeters to observe phase transformations—researchers can accurately provide information on changes in enthalpy. These measurements have intrinsic interest because they allow investigation into reaction mechanisms and the thermodynamic properties of materials. In addition, calorimetry is an essential part of quality control in industries, where it ensures that materials meet a specific thermal performance requirement, thus making it an essential part in research studies as well as application. Other than application, calorimetry provides enormous amounts of information about the kinetic behaviors of materials. For instance, the analysis of temperature-time curves measured by constant rate thermal analysis (CRTA) can determine the kinetic models for the thermal degradation processes. As a recent article has demonstrated, these curve shapes are directly associated with the reaction mechanisms behind, for instance, phase boundary reactions or random scission effects in materials such as polytetrafluoroethylene (PTFE) and poly(vinyl chloride) (PVC) (Luque et al., 2011).

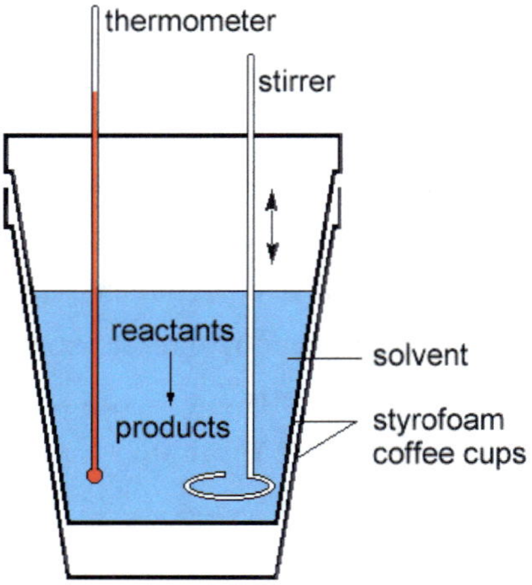

Figure 4.1. *Understanding calorimetry in chemistry.*

Source: https://www2.chem.wisc.edu/deptfiles/genchem/netorial/modules/thermodynamics/chemical/coffee.gif.

This relationship makes it possible for scientists to separate the kinetic model functioning on one CRTA curve, hence obtaining more data about material behavior under thermal stress. Calorimetry is thus a good method both for research studies and development of materials with optimized thermal properties. The use of calorimetry extends to structural analysis outside thermal analysis. It is a fundamental method for the identification of weak points in inorganic compounds, thereby allowing room for advancement in the creation of stronger and tougher compounds. An example is that a PhD thesis explored various methods, including calorimetry, to quantify structural and mechanical properties of inorganic binders (Wasserbauer et al., 2013). With the use of this technique in conjunction with other techniques such as nanoindentation and ultrasonic testing, scientists are capable of gaining a comprehensive material characterization. The information gained not only yields greater knowledge about the properties of a material but also aids in the creation of improved materials tailored to a particular application. Therefore, calorimetry remains an integral component of experimental methods, bridging the gap between prediction and reality in material science (Figure 4.1).

Definition and Importance of Calorimetry

Calorimetry, simply put, is a quantitative measure of the transfer of heat during chemical reactions, physical transformations, or changes of state. Calorimetry is an essential scientific instrument to comprehend macro and micro-level thermodynamic properties. With the aid of various types of calorimeters—such as bomb calorimeters for combustion reactions or differential scanning calorimeters for the investigation of phase transitions—scientists can quantitatively measure the heat that is absorbed in or released in processes. Its ability to produce reliable thermodynamic data renders calorimetry highly applicable in chemistry, physics, and biology. Secondly, the technique helps explain the interactions and stability of molecular forms, enhancing the prediction of behavior in a broad array of environments. Beyond just simple scientific observation, the pertinence of calorimetry extends to pharmaceuticals, materials science, and environmental study. For instance, calorimetric data can be employed to direct the design of more effective drugs based on the solubility and stability of substances under different conditions. Calorimetry also plays a key role in the investigation of the energetics of biological systems, such as protein folding and enzyme kinetics. Microcalorimetry has been employed recently to demonstrate how the external environment, as solvent composition, influences the stability of biological macromolecules. Interestingly, research revealed that acetonitrile presence affects hen lysozyme denaturation temperature, thereby emphasizing the empirical relevance of calorimetric research to protein stability (Kovriguine et al., 2000). Calorimetry is important in chemistry for the determination of reaction enthalpies, and it provides crucial information on bond energies and reaction pathways. Using precise calorimetric measurements, scientists are able to derive valuable Bond Dissociation Energy (BDE) values that are reflective of molecular bond stability. For example, research based on calorimetry of organometallic complexes has revealed how the liberated energy in specific reactions can be substituent group-dependent, showing the inseparable connection between molecular structure and thermodynamic characteristics.

These results are invaluable in optimizing synthetic procedures and adjusting reaction conditions in many chemical transformations (González et al., 2018). Overall, calorimetry is a cornerstone of experimental methodology, gaining greater insight into energy change in advanced systems and stimulating innovation in a wide range of scientific disciplines (Figure 4.2).

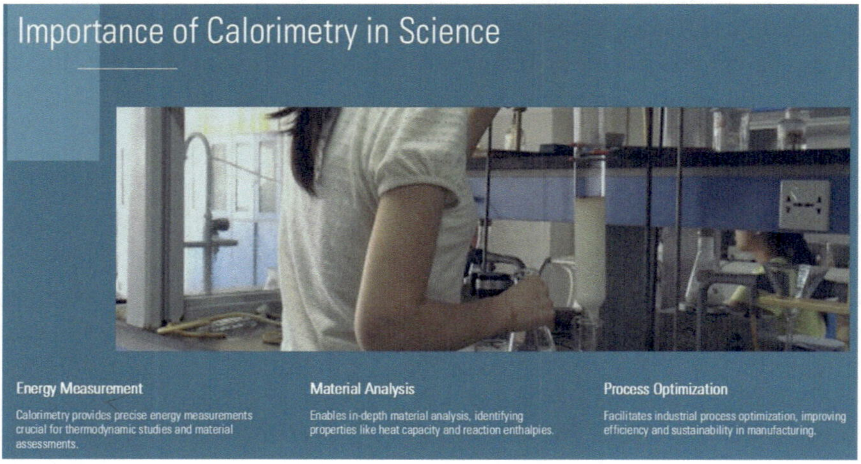

Figure 4.2. *Importance of calorimetry.*

Source: https://www.slideteam.net/media/catalog/product/cache/1280x720/u/n/understanding_calorimetric_methods_techniques_applications_and_benefits_ppt_powerpoint_st_ai_slide03.jpg.

Historical Background and Development

The origin of calorimetry in historical context is well-founded in the evolution of thermodynamic principles. The initial experiments of calorimetry were directly rooted from the works of leading scientists such as Antoine Lavoisier and Joseph Louis Gay-Lussac between the late 18th century and early 19th century. Their extensive investigation of heat transfer and reaction energetics developed key principles such as specific heat and latent heat. In the 19th century, technological advances allowed for accurate measurements and the development of the first calorimeters. These devices transformed scientists' measurement of the heat exchange in chemical reactions, the structure required for understanding energy transformations in general. This laid the foundation for additional research that eventually led to the invention of current experimental methods used in calorimetry as well as expanding its uses in various scientific fields today. With the advent of calorimetry, the 20th century witnessed significant advancement in both the techniques and theory employed within the discipline. The introduction of advanced calorimetric techniques, particularly differential scanning calorimetry (DSC), transformed thermal analysis of material properties. By precisely and repeatable measurement of heat flow, DSC enabled scientists to measure subtle thermal transitions that are critical to the characterization of phase transitions in materials. Besides this, sophisticated computational models that were created for utilization with experimental data allowed researchers to model and predict calorimetric behavior. Technological advances and instrumentation broadened the uses of calorimetry in material science, biochemistry, and pharmaceuticals. This union of modern technology and traditional procedure represents a significant step in the accuracy and trustworthiness of calorimetric data. Over the past decade, the changing face of calorimetry has also been driven by international collaborative research and technological advances, e.g., in the International Linear Collider (ILC) project. The ILC will push our knowledge of particle physics, which includes calorimetric detection of energy released in high-energy particle collisions, that plays a role in the physics of the Higgs boson and beyond (Barish et al., 2013). In addition, developments in the study of neutrinos highlight the importance of calorimetric techniques in the detection and characterization of neutrinos, such as their mass and oscillation properties (Gil-Botella et al., 2015). These activities highlight the growing importance of calorimetry not just in physics and chemistry but even in cross-disciplinary use. Generally, the historical trajectory of calorimetry shows an interaction between experimentation and theory that remains an impetus to scientific research.

Applications in Various Scientific Fields

Calorimetry and experimental methods find far-reaching applications in scientific fields such as biomedical science and materials science. Nanomagnetic hyperthermia (NMH), in particular, is a potential cancer treatment method. With precision radio frequency circuits, researchers can measure heat absorption in therapy without using traditional invasive calorimetry methods. According to this new technique, which relies on measuring variations of resonant circuit quality factor, power dissipated can be continuously monitored in hyperthermia therapy. Not only does this method avoid the inaccuracy in conventional calorimetry but also integrates very well with magnetic resonance imaging (MRI) protocols, thereby enhancing the overall efficiency of cancer therapy programs. Certification of this method in comparison to conventional calorimetric tests speaks to its efficacy and potential for broader clinical application (Gresits et al., 2018). In material science, calorimetric techniques are necessary to analyze the thermal properties of composite materials, particularly those reinforced with fibers. Composite materials are notable for their high strength-to-weight ratio and an unprecedented degree of durability, a quality that makes them ideal candidates for aerospace engineering. The ability to quantitatively measure the heat capacities and thermal transitions of such materials using calorimetry not only contributes to the design effort but to quality control in production as well. Such developments over the mid-1960s have witnessed new applications in various aerospace components, where material performance is of paramount concern. Improving calorimetric methods keeps increasingly revealing the thermomechanical behavior of composite materials, ultimately culminating in enhanced durability and damage resistance in high-performance applications where they are most required (Ansell et al., 2025). Furthermore, incorporation of calorimetric methods in interdisciplinary research fuels innovation in the sciences. Scientists can explore new areas of energy storage systems such as batteries and supercapacitors by utilizing calorimetric data, where thermal management plays a very significant role. Understanding the heat generation and dissipation characteristics in such systems is crucial in order to ensure maximum efficiency and safety. Ultimately, calorimetry is an important experimental tool—not simply to measure thermal properties but also to put new technologies into widespread application across many fields. Since the demand for more sustainable and efficient technology is on the rise, the role of calorimetric methods will remain crucial in advancing basic research and applied technology for contemporary science.

4.1. BASIC PRINCIPLES OF CALORIMETRY

The foundation of calorimetry revolves around heat transfer measurement during chemical processes or physical changes. Essentially, calorimetry serves to quantify the energy transformations that are implicated in such processes so that scientists can effectively examine thermodynamic qualities. The most typical calorimetric experiments utilize the concept of a calorimeter, an apparatus that is utilized to isolate the system from its surroundings so measurements can be as accurate as possible. By controlling variables such as pressure and temperature exactly, scientists can extrapolate crucial information regarding enthalpy changes, which are of significant interest across many disciplines, from chemistry to materials science. Knowing these energy interactions is significant in predicting the state of compounds under changing conditions, particularly when testing combustible materials, where calorimetry provides valuable data to evaluate safety and performance. In calorimetry, a number of methods for measurement have been developed, each suited for particular purposes. A specific example is differential scanning calorimetry (DSC), which is commonly used to examine phase change, e.g., melting and crystallization, of materials. With this technique, scientists are able to view the manner in which an object takes in and gives out heat so that they can get a clear picture of its thermodynamic properties. Bomb calorimetry, on the other hand, is primarily applied in combustion studies, where heat produced when a sample is combusted is measured in a sealed container. Bomb calorimetry is particularly crucial in ascertaining the calorific value of fuels, which directly impacts energy production and efficiency. The calorimetric method of choice is greatly based on the sample type and nature of information to be collected, thereby indicating the importance of experimental design tailored to the needs of the calorimetric study. Lastly, calorimetric data should be handled with care and rigorously evaluated in line with proper thermodynamic knowledge. Calorimetric experiments results may be influenced by numerous factors such as sample size, reaction kinetics, and conditions. Hence, the variables have to be accounted for by the scientists during data analysis to make it both reliable and valid. For instance, when applying calorimeter methods in the research of materials like wood, which undergoes complex thermal decomposition, high-class analysis must be employed to identify good fire retardants and improve material safety. Recent developments in calorimetric techniques, including use of nanocomposites and fire-retardant formulations (Hull et al., 2013; Iglauer et al., 2013), demonstrate the immense variety of applications of calorimetric concepts in fields and indicate their importance in the development of green and safe materials (Figure 4.3).

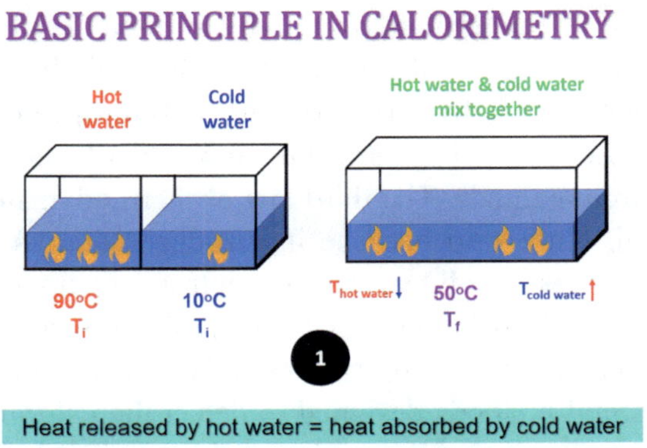

Figure 4.3. *Basic principle of calorimetry.*

Source: https://i.ytimg.com/vi/3tMNrPKvLGI/maxresdefault.jpg.

4.1.1. Laws of Thermodynamics

The thermodynamic laws form a fundamental framework that governs energy transfer and energy conversion processes in physical systems. The first law is also referred to as the law of conservation of energy, whereby energy cannot be created or destroyed but can be transferred from one form of energy to another. This is of prime importance in calorimetry, which is a technique for measuring heat transfers in chemical reactions or physical changes. By tracking these energy changes, scientists are able to estimate the heat capacities and enthalpy changes of various materials, thereby providing them with vital information about their heat behavior. Secondly, having the understanding of this law, scientists can predict the behavior of the system when put in various conditions, which is a critical aspect in experimental procedures that analyze the thermodynamic paths of matter.

These affirm the central role of the first law in connecting theoretical ideas with experimental calorimetric techniques. The second law of thermodynamics introduces yet another fundamental principle to explain entropy, a measure of the extent of randomness or disorder in a system. This legislation states that, in any process of energy transfer, the combined entropy of a closed system never diminishes with time; instead, it remains constant in the case of reversible processes or increases in the case of irreversible processes. In calorimetry, this concept sheds light on the spontaneity and feasibility of chemical reactions. For instance, the spontaneity of a reaction may be discerned from the accompanying changes in entropy so that chemists may evaluate reaction conditions for favorable outcomes. Recent work has highlighted the significance of entropy change for network-forming materials, which often exhibit characteristic calorimetric behavior that could possibly reflect hidden thermodynamic principles. This connection between calorimetry and entropy provides additional understanding of how energy disperses at the molecular level, giving substance to the field's research methods. The third law of thermodynamics also extends the definition of absolute zero because it introduces that as the temperature of a system approaches absolute zero, the entropy approaches a minimum constant. This act has deep implications for the theoretical underpinnings of calorimetry, particularly for phase transitions and heat behavior of materials. For example, experimental methods designed to detect the calorimetric properties of materials at lower temperatures can detect phase transitions not apparent at higher temperatures. Glassy relaxation processes, like those handling network glasses, are likely to be interested in these low-temperature processes and how they relate to the concepts of thermodynamics. Besides, measurements from calorimetric observations close to absolute zero temperatures help further refine material behavior models and improve prediction processes for thermal properties, and therefore validate the third law in contemporary thermodynamic research.

4.1.2. Heat Transfer Mechanisms

Mechanisms of heat transfer are critical to calorimetry theory and practice as well as applications in experimental techniques. They consist of three prominent modes: conduction, convection, and radiation, each linked to certain processes. Conduction occurs through the interaction of molecules in solids, in which the heat is delivered from the warmer regions to colder regions through direct contact. It makes up an important segment of experiments with solids because it regulates how quickly and effectively temperature equalization happens in a material. The efficiency of conduction can be a function of a number of variables including material conductivity and thermal gradient. Of greater concern is the fact that the significance of precise temperature-time profiles in thermal investigation cannot be overstressed, particularly in methods such as constant rate thermal analysis (CRTA) where kinetic model building is enabled by the precise understanding of heat transfer in materials (Luque et al., 2011). Convection, the second mechanism of heat transfer necessary for operation, is achieved through motion of liquids where warm areas rise and cool areas descend, producing a cycle of circulation. This is most critical in interactions between liquid and gas phases and is greatly responsible for results of calorimetric tests. For instance, properties of the fluid being tested, like viscosity and thermal properties, may be factors affecting efficiency of heat transfer during testing. It is of major significance to comprehend convection while designing experiments grounded on fluid dynamics since convection determines thermal homogeneity and stability of samples directly. In most experimental conditions, such as tests used to quantify thermal degradation and combustion reactions, such as that found in wood materials, the heat transfer properties can determine the outcome of testing protocols as well as interpretation of data (Iglauer et al., 2013). Radiation, the third process, is the transfer of heat by way of electromagnetic waves, mostly in situations where bodies are not in contact with one another. Radiation is extremely crucial in hot climates where both convection and conduction cannot occur. In calorimetry, the effect of radiation needs to be adequately managed in order to obtain precise readings because it can contribute to error unless adequately accounted for. For example, in thermal analysis of materials like wood, the radiation coming out on decomposition can influence the findings. Adding shielding material, experimental setup alterations, and taking environmental factors into account are therefore significant. In general, a general knowledge of these heat transfer processes not only directs the design and execution of calorimetric measurements but also enhances the quality of results obtained in research on the properties of materials in thermal environments (Iglauer et al., 2013) (Figure 4.4).

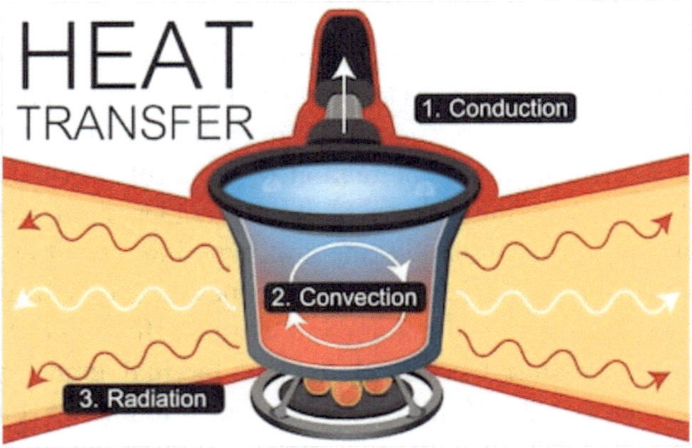

Figure 4.4. *Heat transfer mechanisms.*

Source: https://www.researchgate.net/publication/343084958/figure/fig2/AS:915518303264774@1595287981934/An-illustration-of-the-primary-mechanisms-of-heat-transfer-conduction-convection-and.ppm.

4.1.3. Measurement Units and Conversions

In calorimetry and experimentation, precision units of measurement and conversion are very important to ensuring accuracy and uniformity in scientific research. Different branches of science use different units of measurement, and conversion from one unit to another, e.g., joules to calories or grams to kilograms, requires a proper knowledge of conversion factors. These characteristics not only simplify calculations but also provide uniform results in different studies and experiments. Adherence to a measurement system on a strict basis allows scientists to compare results, verify experimental results, and build a coherent picture of energy behavior during chemical reactions, as achieved through calorimetric experiments. A complete understanding of the units of measurement thus acts as the foundation of credible data, which supports the integrity of scientific methods. Furthermore, the importance of accurate conversion is emphasized in experiments that determine thermal characteristics by techniques such as differential scanning calorimetry (DSC). In this technique, widely applied to assess the thermal behaviors of compounds, the measurement units in heat flow and temperature changes must be carefully considered. For instance, conversion of Celsius to Kelvin or vice versa is crucial in avoiding inconsistency in interpreting thermodynamic characteristics. In practical contexts, investigators may examine such parameters as the melting capacity of various anti-icers, where measurement with accuracy and unit conversions directly impact the determination of material effectiveness (as evidenced by results reviewed based on statistical framework) (Alnuaimi et al., 2019). Understanding unit conversions, thus, ensures results that are credible and apt for use in practical contexts, in support of the universal applicability of experiments conducted in the field. The consequences of measurement units and conversion are not limited to specific experiments but reach data compilation and synthesis of larger scientific endeavors. For instance, in sophisticated experiments, for instance, the ones that seek to comprehend neutrino behavior using detectors like the ICANOE, precise measurement of momentum and energy becomes paramount (Collaboration et al., 2000). Unit transfers must be managed with expertise on various parameters to maintain the soundness of the measurements of oscillations and to test theoretical predictions against empirical findings. While scientists collaborate across disciplines, the ability to harmonize units derived from various scientific paradigms allows researchers to develop a more interconnected body of knowledge. Consequently, the supportive role of measurement units and unit conversions facilitates the collaborative scientific inquiry of the contemporary period and a collective approach to examine complicated phenomena across disciplines.

4.2. TYPES OF CALORIMETERS

The calorimetry sector is diverse and is typified by different calorimeters, each designed to fulfill specific experimental needs. The two most common types are the electromagnetic calorimeters (ECAL) and the hadronic calorimeters (HCAL). ECALs optimally identify the energy of electrons and photons through electromagnetic processes and have materials that provide good resolution for these particles. Conversely, HCALs are intended to absorb hadron energy, which engages through multiple modes. The divergent operational methodologies of the two classes of calorimeters highlight the significance of material selection and design, optimized to provide high efficiency in absorption and measurement. The difference is one of vital significance in high-energy particle physics, wherein reliable detection plays a pivotal role in data interpretation and experimental verification. In addition to these standard forms, new developments like dual-readout calorimetry have appeared to advance the measurement in capabilities. This technique overcomes inherent limitations with hadronic energy detection, as standard calorimeters find it difficult to provide accurate readings. Through the use of two separate readout systems with one being sensitive to electromagnetic showers and the other to hadronic interactions, researchers can minimize inconsistencies in the measurement of energy (Lee et al., 2017). This approach improves not only the resolution for energy measurements of the determinations but also allows discrimination among different particle types better. Dual-readout systems, therefore, provide a significant breakthrough, which combines the beneficial features of ECAL and HCAL with simultaneously reducing the issue of high-energy hadrons. Calorimetric approaches continue to be on the rise in terms of progress as dual-readout systems are refining with the rising finesse. Dual-readout systems have proven useful in high-energy physics experiments. Modern improvements highlight the way this technique has now become pivotal to measuring the properties of hadrons and hadron jets, with huge advances over earlier calorimeters (Lee et al., 2017). Improved performance illuminates fine details in particle interactions and patterns that help physicists pursue unsolved questions at the high-energy frontier with greater precision. With advancements in calorimetry methods, continued evaluation of their performance and limitations will always be critical, resulting in future breakthroughs that aid us in understanding more about basic physics concepts.

4.2.1. Constant Pressure Calorimeters

Constant pressure calorimeters are of critical significance in experimental thermodynamics, particularly in the measurement of heat capacities of various substances. The instruments maintain constant pressure during measurement process, which allows for accurate measurement of thermal properties within a controlled environment. The continuous pressure calorimetry is especially useful in the investigation of phase transformations and reactions at atmospheric pressures, under which changes in pressure can lead to erroneous results. With scientists increasingly exploring the complexities of thermal behaviors, such calorimeters provide a uniform approach that can enhance the accuracy and reproducibility of experiments. Their accuracy in measurements they make ensures that theoretical models are able to be stringently tested, promoting the field's understanding of thermodynamic principles. A key innovation for the field of calorimetry is the development of microfabricated constant pressure calorimeters, or μ-calorimeters. These devices utilize high-performance materials such as low-stress silicon nitride to minimize thermal mass, enhancing sensitivity to measurement. With the capability to analyze tiny liquid samples as little as 200 nanoliters, μ-calorimeters are capable of performing a variety of analyses including differential scanning calorimetry and thermal wave analysis. Integration of a heater and temperature sensor in these instruments allows simultaneous analysis of thermal diffusivity along with specific heat, hence expanding the range of thermodynamic research. These advances are testimonies to the ongoing evolution of calorimetric technology, and they open up new avenues of research across fields, including biochemical and materials science (Arpaci et al., 2014). Moreover, the applications of constant pressure calorimeters extend beyond traditional thermal analysis and find growing applications in high-energy physics. The Particle Flow Approach, the new paradigm for particle-level event reconstruction in particle physics, highlights the need for extremely granular calorimeters, which are vital for precision measurements in future collider experiments. Experimental and theoretical studies, particularly from collaborations like CALICE, have found these calorimeter techniques successful in probing detector concepts. Prototypes subjected to beam tests were not only resilient in performance but also provided illuminations such as never before on hadronic cascades. This union of calorimetry with advanced particle detectors unmistakably demonstrates the adaptability and applicability of constant pressure calorimeters to cutting-edge scientific research, broadening our horizons for basic physical phenomena (Kawagoe et al., 2015) (Figure 4.5).

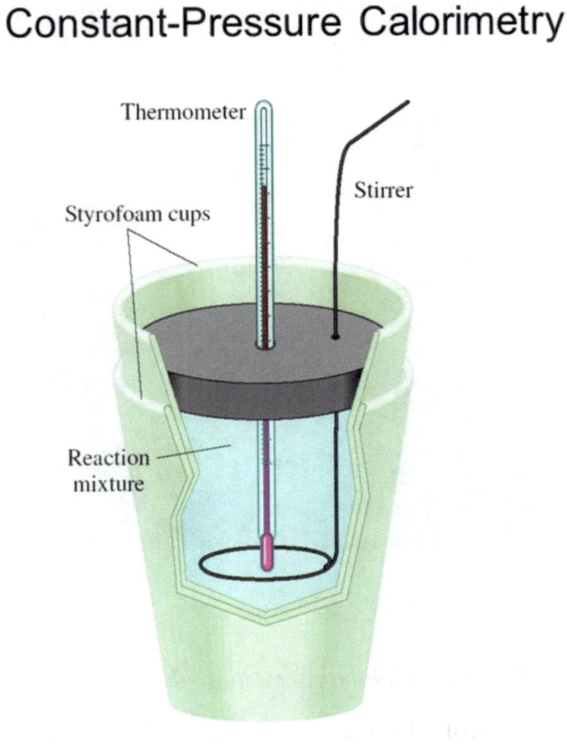

Figure 4.5. *Constant pressure calorimeters.*

Source: https://blogger.googleusercontent.com/img/b/R29vZ2xl/AVvXsEgNeyzN-Z7s1R2wqgXZP7F4ptS0WH5ajymUN34MQgRaUxGQlD5SSitKsafBZdspgvuN-8DzI_WTOSYhPzXpuA0ui91L671MJA7aIH_huWhEVgfLTjdovvQe48YOo9tb-8ceucWaZA9tE5PsRvD/s1600/11.jpg.

4.2.2. Bomb Calorimeters

Bomb calorimeter is a vital instrument utilized in the process of calorimetry for measuring the heat of combustion of compounds with high accuracy. A steel vessel, or bomb, is incorporated in the bomb calorimeter that holds the sample to be tested as well as an oxygen atmosphere. When the sample is ignited, combustion is generated and energy is released in the form of heat. This heat is subsequently absorbed by the surrounding water, whose temperature increase is quantified to find the heat of combustion of the sample. Not only does this apparatus give a highly accurate measurement of energy content, but it also sets up a valuable relationship between released energy and chemical composition of fuels, which can be invaluable in applications ranging from industrial energy production to ecological research. One of the distinguishing strengths in using bomb calorimeters is that they can give repeatable and consistent measurements, which is of paramount importance in scientific testing.

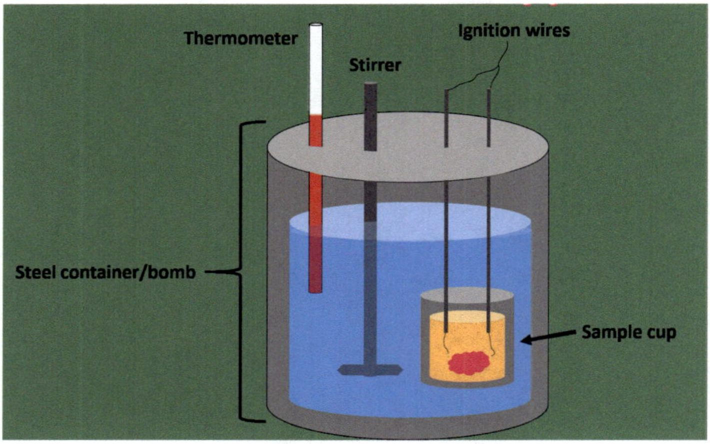

Figure 4.6. *Bomb calorimeters.*

Source: https://d20khd7ddkh5ls.cloudfront.net/bomb_calorimeter_apparatus.png.

The extremely controlled setup of the bomb calorimeter represses outside parameters, hence ensuring that only the combustion properties of the sample are represented by the measurements. This consistency is essential when assessing different fuel materials in solid, liquid, and gas forms. This kind of strategy carries far-reaching consequences beyond mere academic interest; it is especially relevant in the development and maximization of alternative sources of energy. By determining the heat of combustion values, researchers can fine-tune the efficiency of waste-sourced fuels, for example, as observed in pyrolysis activities, which converts organic waste into efficient energy products (Bulmău et al., 2019). Moreover, bomb calorimetry plays a pivotal role in enhancing the knowledge of processes of energy transformation in various applications, from hydrocarbon combustion to synthetic fuel analysis. Quantities obtained by bomb calorimeters can direct the creation of systems that are capable of optimizing fuel performance, particularly in pioneering technologies that utilize renewable resources. For instance, through the use of feedback control systems in combination with calorimetric analysis, researchers can produce real-time energy output measures, which encourage greener methods of producing energy (Chaturvedi et al., 2025). As the globe becomes increasingly conscious of the need for greener sources of energy, the role played by bomb calorimeters in experimental research cannot be substituted, giving way to effective analysis and improved energy conversion methods (Figure 4.6).

4.2.3. Differential Scanning Calorimeters (DSCs)

DSCs are highly relevant calorimetry devices which allow the investigation of thermal properties of materials. DSCs provide information regarding phase transitions, e.g., melting, crystallization, and glass transition, by quantifying heat flow to and from a specimen under a set temperature program. The method relies on comparing the heat flow of an example to a reference substance under identical conditions, thus allowing precise determination of thermal events. This capability makes DSC a valuable tool in numerous research applications, e.g., polymer science, pharmaceutics, and food technology, where behavior of materials when the temperature is changed should be known in order to maximize the process and product formulation. Recent advancements in the technology of DSC have enabled more sophisticated analysis through the development of microcalorimeters such as the μ-calorimeter, which is more sensitive and has lower thermal mass. The on-chip heaters and sensors in the microfabricated devices enable scientists to study liquid samples from 200 nanoliters in volume down. These characteristics give researchers greater flexibility and accuracy for measuring such properties as specific heat and thermal diffusivity, which are important in separating the unique characteristics of complex materials. Its ability to perform both direct current (DC) and alternating current (AC) calorimetry extends the use of thermal analysis and enables the study of dynamic processes in a broad range of applications, which underscores the significance of DSCs in contemporary experimental techniques (Arpaci et al., 2014). The significance of differential scanning calorimetry goes beyond material science and engineering but also includes biological researches and specifically the comprehension of thermodynamic responses. With regard to homoeothermic animals, calorimetric techniques can measure the energetic expense of acclimatization between varying environmental conditions. By measuring thermal energetics quantitatively, DSCs can illustrate mechanisms by which animals acclimatize physiologically to environments, with emphasis on acclimatization mechanisms. This research into animal energetics not only serves as a focus of biological science but gives an insight into evolution adaptation as animals face shifting climates and environments. Consequently, applications of DSCs in the study of biological processes highlight their potential, making them important tools in life and material sciences (DeShazer et al., 2009) (Figure 4.7).

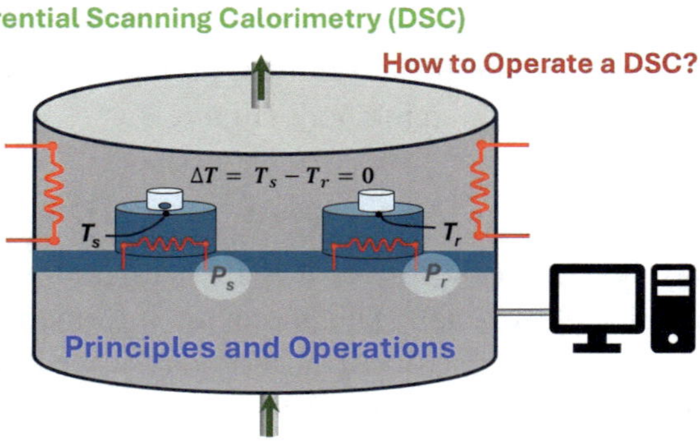

Figure 4.7. *Differential scanning calorimeters.*

Source: https://i.ytimg.com/vi/0os3qzVoXkQ/maxresdefault.jpg.

4.3. EXPERIMENTAL METHODS IN CALORIMETRY

Experimental techniques in calorimetry are important in accurate measurement of the energy changes associated with chemical reactions or physical changes. These methods are largely based on calorimeters, which are classified into types based on their structures and operating principles. For example, bomb calorimeters are largely applied in constant volume reactions to allow precise measurement of the liberation of heat in combustion processes, while coffee cup calorimeters can be useful in studying reactions under constant pressure. The accuracy of each method relies mainly on control of the environment and complete insulation to avoid heat exchange with the surroundings. Effective calorimetry thus requires sophisticated instrumentation, as well as careful experimental technique to produce reliable data, that are necessary for the interpretation of thermodynamic properties in a wide range of applications from materials science to biochemistry. With the investigation of more advanced experimental techniques in calorimetry, researchers are further incorporating cutting-edge technology, such as digital signal processing techniques. For instance, the application of $LaBr_3$:Ce detectors in gamma-ray spectroscopy presents innovative techniques for high-rate measurements that can be exploited in rapid calorimetric assessments under high count regimes. These developments make the efficient enhancement of resolution achievable at count rates greater than 1 MHz, further optimizing calorimetric data extraction and analysis.

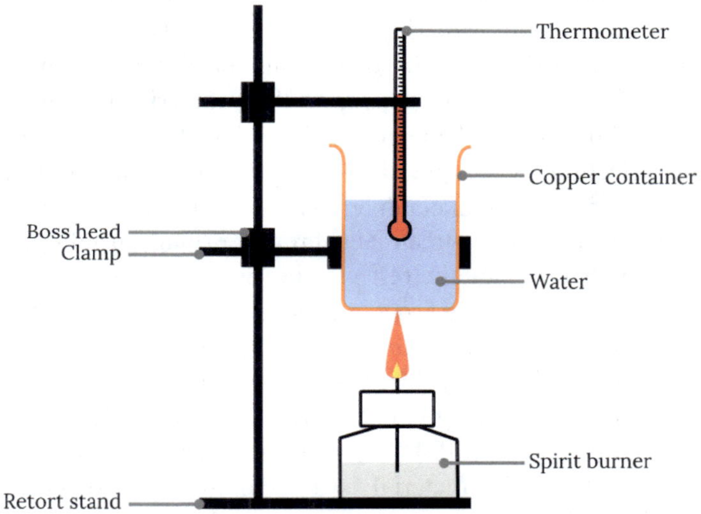

Figure 4.8. *Experimental methods in calorimetry.*

Source: https://www.learnable.education/wp-content/uploads/2020/05/Year-11-Chemistry-Practical-Calorimetry-experimental-setup-1.png.

The inclusion of pile-up correction methods further enhanced performance and effectively dealt with overlap problems of signals that can confound typical calorimetric measurement. This shows how detector development can revolutionize calorimetric methods to enable greater accuracy when measuring energy transitions. In addition, overlap between calorimetry and materials characterization strategies is becoming progressively vital, especially when it comes to the fire characteristics of organic materials. Wood thermal degradation studies, highlighting the need for effective fire retardants, rely on calorimetric measurement in conjunction with thermogravimetry. Cone calorimetry, specifically, allows scientists to determine heat release rates and ignition behavior of various wood treatments. Analysis of ignition resistance and integrity strategies for individual wood products is helpful towards sustainable building. Sustained development of new experimental techniques emphasizes the decisive role of calorimetry in theoretical calculations as well as practice in material sciences and safety engineering, as indicated in the recent achievements in the measurement of wood products flammability (Iglauer et al., 2013) (Figure 4.8).

4.3.1. Sample Preparation Techniques

Effective sample preparation methods are necessary for precise calorimetric measurements and overall experimental outcomes. These techniques have a significant effect on the accuracy of information derived from calorimetry through the achievement of consistent, well-characterized samples that are representative of the material under consideration. One common technique involves the selection of sample size and geometry appropriately because these parameters affect thermal conductivity and heat transfer rates. Moreover, the sample can also need pre-treatment through drying, grinding, or sieving to reach a state that accurately reflects the properties of the material for calorimetric analysis. Applying isoconversional methods of thermal analysis allows researchers to conclude stable parameters of materials under varying heating rates, which ultimately leads to improved predictive simulation of material behavior under varying temperatures, an element developed in depth through the various works on PMMA composites with silica (Čechová et al., 2012). Homogeneity of the material is a factor that cannot be overlooked when conducting calorimetric experiments as it directly affects the thermal behavior of the materials. In calorimetric sample preparation, homogeneity of composition is a requirement to prevent heat capacity disparities that could be caused by phase separation or the existence of contaminants. Techniques such as stringent mixing and precise partitioning by particle size ensure homogeneity. Novel techniques involving mechanical homogenization and cryogenic milling have also been found extremely promising in the achievement of homogeneous material properties. The importance of rigorous sample preparation methods is also evident in research on inorganic materials, where determination of the weakest structural phases is essential for maximizing mechanical properties, as pointed out in research studies probing the strength of cementitious material through calorimetric approaches (Wasserbauer et al., 2013). In addition to physical conditioning, the controlled environment under which the samples are prepared also has a significant role to play in their ultimate performance in calorimetric tests. Preparation of the samples needs to be conducted under specified conditions in order to reduce exposure to moisture, impurities, and temperature fluctuations that may alter the properties of the samples. Working under inert atmospheres or vacuum conditions can dampen oxidation or water intake, which is particularly critical with reactive materials. Besides, precise thermal history prior to calorimetric measurement enhances reproducibility to enable scientists to uniformly compare their results. Such elaborate preparation protocols not only enhance the accuracy of calorimetric measurements but also form the foundation of successful experimental analysis in various applications, evidencing interaction between sample preparation techniques and accurate calorimetry.

4.3.2. Calibration Procedures

Calibration procedures in calorimetry are critical in making precise and dependable measurements. Such processes would typically be the methodical comparison of a calorimeters reading with standard reference material or known standards. This allows for scientists to define the accuracy of the calorimeters output in an attempt to minimize deviations that are brought about by environmental factors or equipment malfunction. For calorimetric techniques to give credible data regarding metabolic rate or thermoregulation, there must be rigorous calibration conducted on a regular basis. Notably, the integration of extremely fine-grained detectors, such as those presented in new advances in particle flow calorimeters, demonstrates the complexity and need for exact calibration in experimental physics. Through clarified measuring procedures, researchers can safely interpret calorimetrically relevant information to their work, hence enhancing scientific understanding of thermodynamic behaviors in various systems (Kawagoe et al., 2015). Apart from routine calibration operations, new techniques have been created to improve calorimetric measurement accuracy. This entails the application of advanced computer software algorithms that check for discrepancies and compensate for systematic errors. Routine maintenance and performance testing of calorimetric instruments as well guarantees optimal instrument performance. For instance, when defining subjects' energy expenditure when using water cooling garments (WCG) in experimental designs, precise calibration guarantees that data points towards accurate metabolic rates and heat production (Annis et al., 2025). Calibration procedures like these are not only essential to provide sound data but also to enable comparisons across various experimental designs. As such, researchers are able to gain better insights into the intricacies of biological thermoregulation and energy processes in different environments. In the end, precise calibration procedures are the key to reliable calorimetric study and analysis. From enormous scale experiments in controlled environments to intricate studies on human metabolism, precision in calorimetry is of utmost concern. Utilizing standard calibration procedures with the aid of modern technology, researchers are able to enhance the fidelity of calorimetric data. With advancing discipline, particularly with new advancements in particle physics and calorimetry, following tight calibration standards will remain essential. This commitment to precision not only enhances the validity of scientific findings but also fosters an improved appreciation of calorimetric techniques important for diverse uses across disciplines (Annis et al., 2025; Kawagoe et al., 2015) (Figure 4.9).

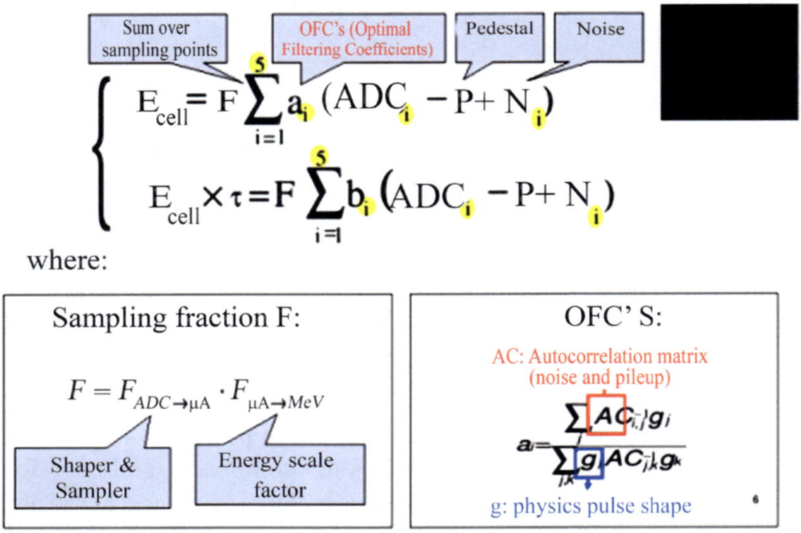

Figure 4.9. *Calorimeter calibration.*

Source: https://www.physics.smu.edu/~dgoldin/LArCalo/Calibration/eq1.png.

4.3.3. Data Collection and Analysis

Data collection and analysis is the pillar of calorimetry as it decides whether experimental data is reliable and accurate. Quantification of heat transfer in physical or chemical transformations within calorimetry is an important aspect, and hence a robust methodology needs to be set up. Choosing the equipment, sample preparation, and conditions for conducting experiments can significantly influence resulting data. Augmentation of experimental methods by professionals has verified that rigorous studies of the phase equilibria and the thermodynamic properties are crucial. The history of the period during which research into materials such as zirconium and hafnium carbonitrides was being carried out shows how knowledge at the grass-root level impacts contemporary research into their behavior and properties. The continuity and continuity of such methods, thus, are crucial for experimental verification of future theories (Hong et al., 2019). With the advancement of data gathering, more precisely with the integration of new techniques, researchers are provided with additional analytical capability. Technologies such as electrical pulse heating and laser melting are advanced methods of synthesizing and studying samples, offering the gaps in previous methods. These findings offer invaluable information that facilitate strict testing of theoretical predictions regarding materials such as hafnium carbonitrides, which have complex phase equilibria and thermodynamic properties. The efficiency of calorimetry data collection is carried forward to numerous qualitative and quantitative analysis aspects. With computational evidence complementing experimental understanding, the dependency between empirical and theoretical models is even more significant, further enhancing the need for data analysis rigor (Hong et al., 2019). Besides, data validation for calorimetric data involves full understanding of statistical methods in order to ascertain that data collected is not only properly interpreted but also suitably reported. Such analysis precision facilitates differentiation among trends and establishing correlation between varied thermodynamic concepts. In experimentation procedures, transparency and reproducibility of data are of highest priority, calling for well-documented protocols and peer assistance. The ability to handle large data sets combined with advanced computational techniques enhances the accuracy of prediction of material behavior under varying thermodynamic conditions. Ongoing research thus introduces a continuous flow of information that not only refines our understanding of calorimetric concepts but also provides applied insights in materials science.

4.4. FACTORS AFFECTING CALORIMETRIC MEASUREMENTS

The accuracy of calorimetric measurements is highly dependent on a broad range of external and internal factors that can skew results and generate inconsistent data. Thermal conductivity in materials that are applied in the measurement is a critical factor that affects heat transfer inside the calorimeter. For instance, the ambient temperature and the calibration standards must be properly controlled in order to minimize deviations. Experimental conditions such as room temperature or pressure variations can create adverse effects on precise measurements of heat capacities or enthalpy changes. To counteract such deficiencies, calorimeters must be calibrated before experiments are performed so that systematic errors will be calibrated out and corrected. This is essential in the achievement of true calorimetric values, especially in high-end research founded on high-pressure or high-temperature conditions. The chemical stability of the measured compounds also decides the accuracy of calorimetric data. Compounds that decompose or react while going through calorimetric measurement can potentially provide false readings as they tend to drastically alter the pattern of heat transfer. For example, in experiments conducted on complicated materials such as compounds of zirconium and hafnium, parameters such as phase equilibria and melting points must be precisely taken into account, since these will directly influence calorimetric measurements obtained through experiment. As Hong et al. (2019) further highlights, there should be a proper understanding of material properties, such as their thermaldynamics under varying conditions, to ensure effective calorimetric measurement. The ability to predict the performance of these materials under thermal stress not only enhances data quality but also allows researchers to make correlated optimizations in experimental configurations. Instrument selection and operating protocols are also key to calorimetric measurement precision. The type of calorimeter used, differential scanning calorimeter or bomb calorimeter, can introduce disparities in measurement accuracy based on the material and involved energy changes. As seen in concrete mix design research, workability and hydration behavior can easily affect calorimetric tests, which reflects the importance of mindful experimental planning. By using additives like calcium sulfate or lime, experimenters can manage factors like setting time and heat release, thus demonstrating the power of experimental conditions to shape calorimetric tests (Belie et al., 2016). Hence, full control over the experimental conditions as well as the materials utilized is necessary to achieve optimal calorimetric analysis, further maintaining the importance of systematic methodology in scientific experimentation (Figure 4.10).

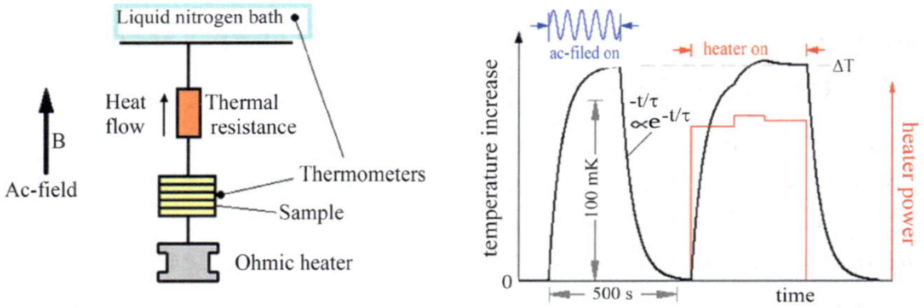

Figure 4.10. *Factors affecting calorimetric measurements.*

Source: https://www.researchgate.net/publication/231137378/figure/fig1/AS:388888964157444@1469729762568/Principle-of-the-calorimetric-method-temperature-trace-and-calibration-heater-power-in-a.png.

4.4.1. Environmental Conditions

Environmental conditions definition is crucial in calorimetry research and experimentation methods, particularly while investigating the thermal behavior of materials like wood. The combustion and thermal degradation of wood under varying temperature and humidity conditions can have a cataclysmic effect upon its properties. When wood is being heated, gases, vapors, and char are released from it, depending on particular environmental conditions. Thermogravimetric analysis and cone calorimetry are just some of the techniques employed to study such variations, typically yielding results that reflect the nuances of temperature, heating rate, and sample mass. A thorough understanding of the interactions is essential to ensure the optimal material safety and performance under fire-prone environments. Custom-designed experiments based on specific environmental conditions, therefore, become essential in predictive modeling of wood behavior when it is being heated, thereby leading to the development of effective fire-retardant materials (Iglauer et al., 2013).

In addition, the ambient environmental conditions under which calorimetric tests are conducted can result in quite large deviations within the wood material's combustion profile. The wood's dimensional characteristics, surface area, and inherent moisture content can modify its behavior when exposed to heat. For instance, research has illustrated that higher humidity tends to decrease resistance to ignition and change the heat release rate. Through the use of sophisticated flammability analysis methods, scientists are able to understand through what mechanisms wood degrades and evolves through its different thermal phases. It is necessary to understand that environmental conditions are not background factors; instead, they play an active role in determining experimental results. Thus, researchers seeking accurate results in calorimetry must carefully control and measure environmental conditions when testing so results are consistent and meaningful in various contexts (Hull et al., 2013).

As far as environmental conditions and impacts on calorimetry are concerned, it is equally important to explore new technologies that enhance the safety of combustion along with efficiency. Nano-scale fire retardant design work is an exciting area of research in this context. These products, intended to increase the fire resistance of wood, rely on deep understanding of the effect of environmental conditions on material response. By taking char formation and wood encapsulation through various chemical reactions into account, researchers can design materials that resist very high temperatures without deformation. Experimental methods for the formulation testing with silicon, nitrogen, and phosphorus, or new polymer-based composites, will have to be designed carefully to consider the effect of environmental conditions. Therefore, build-up of our knowledge regarding the interaction of these conditions with calorimetric analysis will not only increase scientific knowledge but pave the way towards safer, more environmentally friendly use of wood in construction and other applications (Iglauer et al., 2013).

4.4.2. Sample Purity and Composition

Sample composition and purity evaluation is an integral part of calorimetry and experimental practice, considering the effect impurities may have on thermal material properties. Sample purity ensures that detected thermal transitions are solely reflective of the samples intrinsic properties and not masked by contaminants. For instance, while analyzing the miscibility of dipropylene glycol propyl ether (DPnP) with water, differential scanning calorimetry (DSC) is an effective technique to analyze the purity of the concerned components. With careful inspection of the thermal transitions, researchers can see departures from the expected values, which may indicate the presence of impurities or varying compositions disrupting the accuracy of calorimetric measurements (Góralski et al., 2015). Thus, high purity of samples is essential to gain reproducible and reliable calorimetric data. Not only is the sample composition advantageous for enhancing calorimetric data interpretation, but it also aids in the prediction of thermodynamic behavior at varying conditions.

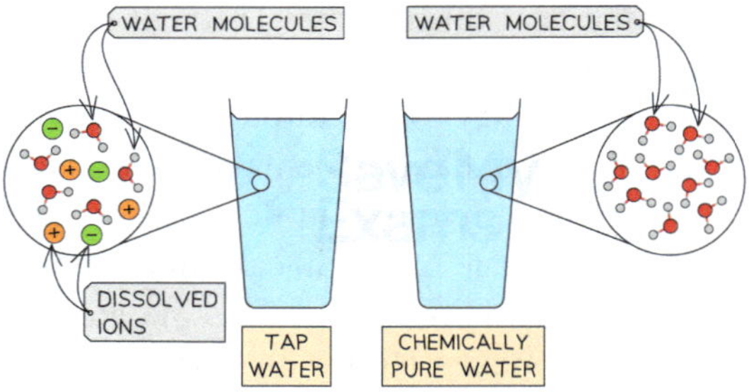

Figure 4.11. *Sample purity and composition.*

Source: https://cdn.savemyexams.com/uploads/2024/03/pure-v-impure.png.

Microcalorimetry was applied in the study of hen lysozyme with acetonitrile (MeCN) to determine denaturational heat capacity change at various pH values. The reliability of such measurements relies on the sample composition, as molarity of the organic additive will change conditions of equilibrium and thermodynamic properties. As already noted, MeCN concentrations are dependent on lysozyme stability, and thus the importance of sample composition to calorimetric measurements (Kovriguine et al., 2000). Therefore, detailed compositional analysis not only leads to better thermal characterizations but also reveals information on the biophysical consequences of solute interactions at nanoscale sizes. Lastly, interaction between sample composition and purity highlights the significance of strict methodological protocols in calorimetric analysis. Thermal transition complication is better appreciated when samples are adequately characterized, paving the way to finer insights into the structure and energetics of the materials under study. Sample purity fluctuations will hide vital information, while an informed understanding of composition allows experimental conditions to be managed so that specific thermal effects may be isolated. Aiming for precise characterization techniques within calorimetry thus ensures data obtained are representative of actual physical behavior, allowing broader use in material science, biochemistry, and countless industrial processes. Here, composition and purity take center stage in seeking reliable calorimetric analysis and application (Figure 4.11).

4.4.3. Instrumental Limitations

Instrumental limitations are a crucial part of calorimetry and experimental methodology since they play a significant role in the accuracy and reliability of data obtained by thermal analysis. One such major challenge is presented by the intrinsic measurement limitations of tools themselves, which may have inherent systematic error or may not be sensitive enough at extreme high temperatures. For example, traditional calorimetric procedures may lack precision when working under extreme high-temperature conditions or undergoing rapid temperature fluctuations, leading to likely inconsistencies in the results. Besides, baseline drift and thermal lag will influence measurement accuracy under examination of phase transitions or reaction kinetics. Such limitations demand proper calibration and testing of equipment before experimentation to ensure that the data represents actual thermal behavior because any inaccuracy might skew results and inferences in a study. Additionally, the nature of material to be studied might introduce further instrumental limitations in calorimetric experiments. Some materials, particularly amorphous ones like glasses, can complicate the analysis due to their inherent thermal behavior and reaction to variable conditions. According to the research on glass aging and crystallization, the mechanical response measured in calorimetric tests may be significantly affected by parameters like atomic diffusion and viscosity, which vary with time and thermals (Cerullo et al., 2015). Therefore, it is not only to measure the thermal response accurately but also to learn about the effect of material properties on the same. All this calls for an intimate understanding of both sample and equipment properties, ultimately counseling researchers on selecting the proper means in order not to make mistakes. Moreover, the application of new materials and experimental techniques can also result in instrumental constraints. For instance, in the application of advanced materials such as zeolites in drug entrapment, researchers will need to overcome characterization challenges such as those from composite systems. Techniques such as differential scanning calorimetry (DSC), while effective, might find themselves limited by the extent to which they can differentiate the intricate interactions between the zeolite matrix and the loading drug molecules (Abadeh et al., 2020). Variation in structural integrity and loading efficiency within the zeolite matrix can make calorimetric data interpretation problematic, bearing in mind difficulties in measuring low aqueous solubility compounds like curcumin. Thus, effective experimental designs must integrate a combination of analytical approaches, resolving the strengths and weaknesses of each with the aim of efficient, reproducible results in calorimetry and related allied experimental procedures.

4.5. CASE STUDIES IN CALORIMETRY

Calorimetry is now an essential method to comprehend the thermodynamic properties of materials using thorough case studies that emphasize various experimental methods. One such case is the thermal degradation reaction in which CRTA is applied in the study of the kinetic models that govern the reaction. One in-depth research was centered on the analysis by researchers of the temperature-time curves produced during degradation for a variety of polymers, i.e., polytetrafluoroethylene (PTFE) and poly(1,4-butylene terephthalate) (PBT). The study emphasized the role of reaction mechanisms, meaning that the characteristic temperature profile shapes are directly associated with different degradation mechanisms, such as phase boundary and random scission mechanisms. These results, validating the validity of CRTA in assessing actual kinetic models, not only enrich our understanding of polymer chemistry but also show the actual material science applications of calorimetry (Luque et al., 2011). Furthermore, digital signal processing techniques have improved the accuracy of high-frequency measurements considerably. An elite case study examined the performance possibilities of $LaBr_3$:Ce detectors for gamma-ray spectroscopy at high count rate settings, i.e., how the detectors handle energy resolution and photo peak efficiency with high count levels. Scholars employed a sophisticated setup featuring a 3x3 detector for experiments that explained the relationship between count rate and performance to prove gamma-ray spectroscopy could be effectively applied in count rates over 1 MHz. The introduction of a novel pile-up correction method enabled better precision, enabling scientists to reverse losses in efficiency and continue calibrating calorimetric measurements. These enhancements demonstrate the ongoing advancement of calorimetry through technology, turning it into a core methodology in thermal phenomenon research (Löher et al., 2012). The examination of calorimetry case studies not only highlights the variety of experimental methods but also emphasizes their utility across a wide range of scientific fields. From nuclear spectroscopy to polymer degradation, calorimetric techniques yield valuable insights into the thermal properties and behaviors of complex systems. These studies demonstrate that effective calorimetry requires astringent experimental design, including the application of appropriate measurement procedures and analysis tools. Furthermore, the ever-evolving nature of calorimetry, coupled with advances in instrumentation and computer processing, continues to extend the frontiers of thermal analysis. As researchers increasingly utilize more sophisticated techniques and contribute to the growing body of calorimetric data, its impact on industries ranging from materials engineering to nuclear science intensifies in scope, cementing calorimetry's pivotal role in the development of scientific knowledge.

4.5.1. Calorimetry in Food Science

The role of calorimetry in food science is pivotal, particularly in assessing the thermodynamic properties of food ingredients. Calorimetry, or the measurement of heat transfer, enables researchers to establish the energy value of food, a vital parameter for food labeling and meal planning. Through techniques such as differential scanning calorimetry (DSC), researchers can study phase transitions and glass transition temperatures of food components. This is crucial to develop products with desirable textural and sensory attributes. For example, DSC can describe the way ingredients interact with each other, thereby affecting the stability, quality, and shelf-life of food items, which is important for food producers and consumers whose choices are more nutrition information and health-oriented. The integration of calorimetric techniques further advances the understanding of food structure, contributing immensely to food technology and innovation. Beyond conventional calorimetric techniques, advanced techniques such as dynamic mechanical analysis (DMA) and dielectric analysis (DEA) complement the molecular-level knowledge of food structures. These techniques allow mechanical and dielectric properties of food systems to be studied in order to determine how changes in moisture content and temperature affect stability and quality of food. For instance, structural relaxations within the vicinity of the glass transition temperature can be employed to guide the formulation of food models that optimize texture and mouthfeel. A recent study employing the Guggenheim-Anderson-de Boer (GAB) model validated the importance of modeling water sorption isotherms in predicting carbohydrate-protein-lipid system performance under varying conditions, which has a bearing on food product strength and functionality (Auty et al., 2018). This multi-faceted approach stresses integrating multiple calorimetric techniques for comprehensive food analysis. The uses of calorimetry extend beyond product formulation to encompass food safety and preservation habits analysis. Interpretation of heat stability using calorimetric analysis is pivotal in establishing the shelf life of food items, particularly in establishing the effects of addition of different food additives and processes. For instance, studies on the thermooxidative stability of food matrices will assist in understanding the influence of changing environments and additives on food quality during storage periods (Čechová et al., 2012). Food scientists can refine preservation technology through calorimetric principles in a bid to maximize the nutritional value of foods and minimize degradation. Moreover, this understanding makes it easier to have better storage technology, which ultimately results in reduced wastage of foods. Thus, calorimetry is an essential tool in a more complex food world since consumer demands for quality and safety constantly shift (Figure 4.12).

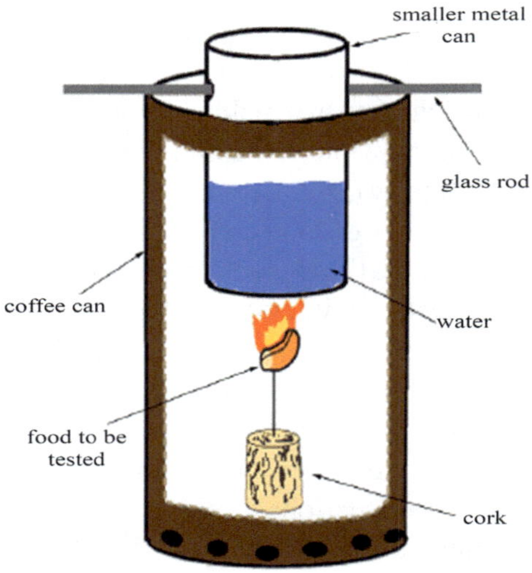

Figure 4.12. *Food calorimetry.*

Source: https://education.mrsec.wisc.edu/wp-content/uploads/sites/283/2017/08/image02Calorimetry.png.

4.5.2. Calorimetry in Material Science

Calorimetry is a significant contributor to material science since it enables researchers to quantify thermal qualities and stability under varied conditions. By measuring heat transfer in physical or chemical changes, calorimetry provides valuable information regarding a material's response to different temperatures. This role is particularly significant in the characterization of polymers, composites, and other novel materials, whose thermal responses are often complex. In studies on poly(methyl methacrylate) (PMMA) composites, for instance, calorimetric techniques such as thermogravimetry (TG) are crucial for evaluating thermal degradation at varying heating rates. These experiments can guide the development of more trustworthy materials by quantifying their performance and lifespan, thereby reducing the potential for failure in actual applications, ultimately for higher material reliability in numerous technological fields (Čechová et al., 2012). Lastly, information regarding heat release rates and ignition resistance, particularly in organic materials like wood, indicates the applicability of calorimetry to materials science. Wood, being an everyday natural material, necessitates thorough investigation of its heat degradation and flammability characteristics for safety reasons. Calorimetric methods, including cone calorimetry, are employed to examine the fire properties of wood on heating, providing data about heat release rates and combustion product formation. Such data are crucial to developing fire-retardant treatments that can improve the materials' properties and safety. By using calorimetry, researchers are able to analyze how treatments and modifications, such as applying nanocomposite coatings, affect wood's resistance to heat so that producers and engineers can have tangible knowledge (Iglauer et al., 2013). Through the integration of calorimetric data into design and evaluation of materials, its application is made possible in a range of applications. The information gleaned through calorimetric examination not only determines the selection of material for a specific application but also assists with the optimization of process conditions. Through knowledge of thermal stability and decomposition behavior, scientists can synthesize optimized materials with enhanced performance characteristics, particularly in composites. Generally, calorimetry is evaluated as a fundamental experimental method in materials science that stimulates innovation and assures the efficacy and safety of products utilized in industries. The continual advancements of calorimetric approaches are expected to give even more details about the thermodynamic properties of emerging materials, thus cementing its future role in research work (Čechová et al., 2012).

4.5.3. Calorimetry in Biochemical Reactions

Calorimetry is a significant technique for analyzing biochemical reactions because it provides scientists with the ability to measure heat developed or absorbed in the process. Calorimetric analysis is essential in the study of reaction mechanisms, enzyme kinetics, and metabolic pathways. By precise measurement of thermodynamic parameters, researchers are able to make important inferences about biomolecular stability, energetics of binding events, and efficacies of substrates. Additionally, enhancements in calorimetric instruments, such as microfluidic calorimeters, have enhanced the capacity to perform high-throughput experiments using small amounts of sample. These developments enable improved investigation of biochemical reactions, opening avenues towards innovation in both academic research and industrial applications. Consequently, calorimetry offers a robust intersection of thermodynamics and biochemistry that enables the investigation of bioenergetics in different biological systems. Recent developments in calorimetry have significantly broadened experimental methods in biochemistry, particularly through microfluidic devices. For instance, high-sensitivity microfluidic calorimeters offer excellent thermal resolution for the examination of very small samples, as evidenced by the ability to characterize reaction heat in 3.5-nL samples at a resolution of 4.2 nW. This enhanced sensitivity is a goldmine for investigating very small thermodynamic changes that accompany biochemical reactions like enzyme catalysis and ligand binding. Besides, they merge soft and hard polymer microfluidics, whose advantage is improved thermal response without the sacrifice of structural strength.

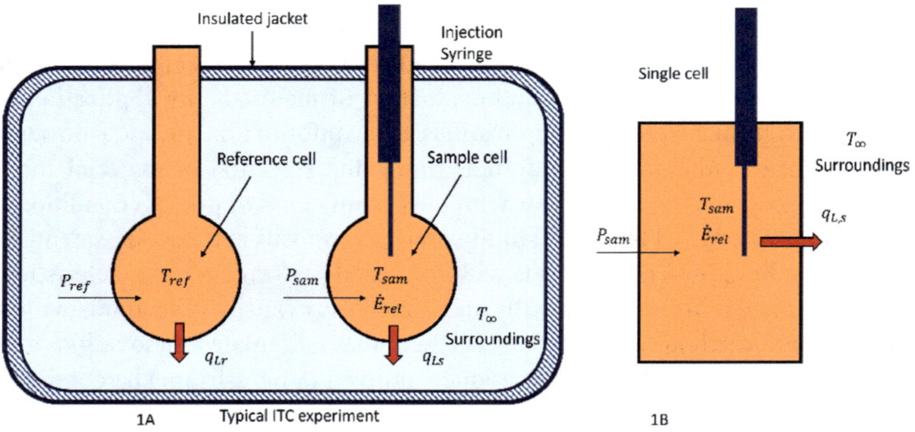

Figure 4.13. *Calorimetric measurements of biological interactions.*

Source: https://www.mdpi.com/entropy/entropy-24-00561/article_deploy/html/images/entropy-24-00561-g001.png.

This is a development that will play a central role in the automation of analyte injection, allowing scientists to successfully carry out many calorimetric procedures. Microfluidic calorimetry is, thus, a groundbreaking development towards achieving high-throughput experimental capability in biochemical science (Axelrod et al., 2009). The use of calorimetry also goes beyond scientific investigation to relevance in practical biochemical industries. In microbial metabolism, for instance, the synergy between calorimetric data and nutrient availability plays a critical role in the growth kinetics of microbes. Amino acids, in particular, have the potential to serve as signaling molecules that affect not only the metabolism but also biofilm development. By examining the extent to which environmental conditions impact microbial reaction calorimetric profiles, researchers are able to elucidate a range of critical aspects of bacterial function, such as virulence and pathogenicity. Indeed, work on *Pseudomonas aeruginosa* l-arginine sensor illustrates the importance of nutrients as regulators of intracellular signal transduction, linking metabolism with phenotypic properties such as biofilm formation. Thus, through the integration of calorimetry and biochemical analysis, scientists can make extremely penetrating observations about the metabolic processes that govern both ecological interactions and clinical outcomes (Cutruzzolà et al., 2018) (Figure 4.13).

4.6. ADVANCES IN CALORIMETRIC TECHNIQUES

Calorimetry has experienced very considerable developments, with significant advances in particle flow calorimetry that has become a revolutionary tool for event reconstruction in high-energy physics. This technique relies on extremely fine calorimeters, meticulously built to offer greater accuracy in the energy deposition measurement from particle interactions. The development and research during the past decade through efforts like CALICE have maintained the efficiency of these large prototypes through beam tests. Notably, the ability to compare test beam data with simulations has offered deeper understandings of hadronic cascades that are crucial in obtaining insight into the underlying physics of particle interactions. These advancements not only enhance the accuracy of experimental results but also offer opportunities for new developments in detector technologies, significantly contributing to probing fundamental particles and interactions in collider experiments (Kawagoe et al., 2015). Parallel to this, isothermal titration calorimetry (ITC) has evolved as a pivotal instrument in biophysical research, particularly in the study of biomolecular interactions. With further research focused on macromolecules, researchers have mastered interpreting ITC data that provide critical information on the formation and dissociation of molecular complexes. Current research has indicated a trend towards linking thermodynamic constants to molecular processes and thus to gain an improved insight into binding mechanisms and kinetics. This has especially proven useful in x-ray crystallography studies, where being able to view individual binding events has allowed for the overall analysis of thermodynamic parameters to be improved. Further, the re-emergence of interest in novel assays combining kinetic and thermodynamic examination of ITC data has made the field relevant once again to offer viable solutions to researchers grappling with complex interactions (Ababou et al., 2016). Calorimetric method advancements underscore the significance of interweaving modern technology with traditional methodology. Particle flow calorimetry is a leaps and bounds advance in experimental physics, while the on-going optimization of isothermal titration calorimetry affirms the value of calorimetry in understanding biological processes at a molecular level. Ongoing improvement in these techniques will seek to uphold the integrity of experimental processes via increasingly precise and reproducible data across a broad scope of scientific disciplines. The combined application of advanced detection tools and refined analysis techniques promises a future of increasing cooperation and imagination and ultimately an enhanced capacity to question and understand intricate systems in particle physics as much as biochemistry.

4.6.1. Innovations in Instrumentation

Calorimetry has been forcefully guided by instrumental advancements, pushing the discipline to previously uncharted regions of precision and application. Previously, it was the trailblazing individuals such as Joseph Black, who first distinguished heat from temperature, who laid the ground for calorimetric developments. The creation of the highly sensitive and automated calorimeters has enhanced the ability to make measurements to astounding precision. This has also widened the application areas of calorimetry so it can be applied across various disciplines, including thermodynamics and materials science, particularly for the study of phase transitions and thermal stability. The technological shift from manual to automatic systems is representative of a larger trend in the scientific community, where increased efficiency and reliability of data are paramount to tightly controlling experimental parameters and results, as is witnessed in the pioneering work of Glasgow's successful scientific legacy (Johnston et al., 2006). In present studies, the integration of sensor-driven monitoring systems in calorimetry showcases innovation in instrumentation. New methods allow for observation of temperature fluctuations and heat transport in real time, increasing the level of information gathering. These feats surpass those of traditional calorimetry, employing principles in sensor science and data analysis to enhance experimental strategies in numerous applications. The creation of automated methods for activity observation and wellness monitoring, for instance, employs calorimetric information to inform rehabilitation and health strategies. This transformation emphasizes the importance of interdisciplinary approaches, as transferred knowledge from associated fields such as algorithmic data analysis assists in building calorimetric practice. Illustrations of such advances evidence the interdependence of today's scientific research and the importance of adaptive methods in addressing the demands of ongoing research (Achumba et al., 2013). The implications of these advances are profound, for they not only improve experimental accuracy but also reconstitute research capacity within calorimetry. Encouraging intelligent instrumentation encourages interdisciplinarity, whereby researchers are able to deal with complicated problems using cutting-edge analysis capabilities. Real-time calorimetry in pharmaceuticals, for example, enables one to properly assess drug interaction and stability, which is essential when developing effective therapeutic interventions. Additionally, the challenges of calibrating and validating new instrumentation put a high value on ongoing method development refinement and the importance of user engagement with new technology. With calorimetry as a science evolving further, the commitment to developing innovative instrumentation will remain crucial in addressing upcoming scientific issues and ensuring robust experimental results (Johnston et al., 2006).

4.6.2. Integration with Other Analytical Methods

The interaction between calorimetry and other analysis methods is the foundation for enhancing the validity and effectiveness of thermal analysis in material science. When calorimetry is combined with techniques such as thermogravimetric analysis (TGA) and differential scanning calorimetry (DSC), scientists are able to derive important data on the thermal stability and degradation behavior of materials. For instance, TGA provides quantitative data on mass loss as a function of temperature or time and can be applied to identify thermal degradation pathways in composite materials. Through simultaneous combination of TGA with calorimetric techniques, better insight into the material performance with enhanced predictability of long-term stability and potential applications is achievable. This integration also allows validation of experimental data, where researchers can cross-check results and draw cogent conclusions regarding material safety and stability. Further, integration of calorimetry with spectroscopic techniques such as infrared spectroscopy (IR) and nuclear magnetic resonance (NMR) gives added analytical value to calorimetric measurements. This synergistic method not only enables examination of thermal properties but also the sensing of chemical changes occurring during thermals. For example, IR coupled with calorimetric measurement enables exact functional groups transforming while thermal breakdown occurs, thereby correlating thermal properties with structural quality. The capacity for analyzing chemical change alongside thermal response allows for greater interpretative development of data, hence enhancing the understanding of the processes determining material stability. This interdisciplinarity, seen in the synergy between different analytical approaches, allows scientists to build comprehensive profiles of material performance under multiple environmental conditions, leading to potential revolutions in material science. Furthermore, using a multi-disciplinary analytical technique can reveal latent interactions in complex material systems, as demonstrated in studies of polymer composites. Coupling calorimetry with X-ray diffraction (XRD) and transmission electron microscopy (TEM) in this case can reveal dominant information on microstructural development during thermal processing. For example, the incorporation of magnesium-aluminum layered double hydroxides (MgAl-LDH) and ammonium polyphosphate (APP) into polystyrene (PS) matrices has also shown significant improvement in thermal stability and fire performance when collectively assessed by TGA and calorimetry. By the analysis of structural morphologies together with thermal dynamics, researchers can discover the synergistic effects towards enhanced material performance, as illustrated in a study where enhanced silica particle size affects the stability of PMMA composites (Čechová et al., 2012; Kandare et al., 2008). The pairing explains the complex relationship between thermal behavior and material composition, facilitating the development of highly advanced materials tailored for applications.

4.6.3. Future Trends in Calorimetry Research

The future of calorimetry research is full of promise driven by the incorporation of new technologies and interdisciplinarity. As the field continues to develop, researchers are turning to combining calorimetry with other types of research, such as molecular modeling and advanced spectroscopic methods. By doing so, it becomes possible to study thermodynamic properties in great detail on the molecular level, gaining further insights into intricate systems and materials. The potential for high-throughput calorimetric screening is highly promising, making it possible for researchers to rapidly determine a large set of compounds or reactions and therefore accelerate the discovery process in materials science and pharmaceuticals. New computational methods will also be necessary, making better predictions from calorimetric data, ultimately leading to more rational experimental design and improved data interpretation. In addition, miniaturized calorimetric devices represent a significant direction that has the ability to fundamentally change experimental methods in calorimetry. Micro-scale calorimeters which are more efficient and smaller are becoming increasingly available, allowing for workers to conduct research with samples that require fewer materials. Not only does the trend conserve precious resources, especially in drug discovery and in environmental research, but it also allows research on rare or expensive compounds. Also, microfluidic advances can enable simultaneous real-time calorimetric measurements along with other analytical methodologies, giving rise to an entirely integrated comprehension of processes underway. Continual progress along these directions further highlights the necessity for expanding calorimetric techniques to more generic scientific studies and for catalyzing interdisciplinary collaboration between chemists, biologists, and materials scientists to stimulate innovation. Furthermore, the application of artificial intelligence (AI) and machine learning in calorimetry introduces a new field for data analysis and interpretation. Machine learning techniques can easily handle large amounts of data generated by cutting-edge calorimetric techniques and identify patterns and correlations that would otherwise go unnoticed using standard analytical approaches. This would give amazing insights, particularly into areas such as drug discovery, where understanding the thermodynamic profiles of various compounds would inform structure-activity relationship. With growing computational power and acumen, the combination of calorimetric experiments with AI-guided analysis will enable predictive modeling and experimental design optimization, ultimately leading to making research output more efficient. While investigators explore these new trends, the science of calorimetry will be reengineered for sure, in line with present scientific necessity and interdisciplinary demand.

SUMMARY

In combining results from this study, it is apparent that calorimetry and other experimental methods play important roles in advancing material science knowledge. The use of techniques such as differential scanning calorimetry, thermogravimetric analysis, and high imaging techniques is useful in giving information regarding thermal and mechanical properties of inorganic materials. In particular, the dissertation addressed in this essay illustrates how calorimetry can enable the characterization of distinct phases, which is very important in identifying structural vulnerabilities in materials. With the use of a strict systematic approach that pairs theoretical modeling with experimental validation, researchers are in a prime position to draw conclusions not only expanding the knowledge base but also providing the pathway for further material performance and application. The meticulous critique of results, summarized in individual chapters of the dissertation, emphasizes rigorous use of experimental design and data interpretation. Also, the uses of employing calorimetric techniques extend beyond the academic domain and into practical applications, particularly in terms of generating enhanced materials with optimum properties. The chapter presents an enormous step towards the utilization of innovative techniques towards material improvement, as evidenced in the identification of structural defects through the use of calorimetry. The method may be extended to other fields, such as aerospace engineering and materials in buildings, where material integrity and performance are top priority. Specifically, the synergic effects that take place between other additives in calorimetry, as explained in the study, have a direct application to industry stakeholders who wish to optimize material formulation. By filling the theoretical and practical gaps, experimental methods and calorimetry can power the initiation of improved materials that meet evolving demands. Collective research lastly highlights interdisciplinary approaches in calorimetry and experimental methods research. The study of thermal properties within the context of mechanical behaviors provides a better comprehension of the interactions involved between varying material constituents. The synergy seen in studies employing flame retardants, such as those which use MgAl-LDH and ammonium polyphosphate, shows the ability that combinatorial techniques can have to open up developments significantly enhancing material properties. By adopting a broad methodology in which calorimetry is coupled with other experimental procedures, scientists can understand material response better under diverse conditions. As highlighted during the dissertation, the strong synergy of various tools of analysis, if utilized aptly, is the secret to innovation in material science and thus to the ultimate contribution towards improvement for academic scholarship as well as industrial application.

The understanding as well as utilization of calorimetry is the secret in experimental physics, particularly in terms of determining the change in energy during chemical reactions or physical transformations. In a comprehensive review of calorimetric methods, their suitability in the measurement of heat transfers and reaction enthalpies is made evident. Through the application of calorimeters, researchers can accurately monitor temperature changes, thereby learning about reaction mechanisms as well as energy balances. Methods vary from simple coffee cup calorimeters to intricate bomb calorimeters, with varying advantages based on the specific requirements of the experiment. These instruments stress the need for accurate temperature measurement and control, which directly correlate with the reliability of the data acquired. The calorimetric technique thus becomes a significant contributor to the quality of experimental findings, enabling meaningful interpretation of data in materials science and thermodynamics. Apart from traditional calorimetric techniques, novel and innovative experiment techniques are under investigation as increasingly enhanced efforts towards measurement efficiency and accuracy. The trend is supported by recent innovations in calorimeter technology, e.g., the application of hybrid photomultiplier tubes for light yield measurement. One such example is the calibration of lead fluoride (PbF_2) crystals as calorimeters, which has been promising, with an achieved light yield of 1.7–1.9 photoelectrons per MeV. The process of improvement, including optimizing wrapping and coupling techniques, indicates how research methods continue evolving to deal with greater precision standards. Notably, Monte Carlo-based simulations have improved computational efficiency in calorimetry to a great degree, minimizing the time to perform calculations without losing accuracy. Such developments reflect the ever-present drive for perfection in calorimetric techniques, expanding their coverage and applications in various fields in experimental physics (Achenbach et al., 2001). Further, interactions among various calorimetric approaches and numerical modeling show important details on energy interactions on the particle level. In the investigation of reactions like that of hot quasi-projectiles, two approaches with competing philosophies offer complementary perspectives through simulation by different models. A comparison between the two methods of research allows scientists to determine the strengths and limitations inherent in each model. Such observations stress methodological soundness in research design whereby selection is determined by the specific parameters and energies to be investigated. The interplay between numerical simulations and experimental data produces an enhanced picture of reaction dynamics, opening the way for subsequent studies. This integrated approach, combining recognized methodologies with cutting-edge computational methods, witnesses the dynamic nature of calorimetry as a key tool for the scientist's repertoire to understand energy phenomena (Charity et al., 2011).

REFERENCES

1. Ababou, R., Adriaenssens, G., Aguirre, A., Alatorre-Meda, M., Amendola, V., Anglin, J. R., et al. (2016). Applications of isothermal titration calorimetry–The research and technical developments from 2011 to 2015. https://core.ac.uk/download/46564311.pdf (accessed on 4th October 2025).
2. Abadeh, Z. A., Ballirano, P., Santonicola, M. G., Saviano, G., et al. (2020). Curcumin-loaded zeolite as anticancer drug carrier: Effect of curcumin adsorption on zeolite structure. https://core.ac.uk/download/296244739.pdf (accessed on 4th October 2025).
3. Achenbach, P., Baunack, S., Lopes Ginja, A., Grimm, R., et al. (2001). Measurements and simulations of Cherenkov light in lead fluoride crystals. http://arxiv.org/abs/nucl-ex/0108019 (accessed on 4th October 2025).
4. Achumba, I. E., Azzi, D., Bersch, S. D., Khusainov, R., et al. (2013). Real-time human ambulation, activity, and physiological monitoring: Taxonomy of issues, techniques, applications, challenges and limitations. https://core.ac.uk/download/29586954.pdf (accessed on 4th October 2025).
5. Alnuaimi, E., Avalos, B., Ewing, A. C., Havens, K., et al. (2019). Bio-based renewable additives for anti-icing applications (Phase II). https://core.ac.uk/download/185623452.pdf (accessed on 4th October 2025).
6. Annis, J. F., Troutman, S. J., & Webb, P. (2025). Cycles in metabolism and heat loss. https://core.ac.uk/download/pdf/42896135.pdf (accessed on 4th October 2025).
7. Ansell, G. S., Loewy, R. G., & Wiberly, S. E. (2025). Composite Structural Materials. https://core.ac.uk/download/pdf/42847094.pdf (accessed on 4th October 2025).
8. Arpaci, V. S., Davaji, B., Lee, C. H., Bak, H. J., & Chang, W. J. (2014). A novel on-chip three-dimensional micromachined calorimeter with fully enclosed and suspended thin-film chamber for thermal characterization of liquid samples. https://core.ac.uk/download/213081950.pdf (accessed on 4th October 2025).
9. Auty, M. A., Lim, A. S. L., Maidannyk, V., Roos, Y. H., et al. (2018). Effects of lipids on the water sorption, glass transition and structural strength of carbohydrate-protein systems. https://core.ac.uk/download/287655076.pdf (accessed on 4th October 2025).
10. Axelrod, B. W., Johannessen, M., Roukes, M. L., Murphy, K., Recht, D., Thorsen, T., Fon, W., et al. (2009). High-sensitivity microfluidic

calorimeters for biological and chemical applications. https://core.ac.uk/download/4882823.pdf (accessed on 4th October 2025).

11. Barish, B., & Brau, J. E. (2013). The International Linear Collider. http://arxiv.org/abs/1311.3397 (accessed on 4th October 2025).
12. Bossi, A. M., Vestri, A., Castagna, A., Girelli, D., Iacob, E., Andreetto, E., Cenci, L., et al. (2015). Surface plasmon resonance based on molecularly imprinted nanoparticles for the picomolar detection of the iron regulating hormone Hepcidin-25. https://core.ac.uk/download/pdf/81531795.pdf (accessed on 4th October 2025).
13. Bulmău, C., & Ionescu, G. (2019). Estimation of energy potential for solid pyrolysis by-products using analytical methods. https://core.ac.uk/download/322439150.pdf (accessed on 4th October 2025).
14. Casares González, J. Á., Espinet Rubio, P., & Pérez Iglesias, M. (2018). Comparing proteolysis and trans metalation reactions: Microcalorimetric studies of C–Au(I) bonds in [AuRL] complexes. https://core.ac.uk/download/225143515.pdf (accessed on 4th October 2025).
15. Čechová, E. (2012). Thermooxidative stability of PMMA composites. https://core.ac.uk/download/30296004.pdf (accessed on 4th October 2025).
16. Cerullo, G., Ferrante, C., Pogna, E. A. A., Rodríguez-Tinoco, C., et al. (2015). Probing equilibrium glass flow up to exapoise viscosities. https://core.ac.uk/download/55257895.pdf (accessed on 4th October 2025).
17. Charity, R. J., Durand, D., Legouée, E., Vient, E., & Pouthas, J. (2011). Characterization of quasi-projectiles produced in symmetric collisions studied with INDRA: Comparison with models. http://arxiv.org/abs/1203.0232 (accessed on 4th October 2025).
18. Chaturvedi, S. K., & Chegini, H. (2025). On-line measurement of heat of combustion. https://core.ac.uk/download/pdf/42831887.pdf (accessed on 4th October 2025).
19. Criado Luque, J. M., Perejón Pazo, A., Pérez Maqueda, L. A., Sánchez Jiménez, M. A., et al. (2011). Constant rate thermal analysis for thermal stability studies of polymers. https://core.ac.uk/download/187420555.pdf (accessed on 4th October 2025).
20. Cutruzzolà, F., Giardina, G., Janson, G., Leoni, S., et al. (2018). A novel bacterial l-arginine sensor controlling c-di-GMP levels in Pseudomonas aeruginosa. https://core.ac.uk/download/188824095.pdf (accessed on 4th October 2025).
21. De Belie, N., De Maersschalck, J., De Schepper, M., Van Driessche, I., et

al. (2016). Setting control of completely recyclable concrete with slag and aluminate cements. https://core.ac.uk/download/55718446.pdf (accessed on 4th October 2025).

22. DeShazer, J., Ferrell, C., Hillman, P., Nienaber, J., et al. (2009). Measuring energetics of biological processes. https://core.ac.uk/download/38932941.pdf (accessed on 4th October 2025).
23. Gil-Botella, I. (2015). Neutrino physics. http://arxiv.org/abs/1504.03551 (accessed on 4th October 2025).
24. Góralski, Paweł, Piekarski, Henryk, Tkaczyk, Mariola. (2015). Phase behavior and heat capacity of {DPnP + water} mixtures at the temperature range of 273.15–338.15 K. https://core.ac.uk/download/71971908.pdf (accessed on 4th October 2025).
25. Gresits, I., Gyüre-Garami, B., Márkus, B. G., et al. (2018). Non-calorimetric determination of absorbed power during magnetic nanoparticle-based hyperthermia. https://core.ac.uk/download/559100682.pdf (accessed on 4th October 2025).
26. Hong, Q. J., Navrotsky, A., Ushakov, S. V., et al. (2019). *Carbides and Nitrides of Zirconium and Hafnium*. https://core.ac.uk/download/323067876.pdf (accessed on 4th October 2025).
27. Horgan, Graham W., Stubbs, R. J., et al. (2014). An evaluation of the IDEEA™ activity monitor for estimating energy expenditure. https://core.ac.uk/download/19506491.pdf (accessed on 4th October 2025).
28. Hull, T. R., & Lowden, L. (2013). Flammability behaviour of wood and a review of the methods for its reduction. https://core.ac.uk/download/19576180.pdf (accessed on 4th October 2025).
29. Iglauer, S., Lebedev, M., Zhang, Y., Zhou, X., et al. (2013). Flammability behaviour of wood and a review of the methods for its reduction. https://core.ac.uk/download/19576180.pdf (accessed on 4th October 2025).
30. Johnston, S. F. (2006). The physical tourist. Physics in Glasgow: A heritage tour. https://core.ac.uk/download/1393823.pdf (accessed on 4th October 2025).
31. Kandare, E., Nyambo, C., Wang, D., Wilkie, C. A., et al. (2008). Flame-retarded polystyrene: Investigating chemical interactions between ammonium polyphosphate and MgAl layered double hydroxide. https://core.ac.uk/download/213089276.pdf (accessed on 4th October 2025).
32. Kawagoe, K., Pöschl, R., Repond, J., Sefkow, F., et al. (2015). Experimental tests of particle flow calorimetry. https://core.ac.uk/download/211334262.pdf (accessed on 4th October 2025).

33. Kovriguine, E., & Potekhin, S. A. (2000). On the stabilizing action of protein denaturants: Acetonitrile effect on stability of lysozyme in aqueous solutions. https://core.ac.uk/download/67753965.pdf (accessed on 4th October 2025).
34. Lee, S., Livan, M., & Wigmans, R. (2017). Dual-readout calorimetry. http://arxiv.org/abs/1712.05494 (accessed on 4th October 2025).
35. Lee, S., Livan, M., & Wigmans, R. (2017). On the limits of the hadronic energy resolution of calorimeters. http://arxiv.org/abs/1710.10535 (accessed on 4th October 2025).
36. Löher, B., Savran, D., Devanathan, S., Fiori, E., Iltis, A., Kulisek, A., Kurosawa, T., et al. (2012). High count rate γ-ray spectroscopy with $LaBr_3$:Ce scintillation detectors. http://arxiv.org/abs/1202.6495 (accessed on 4th October 2025).
37. Micoulaut, M. (2016). Relaxation and physical aging in network glasses: A review. http://arxiv.org/abs/1605.07403 (accessed on 4th October 2025).
38. Moroni, A., Avdeyev, S. P., Bondorf, J. P., Campi, X., et al. (2003). Experimental signals of phase transition. http://arxiv.org/abs/nucl-ex/0310013 (accessed on 4th October 2025).
39. Reiber, C., & Booker, N. (2011). Event program. https://core.ac.uk/download/72860922.pdf (accessed on 4th October 2025).
40. Rubbia, A. (2000). ICANOE–Imaging and calorimetric neutrino oscillation experiment. https://core.ac.uk/download/25277530.pdf (accessed on 4th October 2025).
41. Wasserbauer, J. (2013). Mechanical properties of microstructural components of inorganic materials. https://core.ac.uk/download/30310832.pdf (accessed on 4th October 2025).

CHAPTER 5

HESS'S LAW AND APPLICATIONS

Contents

Introduction 163
5.1. Fundamental Principles of Hess's Law 167
5.2. Mathematical Representation 172
5.3. Experimental Verification...176
5.4. Applications in Chemistry 181
5.5. Industrial Applications..186
5.6. Educational Importance 190
Summary...194
References...197

INTRODUCTION

The study of Hesss Law is important for the understanding of complicated thermodynamics and chemical reaction dynamics. Lastly, Hess's Law states that the net enthalpy change in a chemical reaction is independent of path, relying on the initial and final states alone. Such a fundamental rule has widespread applicability in numerous scientific as well as industrial processes, particularly in energy systems and renewable technologies. By imposing a consistent pattern of energy transfer, Hesss Law allows chemists and engineers to optimize processes, become environmentally sustainable, and save costs. The significance of the law is evident in new energy technology development, where accurate thermodynamic predictions will govern the reliability and efficiency of a system. With the world embracing renewable energy technology, learning about Hess's Law is increasingly important to developing effective mechanisms for harnessing such energies in the most efficient manner. Moreover, Hess's Law is an important element in the modeling of hybrid renewable energy systems, i.e., PV and H_2 based systems. An example of such application is in the optimization of energy storage mechanisms for countering the uncertainty of renewable resources like solar and wind.

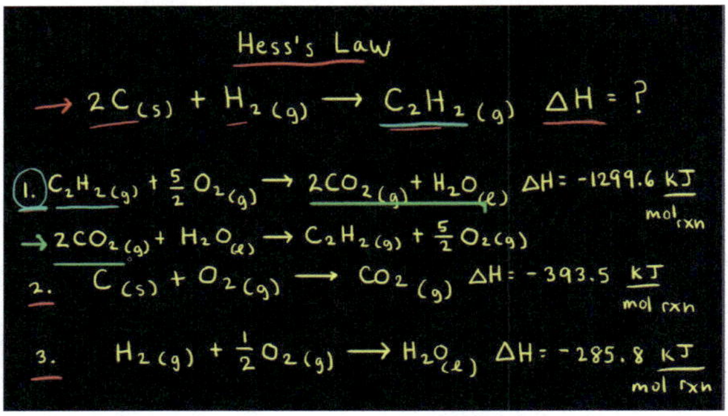

Figure 5.1. *Hess's law.*

Source: https://cdn.kastatic.org/ka_thumbnails_cache/78168b2e-7cad-42fc-8fd8-93c1a3bb8e2b_1280_720_base.png.

By utilizing Hess's Law in energy management solutions, researchers and practitioners can express the exact energy needs of hybrid systems to enable the development of more efficient storage. For instance, the development of a new accurate dynamic system model enables accurate simulations of PV-H_2 systems in the real world, resolving problems related to carbon emissions and energy pricing. Such technological advancements play a crucial role in the ongoing trend towards sustainable energy systems based on renewable energy but still economically sound. The implementation of such principles within hybrid models is a confirmation of the seamless integration of theory and application in modern energy technology. Owing to the growing demand for sustainable forms of energy, the integration of economical energy strategies and principles based on Hess's Law becomes critical for further innovation. Scaling up multi-energy systems facilitates the optimization of the use of resources and the removal of energy losses. The use of this approach, particularly with the inclusion of Smart Energy Hubs (SEHs), makes clear the significant impact of systematic design improvements. The SEHs try to consolidate the variety of energy sources towards optimal performance amidst fluctuating market prices and uncertainty in the demand. Accurate modeling of uncertainty in such frameworks will enhance innovations in energy planning and decision-making processes. As energy planners increasingly use models derived from Hess's Law, the future for the attainment of sustainable energy development becomes brighter. This convergence of theoretical and practical uses ultimately indicates the pertinence and timeliness of Hess's Law towards achieving a sustainable energy future (Figure 5.1).

Definition of Hess's Law

Hess's Law, which is a fundamental concept in the branch of thermochemistry, states that the net change in enthalpy in a chemical reaction is pathway-independent in bringing about the change. According to this law, it does not make a difference whether the reaction occurs in one or multiple steps; the net heat given out or taken in remains the same. For instance, if a reaction can proceed through several steps of intermediate reactions, then the cumulative enthalpy change for such steps will be the same as the enthalpy change for the overall reaction. This implies that chemists can find enthalpy changes for reactions that cannot be readily measured directly, thereby providing more information about energy changes in chemical systems. Applicability of Hess's Law extends beyond the theoretical into the practical realm in reaction engineering and environmental science. Laboratory-wise, scientists can apply this law to determine reaction enthalpies for complex processes, giving significant insight into reaction viability and kinetics. Secondly, Hesss Law plays a central role in the study of calorimetry because it facilitates the calculation of heats of combustion and formation for difficult-to-measure materials. Hess's Law simplifies calculations through the allowance of the combining of known enthalpy changes in different reactions. This is important in developing effective chemical processes. Furthermore, Hess's Law serves as a pedagogical tool, illustrating the conservation of energy concepts and the nature of state functions in thermodynamics. The law also describes not just the relationships between products and reactants but more significantly, the significance of enthalpy being a state function and one that depends solely on the current state of the system and not upon the path to the state. As more is learned about thermodynamics, Hess's Law becomes a universal principle that ties together various fields of energy transformations, confirming the interconnectedness of chemical reactions. The law not only enhances theoretical understanding but also progresses practical skills to be applied in advanced chemical research and manufacturing (Ekambaram, 2012).

Historical Background and Significance

The historical background of Hess's Law traces its roots to the early 19th century when Russian chemist Germain Hess laid down the conditions for energy conservation in terms of the law. According to Hess, he proposed that the total enthalpy change in a reaction is constant irrespective of the path or the number of steps in the reaction. This awareness developed required foundation knowledge that subsequently assisted in developing thermodynamics as a science and an even deeper comprehension of chemical reaction energy conversion. By proving that energy cannot be created or destroyed, Hess's laboratory opened the door to greater innovation in theoretical as opposed to applied chemistry, with energy conservation as a process becoming the very foundation of scientific investigation and industrial application, especially because rising demands for energy efficiency have skyrocketed against the backdrop of the global energy crisis (Malburg et al., 2020).

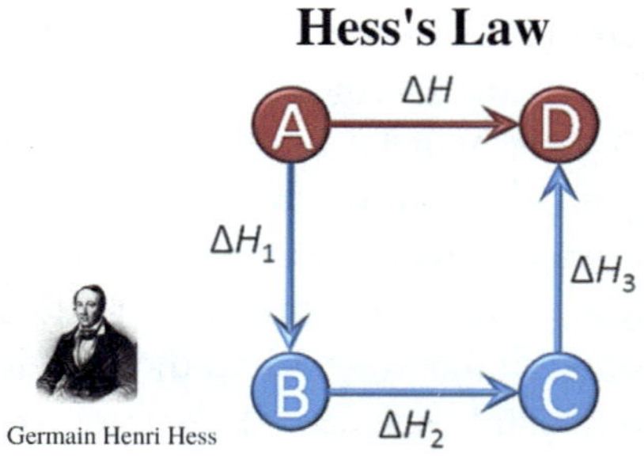

Figure 5.2. *Historical background of Hess's law.*

Source: https://external-content.duckduckgo.com/iu/?u=https%3A%2F%2 Fscienceinfo.com%2 Fwp-content%2Fuploads%2F2022%2F08%2FHesss-Law-.jpg&f=1&nofb=1&ipt=d0f69bf4d49eec2debcfea25fe5c791331d233f27f2a090193f43900899f8bed.

The importance of Hess's Law is not limited to its application in chemistry since it is a central concept in many other applications in several different scientific fields. In everyday situations, including biochemical reactions and engineering design, Hess's Law is used to calculate the enthalpy change in intricate reactions where the measurement might be difficult to perform directly. For instance, in the computation of energy yields in applications of chemical engineering and in gauging a reaction's efficiency, Hess's Law aids scientists and engineers in optimizing conditions for target yields. Its applications also harmonize theoretical insights and practical matters, a reflection of the dependence of energy forces on the practice of sustainability, particularly as societies attempt alternative energy sources as a measure to address climatic threats (Kuikka et al., 2015). Additionally, with the increase in energy demands across the world, it is more critical than ever to grasp Hesss Law in designing sustainable technologies and minimizing the effects of the environment. The drive towards cleaner energy sources, including solar and biofuels, applies the principles set by Hess in determining energy efficiency and viability. Experimental approaches that include the concepts of Hess enable the transition from traditional fossil fuels not only but also promote new modes of energy use and storage, including technological advancements with electric vehicles and green technology. Therefore, Hess's Law continues to be an underlying pillar in education curricula and in applied applications, upholding its ongoing pertinence in building a sustainable energy future against climate change and resource limitations (Malburg et al., 2020) (Figure 5.2).

5.1. FUNDAMENTAL PRINCIPLES OF HESS'S LAW

The fundamental concepts of Hesss Law are rooted in conservation of energy and pointing towards the fact that the net change in enthalpy for a chemical reaction is path independent. Hesss Law enables one to calculate the enthalpy changes for complex reactions as a sum of easier individual steps with known enthalpy changes. This systematic approach not only simplifies the ability to calculate thermodynamic properties but is also crucial in other applied applications. In chemical compound syntheses, for instance, analyzing the individual steps can provide information on the energy dynamics that assist chemists in optimizing the conditions of the reaction. Thus, Hesss Law is a valuable tool in theoretical and applied chemistry alike, helping to give us a better understanding of energy conversion in chemical systems. Another distinguishing characteristic of Hesss Law is that it finds use in determining the enthalpic contribution of a number of reactions when measurement directly isn't feasible. This property is of most benefit in reaction thermochemistry, where enthalpy changes of intermediate species may be deduced from the reactions that they participate. For example, the enthalpy change of formation reactions typically invokes Hesss Law in order to estimate values that are not amenable to direct measurement in laboratory experiments. Also, there are industrial applications for this law in industrial reactions where reaction energetics are necessary for safe and efficient operation. In allowing chemists to build the energy profiles of subsequent steps in a reaction, Hesss Law provides a firm foundation for making informed decisions regarding reaction pathways and energetics. Hess's Law has applications that extend beyond mere calculations; it forms a foundation for better energy systems, especially in the evolving field of sustainable energy. As the transition continues to the integration of sustainable sources, principles encapsulated in Hess's Law can be used in the optimization of energy storage devices that exploit cycles of energy production. For instance, as noted by recent studies, the integration of hydrogen energy storage systems presents the necessity for proper modeling of energy inputs and enthalpic modifications in hybrid systems (Arabkoohsar et al., 2022). By applying Hess's Law principles, researchers can enhance the comprehension of interactions in such systems, thereby enhancing efficiency and minimizing cost (Atteya et al., 2025). Generally speaking, the laws of Hess's Law highlight not only the preservation of energy in chemical reactions but also their widespread application in the creation of technology that attempts to be sustainable.

5.1.1. The Law of Conservation of Energy

The conservation of energy law is a physical principle which states that energy cannot be created or destroyed but can transform from one form to another. This principle is fundamental in explaining many scientific phenomena and forms the basis of significant laws like Hesss Law of thermodynamics. Hess's Law holds that the net enthalpy change in a chemical reaction remains constant, irrespective of whether the reaction is carried out in a single step or several steps. This indicates energy conservation in chemical reactions, reaffirming energy for reaction to be conserved and allows one to determine reaction enthalpies using known reaction values of related reactions. The relationship between energy conservation and Hesss Law thus provides a foundation for thermodynamics and allows us to understand energy transformations in chemical systems. In practical application, the law of conservation of energy is a critical principle for maximizing energy use in most systems, particularly chemical reactions, and the development of energy-conserving systems. For example, electric car battery technology typical of a shift towards sustainable energy practices demonstrates the importance of energy conservation. Modern innovations such as the hybridization of batteries and supercapacitors indicate the ongoing endeavor to store energy in the most optimized way possible while cutting waste. It is precisely such improvements that address the environmental footprint of energy storage devices based on evidence from problems with chemical disposal at lifecycle end for batteries. Thus, understanding the law of conservation of energy not only guides theoretical application but also guides practical endeavors towards producing greener technology that complements sustainable energy goals, such as the case for electric cars (Malburg et al., 2020). Moreover, the law of conservation of energy is directly linked to the thermodynamical discussions at large regarding stability and behavior in systems under perturbation, such as in black hole thermodynamics. Stability of the system can generally be addressed through considerations of energy, equating thermodynamic properties to dynamic behaviors, as in black holes whose differential responses of stability are predicated upon energy characteristics. As explained in more recent literature, the positivity of some energy formulations explains the dynamic stability of those systems and illustrates how fundamental concepts of energy are unchanging across vastly different scientific fields. By the study of such relations, scientists can better observe the way that the conservation of energy not only governs chemical processes but also provides insight into phenomena in the universe, hence bridging gaps between classical laws of energy conservation and emerging principles in theoretical physics (Habisohn et al., 2012) (Figure 5.3).

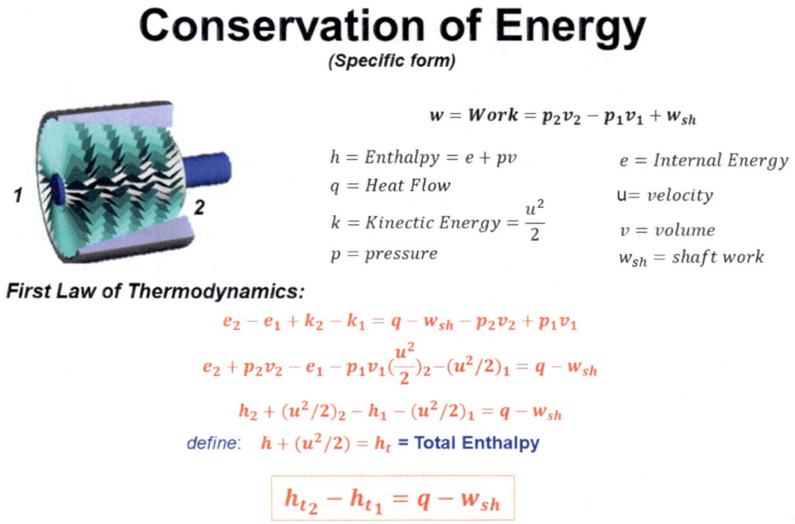

Figure 5.3. *Conservation of energy.*

Source: https://external-content.duckduckgo.com/iu/?u=https%3A%2F%2Fwww1.grc.nasa.gov%2Fwp-content%2Fuploads%2Fconservation-of-energy-1.png&f=1&nofb=1&ipt=985f66809c693adcd45624cbc633f313c9f24c82c705b549df754ba10967d6d7.

5.1.2. Enthalpy Changes in Chemical Reactions

Enthalpy changes play a pivotal role in the calculation of chemical reactions, particularly according to Hesss Law. Essentially, enthalpy is a thermodynamic property of the heat content of a system under constant pressure. The enthalpy change, denoted as ΔH, shows an exothermic or endothermic reaction when there is a transformation of the reactants to the products. The exothermic reactions release heat and hence have a negative ΔH, while endothermic reactions take in heat and have a positive ΔH. Chemists comprehend these changes, and thereby reaction conditions are controlled easily, with maximum yield coupled with the stability of the product. Thus, by including enthalpy changes in thermodynamic calculations, reaction feasibility can be predicted and energy requirements assessed, and thus it becomes a tool of vital importance in theoretical investigations as well as in operational activities of chemistry. Hess's Law that the total enthalpy change of a reaction is independent of the path, is crucial to ascertain the enthalpy change of reactions which are difficult to measure directly. The law allows for ΔH of a desired reaction to be calculated by summing the enthalpy changes of intermediate reactions. For instance, if a reaction can be broken down into a number of stages whose enthalpies of change are known, applying Hesss Law chemists can determine the overall enthalpy change without experimental information for the overall reaction. This is supplemented by models such as Density Functional Theory (DFT), which helps in the calculation of reaction pathways and transition states, further shedding light on enthalpy changes in chemical reactions (Hall et al., 2012). Hence, Hesss Law is a valuable resource for experimental and theoretical chemists alike, allowing for more efficient and accurate energy estimation. The applications of Hesss Law extend beyond academic research to include fields of environmental science and energy exploration. With shifting global needs of energy, there is a requirement for effective and sustainable chemical processes. For instance, the process of converting renewable sources of energy involves the understanding of enthalpy change to develop effective energy storage systems. Hybrid energy systems that combine batteries and hydrogen are an example of such a requirement since they attempt to balance energy storage capacity, efficiency, and sustainability (Guo et al., 2024). Through application of Hesss Law and the use of enthalpy calculations, researchers are able to investigate ideal reaction pathways and energy conversion processes and address the complexity of working with energy sources. This intersection of thermodynamics with useful application increases the importance of enthalpy changes in driving innovation in energy storage and more sustainable practices, pointing to ongoing relevance of Hess's Law in contemporary scientific research.

5.1.3. Relationship Between Hess's Law and Thermodynamics

Hess's Law is a key connection between the thermodynamic laws and the field of chemical reactions, which asserts that the net change in enthalpy during a chemical reaction is independent of the pathway. This basic concept is a foundation of the first law of thermodynamics that energy in isolated systems is preserved and takes it further to state that regardless of whether a reaction is single or multiplicity, net energy change will be the same. Hess's Law thus allows chemists and engineers to estimate changes in enthalpy for reactions that are not easy to measure directly. This theoretical concept is extremely important, particularly in fields like chemical engineering and material science, where understanding energy change is important to developing optimal materials and processes. Hess's Law, therefore, contributes not only to the research of thermodynamics but also to application in various fields. Hess's Law applications transcend pure theoretical background; it is very useful in the investigation of intricate chemical systems.

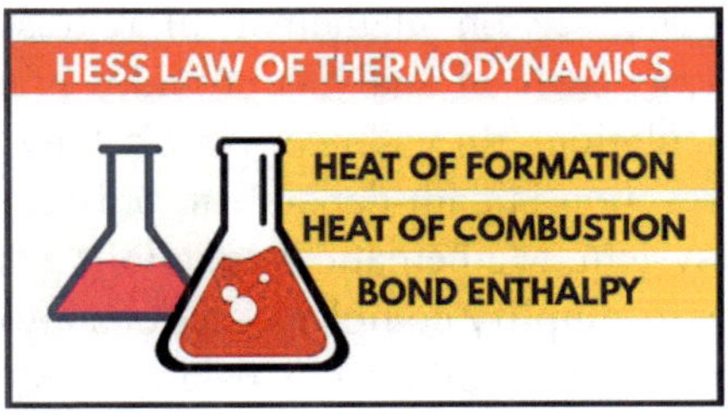

Figure 5.4. *Relationship between Hess's law and thermodynamics.*

Source: https://external-content.duckduckgo.com/iu/?u=https%3A%2F%2Fi.ytimg.com%2Fvi%2F2IsVX_BJgdE%2Fmaxresdefault.jpg&f=1&nofb=1&ipt=106085b5548240e388e9ab256df9ae49329d690925fa6dc1fd557193c911bf34.

Through enabling the calculation of enthalpy changes through addition of known reaction enthalpies, Hess's Law provides the capability to calculate the efficiencies of energy production reactions. For example, in the design of hydrogen-based energy storage systems like those being studied through ongoing research, information about enthalpy changes is highly important. These types of systems must optimize energy input and output for optimum performance and sustainability, and Hess's Law facilitates the evaluation of energy dynamics in such a system. This is the synergy between thermodynamics and the practical uses of energy in the real world, in which Hess's Law serves as a rule of thumb in energy efficiency and technology sustainability. Furthermore, the inclusion of Hess's Law in current energy technologies further establishes its relevance to contemporary energy management and sustainability issues. With continued expansion of the world's energy demands, the maximization of harnessed energy resources is ever more important. Principles of Hess's Laws form a core foundation for simulating integrated energy systems where efficient conversion and storage of energy become paramount. Further innovation in SESs increases the need for a macro-level understanding of energy interaction, which is provided by Hess's Law. As researchers continue to identify alternative methods for the optimization of energy design and operation in response to volatile renewable energy sources, the foundation offered by thermodynamics, encapsulated by Hess's Law, remains pivotal to navigating the complexity of future energy systems. Hence, the interconnection of Hess's Law to thermodynamics is not just conceptual but also instrumental in establishing sustainable energy solutions, accentuating its relevance in theory and practice (Figure 5.4).

5.2. MATHEMATICAL REPRESENTATION

There is a heavy dependence on mathematical representation in thorough analysis of Hesss Law, as such an analysis provides precision and definition to thermochemical calculations. In essence, Hess's Law says that the overall enthalpy change in a reaction is path-dependent, only relying upon the initial and final states of the system. Mathematical equations express this rule simply because they allow one to simplify complex reactions into usable forms. Using stoichiometric coefficients that follow the balanced equation, chemists can consistently relate different reactions and calculate overall enthalpy changes. This approach not only reinforces the fundamental postulates of thermodynamics but also enhances predictive capacity in experimental practice, highlighting the significance of mathematical form in the overall vision of energy conversions in chemical reactions. Mathematical formulation of Hesss Law finds practical application in numerous fields, including energy management systems and sustainable development, where computations of energy must be highly accurate. With increasing interest in renewable energy systems, optimization of energy storage solutions is of utmost priority. Research illustrates the ways in which mathematical modeling facilitates integration of hybrid renewable-hydrogen systems efficiently, with emphasis on the need for such models to optimize design, sizing, and operation strategies. Through the use of equations and algorithms, practitioners can handle uncertainties of renewable energy sources, such as variability and intermittency. For instance, a mathematical model of a dynamic system that considers real-world applications showcases the capability of mathematical representation to utilize energy systems in achieving targets for sustainability and efficiency, as illustrated in existing research (Arabkoohsar et al., 2022). In addition, the success of Hesss Law through mathematical representation goes beyond theoretical studies into tangible solutions for modern energy concerns. Complex computational algorithms and software, as indicated in the topic literature, have enhanced the simulation of energy systems to deliver scalable and affordable energy solutions. Not only does this maximize thermal efficiency, but it reduces the cost and environmental impact in producing energy. For instance, results of optimization analysis indicate considerable energy cost savings and carbon emission reductions with the use of accurate mathematical models for hybrid systems. The coordination of multi-objective optimization methods effectively harmonizes competing interests, showcasing the practical realization of Hesss Law for driving innovation to sustainable energy solutions (Atteya et al., 2025). The link between Hess's Law and mathematical formulation thus underscores its relevance in guiding research and development efforts towards sustainable energy futures.

5.2.1. Formula for Calculating Enthalpy Changes

The calculation of enthalpy changes is the core in the learning of thermodynamics and is most importantly applicable in the case of Hess's Law. Hess's Law defines that the total enthalpy change for a chemical reaction remains unchanged regardless of the number of steps taken to achieve the reaction. This is most vital for the calculation of enthalpy changes in cases where direct measurements are not feasible or even impossible. The change in enthalpy can be determined from the equation $\Delta H = H\_products - H\_reactants$, where ΔH is change in enthalpy and H is the enthalpy of the concerned compound. By applying Hesss Law, complicated reactions can be broken down into simpler reactions with known enthalpy values, thereby allowing us to determine the overall total change in enthalpy of the reaction. Furthermore, Hess's Law may also include using standard enthalpies of formation, which are the enthalpy change when one mole of a compound is formed from its elements under standard conditions. In actual calculations, the enthalpy change in a reaction can be calculated by using tabulated values of standard enthalpy of formation (ΔH_f°) for products and reactants. The formula can be written succinctly as $\Delta H = \Sigma\Delta H\_f^{\circ}(products) - \Sigma\Delta H\_f^{\circ}(reactants)$. This method streamlines the calculation process and allows for a systematic approach to interpreting thermodynamic properties of chemical reactions. By using a large set of data collected from experimentally obtained values, chemists can vastly enhance their capabilities to predict reaction behavior under varying conditions. Besides this, the integration of computational tools such as Density Functional Theory (DFT) has provided new avenues for higher accuracy in the prediction of enthalpy changes. DFT provides strong theories to simulate reaction pathways and determine transition states, particularly in instances involving complex systems not easily accessible through traditional methods. With computer-aided computational chemistry, the scope of Hesss Law has been broadened beyond straightforward stoichiometric reactions to include more complex enzymatic reactions and mechanisms of reactions in different phases of matter. As seen in recent advancements, the application of DFT in studying phenomena offers a stimulating avenue for broadening our understanding of enthalpy changes in different areas, and subsequently to the field of chemical engineering and material science as a whole (Hall et al., 2012; Volodin et al., 2024) (Figure 5.5).

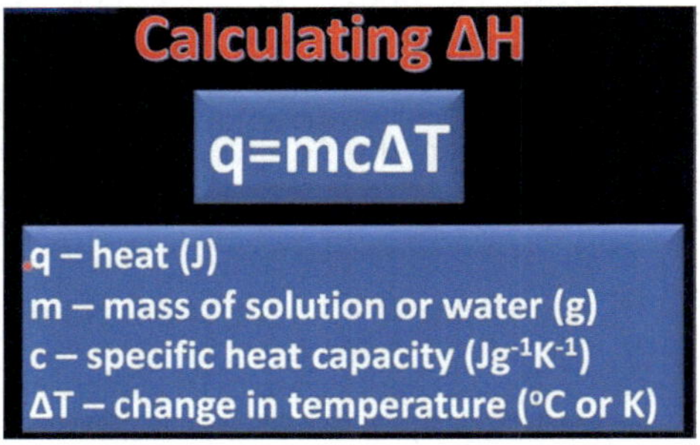

Figure 5.5. *Calculating enthalpy changes.*

Source: https://i.ytimg.com/vi/1eqBG08QEfI/maxresdefault.jpg.

5.2.2. Use of Standard Enthalpies of Formation

The use of standard enthalpies of formation is at the heart of the use of Hess's Law in practical application, since it enables chemists to calculate the enthalpy change of a reaction when measurement by direct methods is not possible. Standard enthalpy of formation is a prescribed value since the heat change resulting from the formation of one mole of a compound from elements in their standard states. This value is a reference point to which the total enthalpy change for a reaction may be obtained by Hess's Law, i.e., the total enthalpy change for a reaction is equal to the sum of enthalpy changes for any set of reactions to the same product. Using tabulated standard enthalpies of formation, scientists are able to compute the enthalpy change in reactions that are complex in nature, thus understanding thermodynamic principles in chemical reactions better. In practical applications, standard enthalpies of formation assist in estimating energy changes that arise in various chemical reactions, particularly in fields such as thermochemistry, environmental science, and food chemistry. For instance, stability of some food emulsions has been studied in terms of their energetic profiles, where enthalpy changes associated with methods of stabilization are compared. Application of standard enthalpies of formation makes it possible for food scientists to design foods to optimize physical stability and reduce oxidation, thus impacting on the quality and shelf life of foods. As new techniques are evolved, having knowledge about such thermal properties can help in developing healthier, stable food systems compatible with market conditions as well as nutritional demands, thus linking chemistry to food technology. Moreover, Hesss Law and the use of standard enthalpies of formation are all a part and parcel of innovation in various industrial operations, far beyond food chemistry to energy production and materials science. For example, in the development of improved catalysts, understanding the enthalpy change accompanying the stabilization of the transition states can lead to further enhanced reaction pathways requiring less input of energy. Novel computational methodologies, such as those studied in studies of Density Functional Theory (DFT), provide the facility to sample energetic landscapes which are vital to optimize the performance of catalysts (Hall et al., 2012). Furthermore, the progressive move towards sustainable processes indicates the importance of ascertaining enthalpy changes in biobased products, as seen in the production of natural emulsifiers with excellent oxidative stability (Schröder et al., 2020). Such uses show the diversity and usefulness of standard enthalpies of formation in driving scientific and industrial advancement (Figure 5.6).

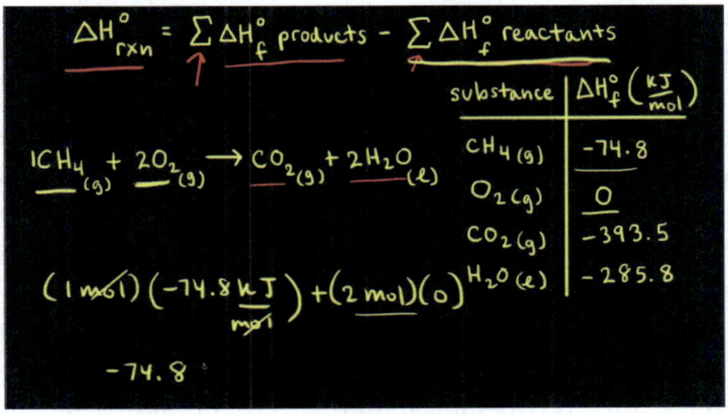

Figure 5.6. *Enthalpy of formation.*

Source: https://cdn.kastatic.org/ka_thumbnails_cache/336116c7-b886-47c7-8d8b-dbc6aab11480_1280_720_base.png.

5.2.3. Application of Hess's Law in Complex Reactions

The application of Hess's Law in complex reactions is an invaluable means for application in thermodynamic calculations to calculate the enthalpy changes in indirect ways. Chemists are able to study multi-step reactions where direct measurements are hard or impossible to carry out, based on this rule. This is particularly applicable in reaction intermediates' situations where the net conversion's enthalpy change is the algebraic sum of reactants and products' heats of formation via various pathways. By measuring the contribution of each step toward the total change in enthalpy, Hess's Law presents a systematic approach to calculating energy that is essential in fields such as organic synthesis and materials science, where thermal property knowledge can mean the difference between process efficiency and yield. In more advanced use, Hess's Law facilitates the ease of scientists' ability to combine varied reactions in order that they may develop energy storage systems that advance beyond traditional methods. This is particularly observed in Hybrid Energy Storage Systems (HESS) which combine various energy storage technology to provide improved performance for a variety of applications, from electric vehicle charging stations to microgrids. These applications rely on precise thermodynamic calculations to estimate the combined components' efficiency and lifespan. Understanding of enthalpy changes has a direct influence on design considerations and working parameters. Thus, applying Hess's Law in the construction of such systems not only makes them more trustworthy but also has a central thermodynamic concept that is necessary for pushing the boundaries of technology in energy storage and management capability (Hassen et al., 2023). Even further, beyond the common chemical reactions, Hess's Law finds applications in cross-domain fields of research, such as material science and environmental sustainability. By the determination of enthalpy changes of various steps in material synthesis or decomposition, Hess's Law enables the development of new materials with suitable thermal properties and thus contributes to green chemistry. Investigations into the energy change in advanced biological systems or polymer processes also rely heavily upon these principles. With scientists extending the limits of material science and energy efficiency, the principles of Hess's Law serve as a firm basis for understanding complex thermodynamic behavior in actual uses, for example, to increase catalyst efficiency or maximize chemical reaction at low temperatures, thus broadening the prospects for environmentally friendly methods for chemical synthesis and utilization (Delande et al., 2024).

5.3. EXPERIMENTAL VERIFICATION

Experimental verification is crucial in verifying theoretical models formulated by Hess's Law and its use in different chemical reactions. Precision in the thermodynamic calculations offered by Hess's Law requires top-notch experimental verification to ascertain that enthalpy change in chemical reactions is precisely estimated. By measurement through calorimetry, researchers can verify the heat released or taken up during a reaction, thus confirming Hesss Law under varied conditions. These experimental findings further enhance our understanding of reaction pathways and thermodynamic behavior, hence enhancing the predictive power of the laws. Furthermore, in view of emerging mobile energy facilities, which are rooted in principles such as Hesss Law, experimental verification becomes even more critical. Creation of reproducible and repeatable results in the laboratory is essential for closure of theoretical predictions to practical applications, thus improving energy management in electronic devices, as reported in Adib et al. (2019) and Malkawi et al. (2018). Experimental verification in the field of energy consumption is complementary to the theoretical foundation of Hess's Law in that it dissipates the complexities associated with various energy conversion devices. For instance, as advancements in electrification of mobile platforms continue, experimental studies' empirical findings can help guide the development of more efficient energy systems. These studies normally investigate the integration of renewable energy sources, which are inherently variable, and hence strong verification methods become crucial to maintaining energy delivery consistency. Further, incorporation of advanced materials and technology in battery systems necessitates experimental verification. As Adib et al. (2019) indicates, effective energy storage will be demonstrated to provide reliability and efficiency that can only be ascertained by systematic testing. Such a bridge between theory and practice eventually enables the transition from laboratory findings to practical use, thus making Hess's Law a more practical choice in modern energy systems. In addition, the exchange between principle and empirical fact serves to indicate the overall importance of experimental confirmation in the advancement of science. Testing and confirmation stringently not only confirm Hess's Law but also encourage innovations such as microgrids and distributed generation systems in general. Such new models leverage the efficiencies that Hess's Law emphasizes and reduce difficulties related to load variance, as in Malkawi et al. (2018). Experimental validation of these systems involves quantitatively measuring the functions of various subsystems under various operating conditions, ensuring the overall reliability and sustainability of energy options. With ongoing studies, the findings of these experimental work will indeed provide the way for further improvements and optimizations, attesting to the use of Hess's Law in the design of new energy systems. This iterative cycle of testing, application, and refining demonstrates the vital function experimentation plays in bridging theory with actual, everyday innovations.

5.3.1. Calorimetry and Its Role in Measuring Enthalpy

An understanding of calorimetry is essential in ascertaining the enthalpy change between chemical reactions, a principle paramount to Hess's Law. Calorimetry is quantitative measurement of heat in a reaction, allowing scientists to determine the enthalpy changes, either endothermic or exothermic. Scientists can have precise measurements in controlled conditions using calorimeters for specific experimental conditions. The first law of thermodynamics dictates that energy cannot be created nor destroyed but only transformed. This principle is the foundation for the operation of a calorimeter, where a change in temperature is measured during a reaction. The data which is generated through calorimetric measurements allows chemists to use Hesss Law to calculate enthalpy changes for reactions which cannot be directly measured. Calorimetry is not merely about the measurement of temperature but also various methods such as bomb calorimetry and differential scanning calorimetry. In bomb calorimetry, for instance, the sample is burned under a sealed container so that scientists can precisely measure released heat. The method is highly utilized upon examining fuel and determining calorific values, which are crucial in energy technology. Differential scanning calorimetry, on the other hand, measures different thermal properties of samples, providing vital information on transition phases, heat capacities, and reaction kinetics. Both techniques are important to a broader understanding of enthalpy change and emphasize the thermodynamic drives behind chemical reactions. Furthermore, these measurements can be used to test or validate theoretical models, extending our conceptualizations of enthalpic relations among chemical systems. In applied contexts, calorimetry is of great importance across a variety of fields, including food science, materials engineering, and environmental science. For example, the ability of measuring enthalpy changes helps to establish food emulsions' stabilization, i.e., in current research where calorimetric measurements provided information on the oxidation stability of emulsions stabilized by new particles (Schröder et al., 2020). Calorimetry's dual function—testing the physical stability and chemical stability of emulsions—illustrates its interdisciplinary strength in uncovering its relevance in creating efficient and sustainable technologies. As the demand for sustainable methods develops further, calorimetry's role in determining enthalpy will also be crucial, leading best practice and encouraging innovations with which to address today's energy and environmental challenges (N/A, 2025) (Figure 5.7).

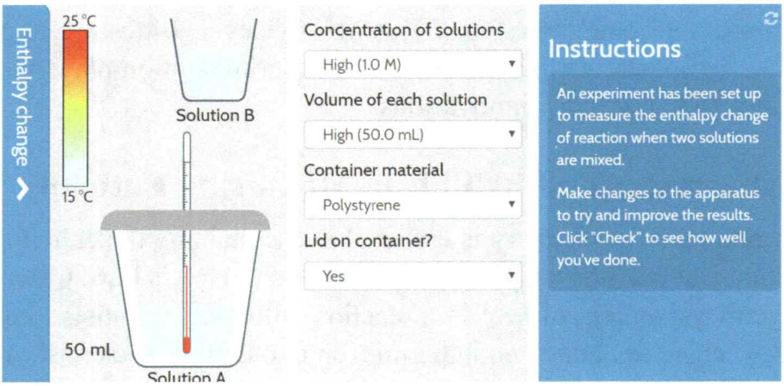

Figure 5.7. *Calorimetry and its role in measuring enthalpy.*

Source: https://cdn.prod.website-files.com/616d929c19707373b131ba49/641967b9616c4a4238cc6352_63e12e34929b0647cc4c845b_screenshot-0%2520(2).webp.

5.3.2. Case Studies Demonstrating Hess's Law

The application of Hess's Law in practical circumstances is brought into perspective using many case studies in chemistry, which demonstrate its applicability. One of the most famous examples is calculating standard enthalpies of formation of complex molecules when measurement is unfeasible or not possible. Hess's Law can be applied by scientists to add the enthalpy changes from a stepwise set of reactions to the synthesis of a target compound. For instance, by observing the reaction of simpler compounds resulting in the production of a high-nitrogen compound, chemists can determine the enthalpy changes without explicit experiment. This case is instrumental in understanding energy content in novel materials, thereby advancing advancements in energetic compounds and corresponding applications in various industrial sectors. Even more demonstrating the relevance of Hess's Law, a second case study displays the significance of prior laboratory instruction in learning chemistry, confirmed through e-learning practices. Through the presentation of students to the background theory of experiments via electronic means, teachers maximize lab time. This is akin to Hess's Law, where the overall enthalpy change of a reaction doesn't matter by which route it takes. In essence, the systematic pre-lab exercises properly prepare students, just as the application of Hess's Law ensures precise energy calculation regardless of experimental pathways. Because students receive preliminary knowledge in advance prior to laboratory experiments, students engage more substantially with experiments, which in turn results in better comprehension of principles of thermodynamics, such as those defined by Hess's Law. In addition, a case study of high-nitrogen compound clearly demonstrates the scope of applications of Hess's Law. Researchers employed density functional theory to study molecular properties and predict the standard enthalpies of formation through isodesmic reactions. Such theory-based grounding allows researchers to validate experimental measurements, thereby allowing for breakthroughs in the synthesis of new compounds. The ability to estimate heats of formation in the condensed phase—like those reactions that lead to such energetic materials—underscores the systematic nature of Hess's Law. This style is indicative of the cooperation between theory and practice of research and development on novel energetic materials and the broader implications of the laws on academic research and the creation of potentially explosive material, affirming the key position of Hess's Law in contemporary chemistry research and applications.

5.3.3. Limitations and Challenges in Experimental Setups

Experimental setups to investigate Hess's Law and its uses have some limitations and challenges, which, in the end, dictate the outcome and interpretation of outcomes. The first challenge is that it is difficult to get precise control over the experimental conditions, such as temperature and pressure, which are required for proper collection of data. Any level of deviation or inaccuracy of these parameters can cause sensational variations in thermodynamic measurement, essentially camouflaging the accuracy of enthalpy calculation based on Hess's Law. With scientists striving for accuracy, the complexity of maintaining such conditions most frequently calls for sophisticated equipment and techniques, which may not always be feasible with budgetary or technical constraints, as seen in similar research activities (Malburg et al., 2020). A second issue commonly encountered in experiment designs is duplicating outcomes. Varying material purity, instrument calibration, and technique consistency can all create variance when experiments are duplicated. In the application of Hess's

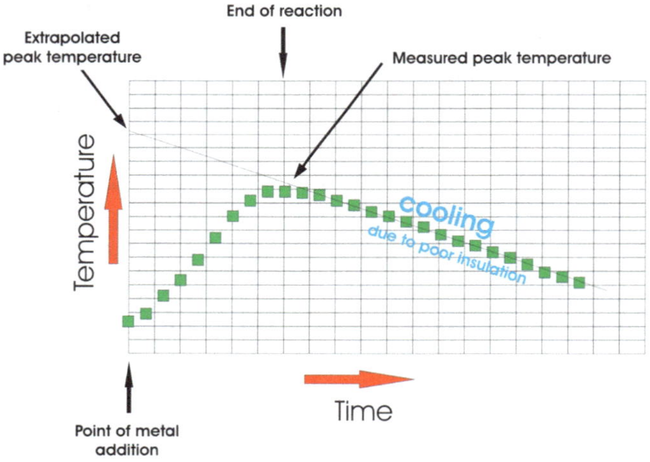

Figure 5.8. *An experimental demonstration of Hess's law.*

Source: https://www.chemedx.org/sites/www.chemedx.org/files/temp_vs_time_graph.png.

Law, such discrepancies can effectively short-circuit the dependability of comparative research based upon consistent energy change values. Researchers in the field should be well aware of these factors and emphasize rigorous standardization procedures to acquire greater reproducibility. Furthermore, merging valuable data from various experimental tests can be hampered due to disparities in experimental conditions, thereby making it difficult to synthesize a uniform picture of the applications of Hess's Law (RAHMAN et al., 2025). Computational and theoretical modeling with experimental configurations is also limited. Although simulations and models can provide valuable information on thermodynamic events, they are often based on assumptions and approximations that may not hold under all circumstances. Ideal gas assumptions, for example, may not hold in certain real-life scenarios, and the resultant inaccuracies in predicting experiment results intended to show Hesss Law could occur. This type of modeling challenges can result in substantial interpretative variations, and it becomes challenging to establish rigorous conclusions for the energy modifications in chemical reactions. This challenge has to be addressed by a fine balance between theoretical models and experimental measurements, ensuring that all analyses are based on robust scientific principles (Malburg et al., 2020) (Figure 5.8).

5.4. APPLICATIONS IN CHEMISTRY

The applications of Hess's Law permeate the different fields of chemistry, providing essential information for calculating energy and reaction mechanisms. In establishing a logical approach to the understanding of enthalpy changes, Hess's Law is a valuable tool in the prediction of reaction energies within theoretical and applied contexts. The rule is particularly helpful when studying complex chemical reactions which could be decomposed into elementary reaction steps. Thus, chemists can successfully calculate total enthalpy changes in a process without needing to carry out each reaction itself, which can be inconvenient or even dangerous. The value of Hess's Law in energy calculation emphasizes its pivotal role within thermochemistry, a driving force toward energy-efficient processes and material development. Apart from energy calculations, Hess's Law is also of immense application in the new field of organic electronics and photonics where new materials are being synthesized for environmental sustainability. The recent work focuses on the exploration of biological materials with outstanding properties to be implemented in organic electronics, illustrating the merits of introducing sustainable methods in chemistry. As highlighted in recent studies, understanding the thermal properties and reaction pathways of such substances is paramount in maximizing their performance in energy utilization. The study of organic and biological substances not only aligns with current environmental issues but also offers new opportunities for technological innovations, integrating the principles of thermodynamics with real-world applications in sophisticated areas of research (Flimelová et al., 2016). In addition, Hess's Law has broader applications than simple chemistry to daily uses such as pharmaceuticals, where understanding reaction enthalpies is essential in the synthesis and development of drugs. With the knowledge of heat change in chemical reactions, chemists are able to design more effective synthesis pathways, with reduced wastage and energy consumption. This maximization is particularly crucial in the pharmaceutical industry, where the safety and effectiveness of chemicals are greatly dependent on proper chemical reactions. As research continues to advance and broaden the horizon in this field, the use of Hess's Law in interpreting and applying thermodynamic concepts is still a vital factor in developing more sustainable and efficient pharmaceutical products. Thus, the application of Hesss Law not just enhances our knowledge about chemical reactions but also has a tremendous contribution towards enhancing scientific endeavors in many areas of application.

5.4.1. Predicting Reaction Outcomes

Predicting the products of reactions is extremely important in thermodynamics and chemical engineering, particularly in the application of Hess's Law. This principle, that the total enthalpy change of a reaction is the sum of the enthalpy changes of individual steps, enables chemists to calculate the energy requirements of chemical reactions without necessarily performing each reaction in an experimental manner. By employing Hesss Law, scientists can simplify the study of complex reaction pathways, thus enabling them to determine the stability of reactants and products based on their enthalpy measurements. This predictive capability is crucial for the design of efficient energy systems, especially as the world moves towards renewable forms of energy, where being able to determine how a reaction would go can be used to maximize processes in hydrogen production and storage (Atteya et al., 2025). For renewable energy, being able to predict reaction outcomes becomes more crucial as we design systems that counterbalance various types of energy. For instance, hydrogen energy storage devices that utilize the electrochemical reaction benefit from a better understanding of reaction routes predicted by Hess's Law. Through correct modeling of these reactions, researchers can establish novel methods of maximizing design and operation to make energy systems efficient in responding to varying demands and secure their efficiency optimized. Such predictive models not only assist in having sustainable energy transitions, but also support the technological innovation required for mass scale solutions in renewable energy, as illustrated through breakthroughs in hybrid renewable-hydrogen systems (FONTANA et al., 2024). Further, the application of Hesss Law in the prediction of reaction yields allows scientists to make more well-informed choices in the design and operation of systems in energy nodes. By the application of various kinds of energy, including electricity, gas, and hydrogen, understanding how thermodynamic properties regulate these reactions allows systems in energy management to operate more efficiently. Automation of energy centers in regulating these can apply predictive models to enhance real-time decision-making. In addition, rational outcome predictability has the ability to address certain energy system uncertainties, a key consideration in our journey towards climate neutrality and sustainability (FONTANA et al., 2024). The continuous development of methodologies to predict reactions not only improves scholarly work but also contributes significantly to the implementation of efficacious solutions towards a more sustainable economy (Figure 5.9).

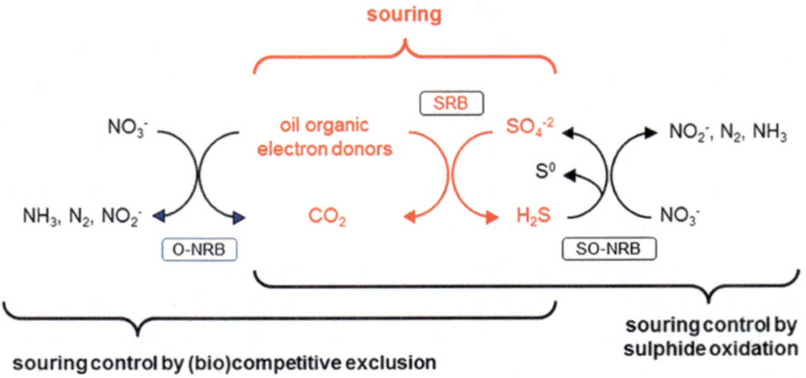

Figure 5.9. *Using Thermodynamics to predict the outcomes.*

Source: https://external-content.duckduckgo.com/iu/?u=https%3A%2F%2Fwww.frontiersin.org%2Ffiles%2FArticles%2F313903%2Ffmicb-08-02575-HTML%2Fimage_m%2Ffmicb-08-02575-g002.jpg&f=1&nofb=1&ipt=73439fc627f6dcde99d45962549355b9aa4a9c28db0f8f15b3fe9070fcdc9ad3.

5.4.2. Calculating Enthalpy Changes for Non-Standard Conditions

Evaluation of enthalpy changes under non-standard conditions helps in understanding chemical reactions under non-laboratory settings, where ideal standard conditions of 25 °C and 1 atm pressure may not be achievable. Hesss Law that the net enthalpy change in a reaction is just the total of changes in enthalpy for individual steps applies in such cases. Environments other than standard involve varying pressures, concentration, and temperature, each having an influence on reaction enthalpy changes. Thermodynamic laws serve as bases for calculation of these changes in terms of adjustable standard enthalpy values using specific equations. A thorough understanding of these principles is necessary for both engineers and chemists, as they are applied in uses ranging from industrial process design to environmental science, where conditions might well deviate far from normal norms. One method of accounting for non-standard conditions applies the Van t Hoff equation, which is useful for determining how much the equilibrium constant of a reaction is affected by temperature. This adjustment also pertains to the enthalpy adjustments, through which one is able to calculate the enthalpy value for desired conditions. The use of equations finds importance specifically when working with reactions involving gases or where concentration shifts might influence the energy profile of systems. Secondly, the use of heat capacities at constant pressure (Cp) or constant volume (Cv) can be utilized to determine enthalpy change in non-standard conditions. By multifying these thermodynamic considerations, an improved approximation of the energy dynamics of a reaction is attainable, crucial to real-world applications such as energy production and pharma, where precise energy calculations are vital. Aside from that, practical uses for the calculation of enthalpy change under non-standard conditions could be in numerous scientific fields, including materials science and environmental chemistry.

For instance, chemical reactions occurring in manufacturing environments typically experience a wide range of temperature and pressure conditions different from standard state conditions. Understanding such changes in enthalpy allows for optimal adjustment of reactions to maximize yields and reduce energy consumption. Additionally, in environmental processes, measurement of enthalpy changes can inform energy conservation strategies as well as the design of sustainable processes. As researchers explore new answers to energy questions, knowledge of the calculations of non-standard enthalpy changes becomes increasingly important to continue advancing the scientific principles and practice of Hess's Law in numerous disparate real-world situations (Figure 5.10).

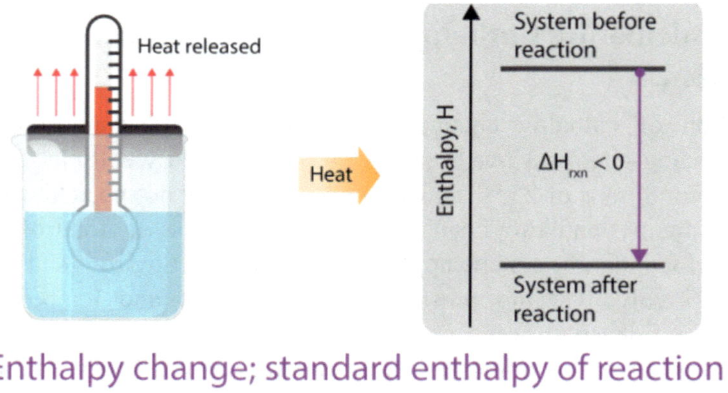

Figure 5.10. *Calculating enthalpy changes for non-standard conditions.*

Source: https://cdn1.byjus.com/wp-content/uploads/2016/11/Enthalpy-Change-Standard-Enthalpy-of-Reaction-.png.

5.4.3. Role in Organic and Inorganic Chemistry

The distinction between organic and inorganic chemistry is an important point to understand in terms of the applications of Hess's Law to many chemical situations. Organic chemistry, being based upon carbon-containing compounds, often involves the complex molecular interactions which are at the core of processes like combustion and synthesis. The application of Hesss Law in this case allows chemists to calculate the net enthalpy change of a reaction by assessing the sum of intermediate reactions, hence facilitating the synthesis of a variety of organic compounds. For instance, the energy changes with the formation and breakdown of organic compounds, such as polymers and medication, are dissected for understanding by Hess's Law, identifying its importance in the development of effective production processes in organic chemistry and the pharmaceutical industry. Conversely, inorganic chemistry is a fairly broad group of compounds that encompasses metals, minerals, and organometallic compounds.

Hess's Law holds good here also, particularly to reactions involving coordination complex and solids. According to Hess's Law, the chemists can evaluate energy changes for a variety of inorganic reactions, a critical parameter in understanding thermodynamic stability and reactivity. Inorganic compounds have tended to show intricate interplay of molecules and atoms that are the foundations for catalysis and materials science. For instance, understanding the enthalpy changes associated with inorganic complex formation can have a huge impact on their use in catalysis and materials construction and thus to green chemistry and sustainable energy applications. Moreover, the boundary area between inorganic and organic chemistry becomes increasingly pertinent when sophisticated materials for electronics and photonics are envisioned as described in recent work.

The dynamics of both organic and inorganic substances can be studied through thermodynamic principles, including those exemplified by Hess's Law. Such a theoretical model allows scientists to analyze the potential use of hybrid organic-inorganic systems in photovoltaic and photonic applications. The investigation of charge transfer mechanisms, for example, can be facilitated through the application of Hess's Law in establishing the energy efficiency of these types of systems in applications. Thus, the integration of inorganic and organic chemistry through reactive pathways enhances our understanding and application of thermodynamic principles, ultimately resulting in new technological solutions and environmental sustainability, as the literature around this research field elucidates (Flimelová et al., 2016; Seton Hall University, 1930).

5.5. INDUSTRIAL APPLICATIONS

Industrial applications of energy storage systems are taking the lead in technological advancements, especially in sustainable solutions to energy. As the world increasingly gravitates towards renewable sources of energy, the necessity of relying on efficient storage systems has been a factor in overcoming the unstable form of solar and wind energy. Hesss Law, under which conservation of energy is a central theme, has a direct bearing on the structure of such systems, enabling engineers to achieve optimum energy transfer efficiency in various applications. The case of supercapacitor and battery developments, for example, demonstrates how principles based on Hesss Law can be used to increase energy storage, with thoughtful design resulting in increased lifespan and reduced environmental footprint, essentially advancing the goals given in today's energy policies (Malburg et al., 2020). This synergy between concepts and application underscores the importance of continued research in energy storage technology in order to achieve sustainable industrial processes. Additionally, hydrogen energy storage systems constitute a revolutionary application of Hesss Law in industry, particularly when combined with renewable energy systems. They have been identified for their environmental advantages and rapid response characteristics, rendering them competent solutions for solving issues of energy intermittency. The advanced dynamics of electrochemical reactions in hydrogen storage highlight Hesss Laws role in preventing system performance improvement to the minimum. New modeling schemes have been established that provide proper sizing and simulation of hybrid renewable-hydrogen systems to facilitate smooth integration into existing energy systems (Atteya et al., 2025). Using such methods, industries can weigh costs against maximized carbon footprints in support of grander sustainability goals. Hydrogen tech advancement is one of the best examples of the common overlap of theoretical thermodynamic processes and real-world applications, advancing progress toward net-zero energy transitions. Hybrid battery-supsca technology is another good example of the use of Hesss Law principles at a commercial level. Researchers have been striving to create hybrid systems that bring together the rapid discharge characteristics of supercapacitors and the energy storage capability of traditional batteries, offering a better energy storage solution for electric vehicles, for example. With the help of computer simulations, engineers proved that such a hybrid system can extend the lifespan of batteries by a considerable margin while also reducing the overall environmental impact associated with the disposal of batteries (Malburg et al., 2020). The findings bring to the fore the importance of further refining energy storage systems to meet new age needs. As industries seek to minimize their footprints, the innovative techniques borrowed from Hesss Law will remain at the forefront of the future of energy storage technology, propelling green operations and efficiency in other industries.

5.5.1. Use in Chemical Manufacturing Processes

The application of Hess's Law in chemical process manufacturing is particularly vital in the optimization of energy efficiency and reduction in the cost of production. By defining the overall enthalpy change in a chemical reaction as the same, regardless of the pathway used, Hess's Law is a means of estimating the changes in heat associated with various processes. This is crucial in industries that utilize exothermic and endothermic reactions. To producers, precise prediction of heat transfer concerning reactions can potentially lead to improved process design such that precise control of conditions can guarantee energy absorption or release, optimizing yield and profit. The connection of Hess's Law with practical applications, particularly with renewable energy and new chemical processes, is very much in line with the growing emphasis on sustainability within processes of manufacture, particularly with regards to evolving environmental laws that dictate the consumption of resources. In line with the foregoing statements, modern chemical manufacturing increasingly applies Hess's Law principles in building more sustainable processes that leave smaller environmental imprints. The principle is particularly relevant in the development of new materials, such as in the development of rechargeable batteries and supercapacitors, where chemical reaction heat management is critical. Effective distribution of heat is crucial in preventing heat degradation and extending the life of the materials. The entry of green energy solutions, as epitomized by the increased focus on green technology, demonstrates how companies are moving away from fossil fuel dependence, creating systems that not only meet energy requirements but also maintain ecological equilibrium. The need to enhance energy storage systems emphasizes the importance of applying Hess's Law in creating cutting-edge, environmentally friendly materials that are both economically and environmentally sustainable in chemical manufacturing, as discussed in scholarly literature on the future of energy (Malburg et al., 2020) and sustainability (Kuikka et al., 2015). Apart from that, Hess's Law also plays an important role in end-on environmental effects analysis and optimization of chemical reactions. With the rising global energy demand, the need for innovative production methods becomes a matter of urgency. Using Hess's Law to measure the energy change of reactions enables manufacturers to assess the effectiveness of their processes, paving the way towards a shift to a sustainable and less resource-intensive system. In a climate-change debate world, measuring the energy conversions achievable using Hess's Law saves money but also appeals to the public-at-large desire for corporate best practice in carbon footprint reduction. Fundamentally, the application of Hess's Law in chemical production processes has a double benefit: advancing technological advances while ensuring environmental stewardship aims, thus becoming increasingly prominent in manufacturing contexts of today and tomorrow (Figure 5.11).

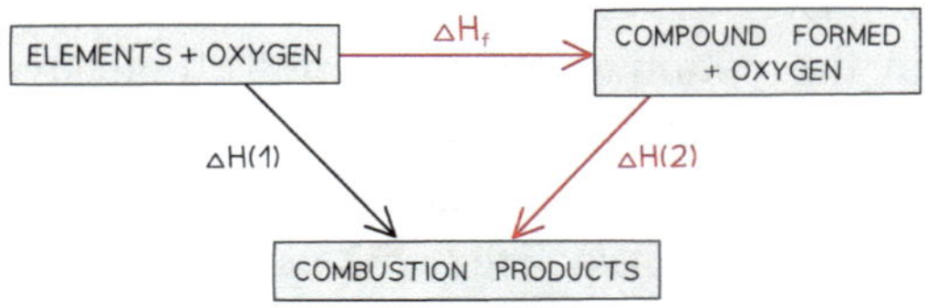

Figure 5.11. *Hess's law use in chemical manufacturing processes.*

Source: https://cdn.savemyexams.com/uploads/2024/01/1-5-chemical-energetics-enthalpy-change-elements-to-products.jpg.

5.5.2. Applications in Energy Production

It is essential to learn about the applications of Hesss Law in energy production since it guarantees efficiency and optimal utilization of resources in energy systems. Hesss Law simply states that the net enthalpy change in a chemical reaction is constant regardless of the path taken from reactants to products. This law has extensive impacts in the energy sectors, particularly when considering renewable energy technology development. For instance, solar energy systems may utilize Hesss Law when designing in order to calculate the energy yield from diverse reaction pathways in solar cells. Greater energy production efficiency is increasingly important with growing global energy demand and reduction in resources, necessitating new initiatives that prioritize sustainable strategies emphasized in cross-disciplinary courses such as those offered in the Bit Bang series of Aalto University focusing on economic, environmental, and societal sustainability in power generation (Kuikka et al., 2015). Within the context of microgrid technology, Hesss Law has the potential to facilitate advancements in renewable energy integration and storage systems. Microgrids, incorporating distributed generation sources such as solar panels and wind turbines, are supported by the thermodynamic postulates set forth by Hesss Law for optimizing energy conversion systems. The integration of energy storage systems, notably lithium-ion batteries and reversible solid oxide fuel cells (RSOCs), plays a vital role in managing the supply of energy and ensuring stability in the operation of microgrid applications. Advanced modeling techniques through the use of tools like MATLAB/Simulink provide researchers the capability to model dynamic interaction as well as predict performance outcomes of these energy systems. As seen in the study of hybrid microgrid architectures, the application of Hesss Law can lead to more effective energy management practices that are adjusted according to changes in demand and availability of renewable resources (Volodin et al., 2024). Exploration of alternative source energy and innovative technological responses supplement Hesss Law utilization in the generation of energy. As energy generation is shifting towards more sustainable processes, the understanding of thermodynamic principles is increasingly important in developing effective ways of energy production. For instance, research has considered the potential for new fuels and developing storage mechanisms, thereby expanding the boundaries of energy generation beyond the conventional. By examining energy transfer processes and applying Hesss Law to different situations, researchers can develop strategies aimed at optimizing performance and reducing costs associated with energy production. With the changing energy landscape, the ability to effectively implement thermodynamic principles will be crucial to establishing a foundation for a sustainable, reliable energy future (Kuikka et al., 2015).

5.5.3. Role in Environmental Chemistry and Sustainability

The intersection of Hesss Law with environmental chemistry and sustainability spotlights the potential to optimize chemical processes with reduced waste energy, towards a sustainable future. Hess's Law, which teaches us that the overall enthalpy change in a chemical reaction is path-independent, is critical to energy management. By accurate computation of reaction energy changes, chemists are able to optimize processes to utilize renewable energy sources, conserving fossil fuels. This aligns with growing emphasis on eco-efficient energy sources, as highlighted in the Bit Bang series addressing sustainable energy technologies (Adib et al., 2019). Renewable mechanisms of chemical reactions not only make way for energy preservation but also promote the production of products which can be used to heal waste and reduce environmental poisons. Lastly, new technologies that involve electrification exemplify the importance of Hesss Law in mobile systems. As energy systems go increasingly electrified, there is a critical need for efficient methods of energy conversion consistent with the precepts of Hesss Law. The transition to electrification, particularly in transportation and portable devices, requires new battery technologies and advanced power electronic converters (Adib et al., 2019). The systems need to be designed keeping in mind the environmental impact through optimal energy efficiency with minimum adverse ecological implications. By the use of Hesss Law in informing energy-related process design and analysis, researchers facilitate the shift towards sustainable energy platforms, thereby assisting in meeting efficiency as well as ecological integrity. The applications of Hesss Law extend past theoretical chemistry, cutting through some of the several different lines of sustainability endeavors. The application of laws in quantifying the enthalpy changes of various chemical reactions can assist in assessing the viability of alternative energy approaches, including focusing on biochemicals and green chemistry. With better understanding of energy dynamics, practitioners are able to promote innovations that not only align with economic viability but also consider social and environmental dimensions of sustainability (Kuikka et al., 2015). This combined philosophy of energy systems and chemical processes will be vital in shaping a sustainable future. While the growing global needs for energy demand continued development, the inclusion of Hess's Law in the discipline of environmental chemistry will be essential in guiding well-informed policies and practices toward sustaining the environment as well as the economy.

5.6. EDUCATIONAL IMPORTANCE

Understanding Hess's Law is paramount to the teaching of chemistry as it is a fundamental concept in describing the changes in energy when referring to chemical reactions. By showing students that the net enthalpy change for a reaction is path-independent, the students gain a very profound understanding of the principles of thermodynamics. This understanding not only facilitates mastery of theoretical principles but also fosters critical thinking and problem-solving abilities through application contexts. Also, while teachers apply Hess's Law in the classroom, they are able to develop a context where students learn to visualize and think about energy transformations, leading to deeper engagement and knowledge retention that are crucial to scientific literacy. The academic relevance of Hess's Law extends beyond the classrooms to research and innovation in most aspects of engineering. For instance, innovations in research work in clean energy technologies presented at conferences, such as the conference organized by Cardiff University, demonstrate how basic principles of thermodynamics, including Hess's Law, underpin modern engineering methods. By adopting these principles, researchers are currently struggling to acquire clean energy technologies, demonstrating the link between theoretical science education and applied sciences. Such links guarantee curriculum relevance, motivating students to appreciate the practical value of the theoretical knowledge they gain, thereby encouraging them to work in fields that contribute to societal development. Furthermore, the application of Hess's Law is fundamentally interconnected with the exigency of international affairs in relation to energy resources, as revealed in literature and research studies regarding renewable energy. The educationally initiated exploration of Hess's Law helps students to critically examine the constraints and potential of current systems for energy, particularly concerning the environmental challenge posed by traditional fossil fuels. By making students learn about the basics of energy conservation, teachers equip them to engage in discussing economic development and environmental sustainability, as in the assignment of coping with the challenges of energy security and recycling accepted in modern literature. Ultimately, this school of thought not only educates the students on the intricacies of thermodynamic principles, but also encourages them to be good-informed advocates of new, sustainable methods in their careers.

5.6.1. Teaching Hess's Law in academic settings

Institutionalization of Hesss Law in school curricula is essential in the enhancement of a heightened level of comprehension of thermodynamics and its applied use in, among other fields, chemistry and environmental studies. Educators may induce increased student interest by employing various pedagogical methods, including inquiry learning and group work. For example, learners can work in groups to design experiments that demonstrate Hesss Law in real-life application, thereby bridging the gap between the abstract knowledge and real-life contexts. Group project usage is in the spirit of interdisciplinary, as exemplified in projects like the Bit Bang series where learners delve into complex issues by using their knowledge to make creative applications (Kuikka et al., 2015). This aligns with the importance of fostering cooperative learning techniques that challenge students to explore and analyze energy concerns based on Hesss Law, which ultimately enhances their problem-solving and critical thinking abilities. Learning institutions must also consider the role that learning Hesss Law plays in addressing global energy demands and sustainability challenges. In-depth understanding of this principle not only helps students make sense of energy conversions but also helps them prepare to contribute to energy resource management and eco-efficient discourse. Courses intended to examine such applications can use case studies and simulations, thus enabling learners to model the implications of Hesss Law in energy production and consumption (Kuikka et al., 2015). Further, by situating Hess's Law within the context of contemporary energy debate, educators can make theoretical concepts relevant to practical realities, equipping future citizens with the information and analytical abilities needed to meet environmental challenges effectively. Lastly, student satisfaction is a rarely discussed but vital consideration in assessing the success of teaching, especially when conveying abstract principles like Hesss Law. Literature indicates that positive educational experiences strongly enhance levels of student satisfaction, which in turn enhances better retention and general academic achievement. To this end, teachers must prioritize creating active learning environments, which could be hands-on experimental laboratory experiences or cooperative learning methods that encourage student involvement and motivation. By accepting various pedagogical approaches, teachers can address multiple learning styles and preferences, ensuring comprehension and retention of fundamental concepts like Hesss Law. This student emphasis not only enhances the learning process but also fosters an academic culture concerned with critical thinking and collaboration, essential traits of future problem solvers and scientists (Figure 5.12).

Figure 5.12. *Teaching Hess's law in academic settings.*

Source: https://i.ytimg.com/vi/Z4ntAj7GcIo/hq720.jpg?sqp=-oaymwEhCK-4FEIIDSFryq4qpAxMIARUAAAAAGAElAADIQj0AgKJD&rs=AOn4CLAX_dxga6e4UuCwuSapcXvXBhBcfQ.

5.6.2. Importance in Understanding Thermodynamic Principles

The understanding of thermodynamic concepts is at the heart of grasping concepts such as Hesss Law, which represents the principle of conservation of energy in chemical reactions. Thermodynamics has recognized laws that regulate the behavior of energy in systems and how reactions occur as well as their feasibility. For instance, how energy is converted and transferred assists chemists and engineers in making processes more efficient so they dissipate minimum energy while maximizing efficiency. The efficiency is particularly vital in domains such as sustainable energy generation where energy planners seek to integrate varied sustainable sources in harmony. With an insight into thermodynamic principles, it is possible to recognize the importance of synergies management of energy, developing new solutions for fulfilling global energy demands in an environmental-friendly manner, as evident from the critical strategies proposed by Arabkoohsar et al. (2022). Thermodynamic principles are particularly important in the context of renewable energy systems. As the world transitions to a net-zero energy economy, it is important to understand such principles as efficient energy storage to offset the unpredictability of renewable resources like solar and wind. Hydrogen energy storage systems are an example, whereby thermodynamic concepts assist in optimizing design and operation strategy. Effective comprehension of thermodynamics enables researchers and engineers to develop precise models of hybrid energy systems, thereby ensuring optimal efficiency in energy conversion and utilization. The process not only facilitates lowering operational costs but also attaining carbon reduction targets. Control over thermodynamic concepts is thus an essential factor in the evolution of sustainable energy technologies as evidenced through learnings derived from Atteya et al. (2025). Furthermore, thermodynamics gives the theoretical basis for Hesss Law, which is that the overall enthalpy change for a chemical reaction is independent of the reaction route. The principle is one of the interdependence of energy states and is a reflection of the importance of understanding energy conversion in chemical reactions. By enabling the calculation of enthalpy changes without the necessity of extensive calorimetric measurement, Hesss Law illustrates the practical application of thermodynamic knowledge in real usage. Researchers and practitioners can rely on this law to effective energy management practice, particularly for hybrid renewable energy system maximization. With continued advancement in such technology, a complete understanding of thermodynamic principles not only facilitates the optimization and operation of such systems, but also enables the pursuit of a sustainable energy future, much like the critical role played by chemistry and thermodynamics in world advancement.

5.6.3. Resources and Tools for Effective Learning

Learning materials and tools that are efficient in communicating complex topics such as Hesss Law and its versatility are required to share complex concepts. The availability of diverse material ranging from textbooks to research papers to interactive models renders students capable of comprehending the intricacies of thermodynamics. Computer-aided platforms and programs within teaching that employ visual aid render learners more capable of observing energy transformations during chemical reactions, thereby enhancing comprehension. Specifically, multidisciplinary teamwork, as evidenced in the case of the Bit Bang series at Aalto University, focuses on interactive learning spaces. Learners participate in critical discourses regarding sustainable energy solutions, hence internalizing scholarly principles through practical application. This multi-faceted approach instils a stronger understanding of scientific principles and stimulates innovative thinking. These tools, when combined, create a learning environment that is conducive to higher-order thinking and problem-solving and reinforces the fundamental principles of Hesss Law. In the pursuit of effective learning, integration of theoretical knowledge and practical tools is critical. Hesss Law, which stipulates that the net enthalpy change in a reaction is independent of the path taken, can be demonstrated using various interactive tools that improve extensive interaction. For instance, machines in which students may simulate different reaction conditions can describe the principles behind enthalpy changes. In addition, practical examples of Hesss Law application through case studies, particularly from energy transition and storage, provide contextual relevance. This single-package strategy of aids and tools not only assists in interpreting abstract notions but also encourages retention and utilization of knowledge in interdisciplinary fields. Secondly, the focus on empowering citizens in energy resource conversations, discussed in the academic review of energy storage systems, necessitates educational resources to drive enlightened participation among learners (Bahloul et al., 2024). Finally, assessing the impact of resources and tools in effective learning means understanding the environment surrounding them, including technological innovation and human needs. Contact with active learning devices such as peer interaction and the provision of feedback mechanisms enables learners to develop critical thinking. Identification and analysis of enablers and barriers within the context of energy storage, as indicated in current literature, show the necessity to get learners ready for knowledge that corresponds to contemporary issues. Through qualitative assessments, instructors can tailor their approach to suit specific learning needs and encourage exploration of solutions that integrate theoretical and practical understanding. The total integration of such tools puts students not only in the position of grasping Hesss Law but also applying it positively to broader environmental and energy contexts (Kuikka et al., 2015).

SUMMARY

The appreciation of Hess's Law and its applications, it is worth drawing attention to the significance of the principle in the broader picture of thermodynamics. Hess's Law not only accounts for the relationship between enthalpy change in chemical reactions but is also applied as a systematic approach to calculating the energy changes associated with complex schemes of reaction. This theoretical accuracy is invaluable to both academic pursuits and practical applications within the pharmaceutical and materials science fields. By allowing chemists to predict reaction behavior without complete experimentation or explicit measurement, Hess's Lawstreamlines processes and enables innovative solutions. The principle's generality makes it a workhorse among the chemist's tools, evidencing its ongoing utility in contemporary scientific pursuit. Furthermore, the implications of Hess's Law extend to practical applications beyond theoretical models into other fields of influence. For instance, in environmental chemistry, information on enthalpic change is crucial in estimating potential energy efficiencies in polluting reactions and resource savings. Similarly, greater competence in thermochemical calculations can hasten the manufacturing of green energy technology. Its use in the lab exercises, as noted by corresponding scientific literature, further solidifies students' grasp of key concepts, bridging the theory-practice divide (Eltsov et al., 2010). These uses of Hess's Law reveal its valuable contribution not only to the understanding of chemical reactions but also to solving real-life issues using the scientific method. Lastly, as the face of education changes, the teaching of Hess's Law and how it applies also must change to develop a stronger foundation among students. Losing student interest and retention in science topics necessitates the reevaluation of existing teaching strategies. The integration of Hess's Law into an active learning curriculum with emphasis on context to the real world can recharge the interest of students (Luca et al., 2014). This approach not only highlights the relatedness of physics and chemistry but also energy transformation across disciplines. Fundamentally, the implication of Hess's Law applications highlights an exigent requirement for new pedagogical approaches that not only transfer knowledge but also inspire future generations to learn and become members of the scientific community.

When the applications of Hess's Law are discussed, it should be known that interest in Hybrid Energy Storage Systems (HESSs) is increasing in all industries. These systems utilize two or more distinct Energy Storage Systems (ESSs) efficiently by capitalizing on their individual strength and merging them to enhance overall efficiency and lifespan. Current studies emphasis that a majority of the research focus primarily on power management and synergy across different storage technologies but fail to emphasize any specific type of ESS. According to literature, Redox-Flow Batteries (RFBs) are distinct because they find excellent application in stationary storage uses with the benefit of hybridization to support features such as availability of peak power and short-term energy supply. The performance metrics driving innovation in this field highlight the transformative power of HESSs as irreplaceable resources in modern energy management practices (Hassen et al., 2023). A second key component of the debate involves optimization issues of applying HESSs, especially in full electric ships. The complexity of designing a HESS is further exacerbated by the necessity of balancing conflicting operating characteristics, such as power and energy density, as well as the process of progressive aging of storage units. A two-stage Pareto-based design optimization approach has been proposed to effectively handle this complexity. The strategy first determines a Pareto front, which represents all optimal configurations with respect to storage capacity satisfying operating constraints. Subsequently, the approach focuses on finding the best configuration from a specific cost function. This method of optimization not only enhances the design process but also reveals the advantages of hybrid storage systems over traditional single-technology systems, thereby indicating their contribution to enhancing energy storage applications (Barrera-Cardenas et al., 2025). Finally, Energy Management Systems (EMS) integration is crucial to optimize HESS operation. A proper EMS allocates power demands among different storage units considering the aging characteristics and constraints of each technology. EMS provides a strategy for guaranteeing that the operating parameters of HESSs are in line with working requirements. Constant research in EMS in hybrid systems means future advancements will increasingly emphasize control algorithms for maximum efficiency and flexibility. The EMS complexity reflects a broader energy systems trend towards greater predictive sophistication, which means that HESS configurations should be capable of dynamically responding to changing energy environments. This focus on EMS will assuredly become the hallmark of energy storage's future and is thus imperative for researchers and practitioners to stay abreast of such developments (Hassen et al., 2023).

REFERENCES

1. Abou-Rachid, H., Dudiy, S., & Goddard, W. A. (2008). Predicting solid-state heats of formation of newly synthesized polynitrogen materials by using quantum mechanical calculations. https://core.ac.uk/download/4880019.pdf (accessed on 4th October 2025).
2. Adib, A., Afridi, K. K., Amirabadi, M., Fateh, R., et al. (2019). E-mobility–Advancements and challenges. https://core.ac.uk/download/287210234.pdf (accessed on 4th October 2025).
3. Anekwe, O. (2007). Satisfaction matters: A comparative study of African American students in education programs within the academy. https://core.ac.uk/download/650404319.pdf (accessed on 4th October 2025).
4. Arabkoohsar, A., Hajizadeh, A., Lasemi, M. A., Mohammadi-Ivatloo, B., et al. (2022). A comprehensive review on optimization challenges of smart energy hubs under uncertainty factors. https://core.ac.uk/download/518026662.pdf (accessed on 4th October 2025).
5. Atteya, A. (2025). Development of a novel platform for the optimal design, sizing and energy management of grid-integrated hybrid photovoltaic-hydrogen energy systems. https://core.ac.uk/download/664729899.pdf (accessed on 4th October 2025).
6. Bahloul, M., Cardo-Miota, J., Efthymiou, V., Khadem, M., et al. (2024). Analysis of barriers and key enablers toward citizen ESS successful integration. https://core.ac.uk/download/613215395.pdf (accessed on 4th October 2025).
7. Barrera-Cardenas, R., D'Arco, S., Piegari, L., Tricoli, P., et al. (2025). Pareto-based design optimisation of hybrid energy storage systems for full-electric vessels. https://core.ac.uk/download/648317398.pdf (accessed on 4th October 2025).
8. Chalmers, J., Goodfellow, C., & Price, G. J. (2006). Using a VLE to enhance a foundation chemistry laboratory module. https://core.ac.uk/download/267013688.pdf (accessed on 4th October 2025).
9. Delande, L. (2024). Sharp spectral gap of adaptive Langevin dynamics. http://arxiv.org/abs/2403.00552 (accessed on 4th October 2025).
10. Ekambaram, S. (2012). General Chemistry. Pearson Education India. http://books.google.com/books?id=LTw8BAAAQBAJ&dq=Hesss+Law+definition+and+applications+in+thermochemistry&hl=&source=gbs_api (accessed on 4th October 2025).
11. Eltsov, S. V., & Vodolazkaya, N. A. (2010). Laboratory exercises on physical and colloid chemistry. https://core.ac.uk/download/46590087.pdf (accessed on 4th October 2025).

12. Fontana, C. (2024). Energy management systems and potential applications of quantum computing in the energy sector. https://core.ac.uk/download/614224283.pdf (accessed on 4th October 2025).
13. Guo, T., Jiang, L., Li, B., Ren, Y., et al. (2024). An overview of application-oriented multifunctional large-scale stationary battery and hydrogen hybrid energy storage system. https://core.ac.uk/download/596422343.pdf (accessed on 4th October 2025).
14. Habisohn, C. X., Zerilli, F. J., Burnett, G. A., Galloway, G. J., Reall, H. S., Racz, I., Corvino, J., et al. (2012). Stability of black holes and black branes. http://arxiv.org/abs/1201.0463 (accessed on 4th October 2025).
15. Hall, M. L. (2012). Density Functional Theory: Development and Applications. https://core.ac.uk/download/161439672.pdf (accessed on 4th October 2025).
16. Hassen, W. F., Molina Gaudo, P., Pettinger, K. H., Poisl, T. M., et al. (2023). Hybrid energy storage systems based on redox-flow batteries: Recent developments, challenges, and future perspectives. https://core.ac.uk/download/572528973.pdf (accessed on 4th October 2025).
17. Kuikka, M., Neuvo, Y., & Ormala, E. (2015). Bit Bang 7: Future of Energy. https://core.ac.uk/download/80716550.pdf (accessed on 4th October 2025).
18. Luca, O. R. (2014). Equilibrium as a conceptual framework for an integrated, synergistic introductory course in chemistry and physics. https://core.ac.uk/download/33113647.pdf (accessed on 4th October 2025).
19. Malburg, L. C. (2020). Analysis of nickel-copper metallic foam supercapacitor for electric vehicles with hybrid battery-supercapacitor energy storage system. https://core.ac.uk/download/344682871.pdf (accessed on 4th October 2025).
20. Malkawi, A. (2018). Single and dual DC buses nanogrids with decentralized control. https://core.ac.uk/download/211521014.pdf (accessed on 4th October 2025).
21. N/A. (1930). Seton Hall College 1930–1931. https://core.ac.uk/download/591698092.pdf (accessed on 4th October 2025).
22. N/A. (2025). Proceedings of the Cardiff University Engineering Research Conference 2023. https://core.ac.uk/download/637921030.pdf (accessed on 4th October 2025).
23. N/A. (2025). Renewable Energy. https://core.ac.uk/download/362227142.pdf (accessed on 4th October 2025).

24. Qadrdan, M., Sami, S., Wu, J., Zhou, Y., et al. (2024). Modelling and simulation of a hydrogen-based energy storage system. https://core.ac.uk/download/613844065.pdf (accessed on 4th October 2025).

25. Rahman, A. (2025). Innovative Control and Energy Management of DC Microgrids with Hybrid Storage: Enabling E-Mobility and Hydrogen Energy Solutions. https://core.ac.uk/download/646503695.pdf (accessed on 4th October 2025).

26. Schröder, A. (2020). Combined Physical and Oxidative Stability of Food Pickering Emulsions. https://core.ac.uk/download/pdf/287783997.pdf (accessed on 4th October 2025).

27. Volodin, J. (2024). Dynamic Optimization of Microgrid with Reversible Fuel Cell and Lithium Ion Battery. https://core.ac.uk/download/646199123.pdf (accessed on 4th October 2025).

CHAPTER 6

ENTROPY AND THE SECOND LAW OF THERMODYNAMICS

Contents

Introduction 201
6.1. Historical Background 206
6.2. Fundamental Concepts of Thermodynamics...210
6.3. The Second Law of Thermodynamics 215
6.4. Entropy in Isolated Systems..220
6.5. Applications of Entropy..225
6.6. Misconceptions about Entropy...230
Summary...234
References...237

INTRODUCTION

Entropy principle forms the core of the second law of thermodynamics, shaping our understanding of energy conversions and the directionality of natural processes. Entropy, commonly known as a measure of randomness or disorder, provides essential information about efficiency in energy exchange in thermodynamic systems. The second law states that, in isolated systems, entropy rises over time, indicating a natural tendency toward equilibrium.

This phenomenon suggests that energy spontaneously diffuses and becomes more spread unless work is performed to maintain things in a state. This general principle therefore not only applies in classical physics but finds extremely significant uses in other fields, like chemistry and cosmology, where maintaining the flow of energy in mind can provide insights into the life of stars or the motion of molecules in a reaction. Entropy is studied beyond the realm of simple thermodynamic systems, and it becomes challenging when dealing with macroscopic bodies where quantum phenomena or gravitational forces are involved. Precisely, whereas the second law holds firmly in equilibrium thermodynamics, its extension to non-equilibrium states proves to be complex. The creation of tools such as an entropy meter is helpful in this sense, allowing for more meaningful characterization of the entropy of non-trivially scalable macroscopic systems. These developments are capable of redefining entropy in terms of equilibrium states and the special behaviors defined by complex interactions and external constraints.

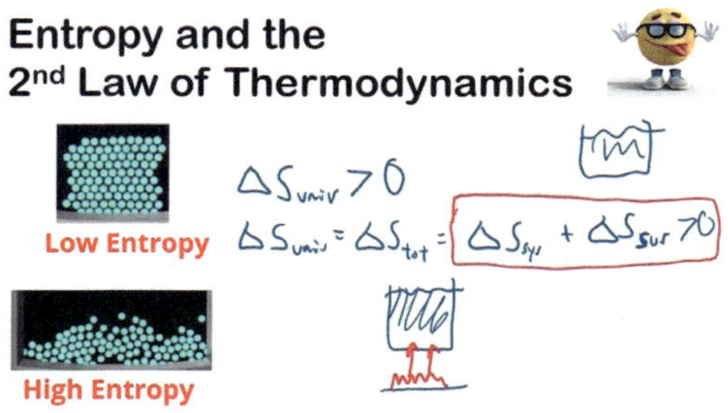

Figure 6.1. *Entropy and the second law of thermodynamics.*

Source: https://i.ytimg.com/vi/Tibi-qLNQa4/maxresdefault.jpg.

By taking up such new approaches, scientists are still expanding the use of thermodynamics so that systems are no longer deemed intractable (Lieb et al., 2014). Furthermore, as our knowledge of thermodynamic theory expands, contemporary theory predicts changes to age-old concepts of entropy, particularly for nonlocal thermodynamic systems. As an example, recent developments in weakly nonlocal thermodynamic theories provide insight into the behavior of entropy across dynamic thermodynamic variables. By incorporating various forms of entropy currents and internal variables referred to as current multipliers, researchers have devised methods to match classical equations like the Guyer-Krumhansl and Ginzburg-Landau equations with modern thermodynamic principles. Increased understanding points out to the intricate relationship between entropy and the second law of thermodynamics as not only being valid for equilibrium systems but also for those involving nonlocality and irreversibility (Van et al., 2003). Through these inquiries, entropy appears as a fundamental variable that not only quantifies disorder but also presents new fronts for enhanced efficiency and sustainability in various branches of science (Figure 6.1).

Definition of Entropy

Entropy, a fundamental principle in thermodynamics, encapsulates the degree of disorder or randomness in a system. In simple terms, entropy quantifies how inaccessible the energy of a system is to perform work, an idea proposed by the Second Law of Thermodynamics. The second law is that the total entropy of an isolated system will increase over time, and that natural processes will tend to reach a state of ultimate randomness. The importance of entropy goes far beyond thermodynamic systems, to many areas of science, including statistical mechanics and information theory, where the word disorder may also be found in anomalous outcomes. One must grasp this multi-faceted definition of entropy to really comprehend its place in the decision of the direction of thermodynamic processes in light of the inevitable trend toward equilibrium. Added to the confusion of comprehending entropy is its extension into phenomena such as black-hole physics, where traditional thermodynamic descriptions of temperature and entropy have been in the limelight. Surface gravity within a black hole can be considered to be a type of temperature, and the surface area of a black hole can be employed as a proxy for entropy, providing analogue underpinnings to challenge orthodox perspectives. The intersection of gravitational physics and classical thermodynamics suggests that entropy is less an outcome of energy dispersal than an inherent characteristic of the spacetime structure. The intersection calls for closer investigation into the conceptual foundation of both fields since it compels scientists to revisit conventional wisdom on how energy, entropy, and the universe itself interact (Curiel et al., 2014). Aside from its basic use in thermodynamics, entropy also raises interesting issues depicted in the Gibbs paradox in situations dealing with gas mixing. Classic thermodynamics contends that mixing two disparate gases will result in higher entropy; yet when the gases are the same, the scenario creates a discontinuous change in entropy that is paradoxical. The solution to this paradox unlocks the field of quantum thermodynamics, wherein the relationships between states and distributions of energy allow for a deeper understanding of entropy's value. In such a context, entropy's operative definition recedes behind mixing ergographs, an indicator of mixed state-accessible work. This quantum explanation not only sheds light on the Gibbs paradox, but also argues in favor of the fact that classical thermodynamic entropy formulations can fall short without the presence of quantum mechanics complications (Allahverdyan et al., 2006).

Overview of the Second Law of Thermodynamics

The Second Law of Thermodynamics identifies a universal principle that determines the direction of physical processes by saying that the total entropy of an insulated system will never reduce over time. This law contrasts strongly with the first law of thermodynamics, which is purely about conservation and energy transformation. While energy can be transferred and converted, the Second Law introduces the concept of entropy, a measurable quantity of disorder or randomness of a system. Entropy is a critical measure of the irreversibility of natural processes, defining that spontaneous processes cause an increase in entropy. This built-in tendency towards more entropy proposes a universal direction toward equilibrium, which effectively delineates our notions of thermodynamic activity for isolated and non-isolated systems. Beyond these theoretical extensions, the Second Law also has broader-reaching effects which impact a variety of scientific investigations in physics, chemistry, and even biology. In thermodynamic theory, phenomena ranging from heat engines to biological processes such as metabolism and evolution are governed by Second Law obeying relationships.

More recently, in particular, research utilizes constructs such as non-equilibrium thermodynamics to investigate entropy interactions within complex systems, and it gains insight into phenomena such as phase transitions and critical points. For instance, advanced theories investigate the contribution made by gradient terms to stress in materials, emphasizing entropy's relevance both in mechanical and thermal contexts. Notably, the employment of mathematical description of phenomena by frameworks like the Cahn-Hilliard and Ginzburg-Landau equations helps to show how thermodynamics transgresses disciplines of science, enhancing our knowledge of the contribution made by entropy to physical systems (Ván et al., 2020). In turn, the Second Law of Thermodynamics compels researchers to recast conventional techniques used in systems that are out of equilibrium. By keeping focus on entropy production, thermodynamic reasoning eliminates the necessity for complex variational methods normally used in material response modeling. The use of constitutive relations that take into account internal variables reflects developments toward attaining compatible models that meet thermodynamic requirements. Here, the operation with entropy flux as a constitutive variable reaffirms an improved understanding of irreversible processes. Elaboration of such concepts serves to emphasize the interdisciplinary character of the Second Law, providing definitive insight into phenomena from materials science to bioregions. As thermodynamics progresses, it forces scientists to keep refining their comprehension of energy movement and the irreversibility of natural processes (Ván et al., 2017).

Importance of Studying Entropy in Physical Sciences

The physical sciences study of entropy is important in that it captures fundamental principles governing energy transformations and the directionality of natural processes. Essentially, entropy is a quantification of randomness or disorder within a system, reflecting the inherent constraints governing energy exchange. It is worth appreciating this concept in order to understand the action of non-isolated and isolated systems, in which entropy always grows according to the Second Law of Thermodynamics. The law identifies that natural phenomena trend towards maximum entropy, reflecting that systems are tending towards equilibrium. Through studying entropy's role in thermodynamic processes, scientists are better able to predict and regulate energy use in various applications, ranging from engineering to environmental studies, to facilitate more efficient and sustainable operations. Further, the topic of entropy has connotations which extend to areas of active systems, where the traditional ideas of thermodynamics must be rethought. As an example, active biological systems that are not in equilibrium challenge conventional perspectives on thermodynamics and require a high-level understanding of entropy production in non-Hamiltonian dynamics.

Figure 6.2. *Entropy in physical sciences.*

Source: https://media.geeksforgeeks.org/wp-content/uploads/20240116161936/Entropy.png.

There are recent works that have described frameworks in stochastic thermodynamics to better treat such systems, with emphasis given to the underlying thermodynamic variables, like work and heat, that govern their dynamics. By extending the definitions of entropy and thermodynamic laws, researchers are acquiring new insights that relate statistical mechanics to biological processes, as seen in studies of the thermodynamics of active matter systems (DeWeese et al., 2017). Such a move not only furthers our understanding of fundamental physical laws but also fuels progress in the areas of biological and materials science. In addition to its underlying role in thermodynamics, the investigation of entropy is at the heart of an education on causal relations in complex systems. Through coupling between subsystems, particularly when driven by a faster, more dynamic counterpart, the emergence of a fruitful thermodynamic arrow of time can be explained. In such situations, the realization of the back-action and the consequences upon the system's entropy should be recognized because the second law can be refuted by non-Hamiltonian effects (Allahverdyan et al., 2007). This correspondence between scales of systems is meant to establish the criticality of a more comprehensive approach towards the study of entropy in physics. Hence, the study of entropy not only improves theoretical models in physics but also provides basic knowledge which is applicable to practical usage, thereby making it a core subject of study across a wide variety of scientific disciplines (Figure 6.2).

6.1. HISTORICAL BACKGROUND

The Second Law of Thermodynamics has a rich, complex history of development of thought on energy and entropy. This axiom, that total entropy in an isolated system will never decrease over time, came to be significant in the 19th century through the work of influential pioneers such as Sadi Carnot. His foundational work laid down the principles of thermodynamic efficiency in engines and opened the way for what came next. Following Carnot, researchers like Rudolf Clausius and Ludwig Boltzmann further established the theory of heat transfer and statistical mechanics, respectively. They proved that entropy was not only a measurable quantity of randomness but an intrinsic characteristic of thermodynamic processes linking microscopic descriptions to macroscopic behavior. The Second Law was thus a cornerstone of physical science that dictated our description of nature and energy conversion. This version of history did not remain constant; it altered significantly with developments in theory and technology. The use of the principles of thermodynamics to broader scientific domains, such as black hole thermodynamics and Hawking radiation, showed that the laws applied even in non-traditional areas. For instance, cosmology, a new domain, necessitated redrafting thermodynamic principles to describe black holes' behavior, leading to new developments in how entropy is found in astrophysical systems. Scientists reinterpreted the effects of entropy with paradigms like the Cahn-Hilliard and Ginzburg-Landau equations, a direct result of the Second Law. This rich texture of historic development intimates the law's foundational role in traditional and modern physics, emphasizing its significance in ongoing scientific research (Ván et al., 2017). Furthermore, the entwinement of entropy with cosmological process is used to illustrate the law's ubiquitous influence. The studies surrounding Hawking radiation have especially highlighted the theoretical and historical significance of entropy in situations ranging from black hole thermodynamics to the early universe. As scientists move further into these areas, they are confronted with the metaphysical issues surrounding information loss and the ultimate fate of energy in the universe. These discussions not only call to mind the conclusions determined by classical thermodynamics but also challenge contemporary theories of physics. This journey represents a critical confrontation with the Second Law, revealing its axiomatic status in physics, and de-mythologizing the delicate interplay between entropy and the intrinsic nature of space and time (Page et al., 2004).

6.1.1. Development of Thermodynamic Principles

The development of thermodynamic principles has been very impressive from classical theories to contemporary methods addressing a larger class of phenomena. Beneath all this change lies the Second Law of Thermodynamics, which basically says that in any enclosed system, the total entropy can never be lower in the future than it was in the past. This law not only tells us about energy transformations but also serves as a sort of directing principle across a wide variety of disciplines such as engineering, biology, and information theory. The implications of entropy and its inexorable increase resonate through much of the sciences, eliciting the irreversibility of real processes. Therefore, the Second Law is not only a constraint but a core concept encouraging improved comprehension of the behavior of complicated systems to guide towards additional theory like the constructal law, which is an independent principle regarding flow dynamics and system design in time (Reis et al., 2006). One of the most important developments in thermodynamics is the formalization of mathematical models which connect abstract principles to real-world applications. Traditional methods, such as variational techniques, have been complemented by techniques that stem from entropy considerations, such as the derivations of the Cahn-Hilliard and Ginzburg-Landau equations. The models illustrate how entropy flux can be rigorously defined based on thermodynamic methods, confirmed by the findings of Lius procedure. These advances allow the compact characterization of thermodynamic processes without necessarily needing only variational methods, which have a tendency to overcomplicate the analysis. The study of the relationships further provides insight into the critical phenomena and phase transitions, demonstrating the relevance of the Second Law in developing modern physical theories (Ván et al., 2017). Moreover, the interdisciplinarity of the principles of thermodynamics has given rise to new applications that extend their use beyond physics into domains like ecology and economics. The constructal theory covered in recent writing demonstrates this evolution by relating the principles of thermodynamics to the dynamics of flow structures within natural and social systems. The constructal law, which states that systems develop to provide simpler access to currents, provides a unique insight into the possibility of thermodynamic laws to be a unifying factor across numerous areas of study. The integration implies a shift towards viewing thermodynamics as a much more expansive theory explaining organization and evolution in numerous systems as opposed to being a set of laws merely explaining heat and work. This ongoing cross-fertilization between thermodynamics and other sciences shows that basic principles continue to power innovation and inform our understanding of complex interactions within nature and society (Reis et al., 2006) (Figure 6.3).

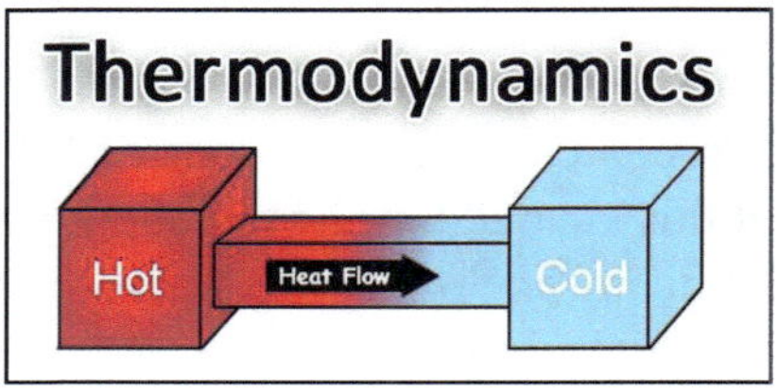

Figure 6.3. *Development of thermodynamic principles.*

Source: https://assets.isu.pub/document-structure/230615083338-bf56db-075d59cf5037c7bdfb91ce68e1/v1/0bbb5c9ec68c34a52bcf36f7fddc6452.jpeg?width=720&quality=85%2C50.

6.1.2. Key Figures in Thermodynamics (e.g., Carnot, Clausius, and Kelvin)

The development of thermodynamics as a science was furthered by the pioneering work of the pioneers such as Sadi Carnot, Rudolf Clausius, and William Thomson (Lord Kelvin). Carnot is generally regarded as the father of thermodynamics and developed the concept of the ideal heat engine in his seminal work, Reflections on the Motive Power of Fire (1824). His findings formed the basis of the knowledge of efficiency in thermal processes, showing that no heat engine can be more efficient than the Carnot engine operating between two heat reservoirs, a simple thermodynamic concept. Carnot's definition actually did hold the principle that a fraction of energy will always be lost to entropy, which later thermodynamic philosophers would explore further in establishing the second law of thermodynamics, which explains the directionality of energy conversion and the inevitable increase of entropy in closed systems. Building on the work of Carnot, Rudolf Clausius further advanced the second law of thermodynamics, significantly expanding the theoretical foundation of entropy as a measure in physics. The definition of Clausius established that during energy transfer, some of the energy always becomes less useful in serving entropy, a measure he defined as a quantity of disorder within a system. His masterful statement that heat cannot spontaneously flow from a body at lower temperature to a body at higher temperature is a cornerstone of current thermodynamic theory and highlights the irreversibility of these processes. This idea is complementary to the notion of the arrow of time, where entropy serves as a measure of times passage in thermodynamic systems and the direct implication of natural phenomena being time-asymmetric, showing the fundamental constraints placed on energy transformations by the second law (Uffink et al., 2001). William Thomson, or Lord Kelvin as he became famously known, was also responsible for establishing the solid foundations of thermodynamics. He invented the absolute scale of temperature, providing a fundamental foundation of measurement of change in energy in thermodynamic processes. Kelvin's work highlighted that the efficiency of heat engines is bound intrinsically by the temperatures of the heat reservoirs, bringing about the definition of the absolute zero point, where thermal motion ceases. This focus on absolute temperature and related equations of efficiency mirrors Carnots concepts in addressing real world efficiencies with regard to real engines. Besides, Kelvins work demonstrates the connection between entropy and energy flow, resulting in a general understanding of the second law of thermodynamics to irreversible processes in physical systems, thus justifying the interconnected interactions between temperature, energy, and entropy in prognosticating the thermodynamic nature of systems (Arnaud et al., 2008) (Figure 6.4).

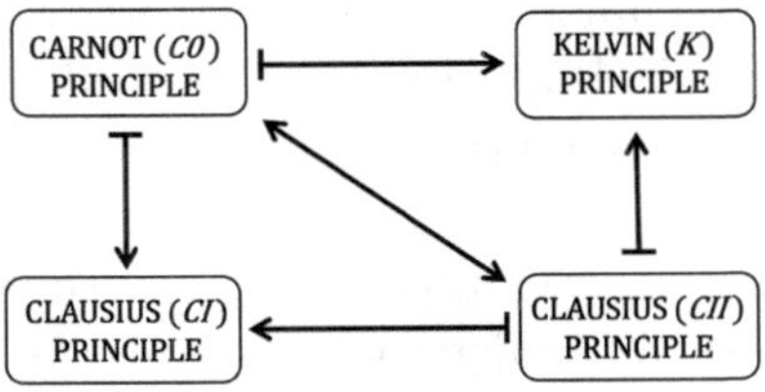

Figure 6.4. *Key figures in thermodynamics (e.g., Carnot, Clausius, and Kelvin).*

Source: https://www.mdpi.com/entropy/entropy-24-00392/article_deploy/html/images/entropy-24-00392-g006.png.

6.1.3. Evolution of the Concept of Entropy

The entropy concept has developed substantially since its development in the 19th century, at first based on the science of heat and work conducted by researchers such as Rudolf Clausius and William Thomson (Lord Kelvin). They developed entropy as a quantification of thermal energy not available for performing work, essentially tying it to the efficiency of heat engines. This initial definition was essentially framed in terms of classical thermodynamics, dealing with closed systems and equilibrium situations. Subsequently, the concept of entropy came to extend to a variety of other phenomena, including those involving irreversible processes. These developments demonstrated that entropy was not only a property of disorder but also a means of characterizing the direction and spontaneity of natural processes, emphasizing the natural tendency toward equilibrium in isolated systems. This foundational shift in perception laid the groundwork for later interpretations and applications of entropy across disciplines. With the advent of statistical mechanics in the late 19th and early 20th centuries, the conceptualization of entropy further transformed, as figures like Ludwig Boltzmann articulated a statistical underpinning to thermodynamic principles. He proposed that entropy could be defined as the number of microscopic states that can correspond to a macroscopic state, introducing the idea of entropy as measuring probabilistic energy distribution between particles. This probabilistic character of entropy was a massive deviation from the classical view and enabled a more refined and logical understanding of thermodynamics that spanned microscopic and macroscopic laws. The notion that entropy is the measure of uncertainty about a systems microstates placed it as a fundamental concept in thermodynamics and one with connections to the concept of information theory, making it connected to concepts such as the Maximum Entropy Production principle discussed in current scientific literature (Brooks et al., 2007). In modern applications, the advancement of entropy continues to echo in many sciences, particularly in nonequilibrium thermodynamics, where its implications are felt across systems. The realization that ecological and biological systems are governed on the basis of entropy production demonstrates its ubiquity. Entropy is purported to be related directly to complex system dynamics by new theories, implying that spatial and temporal orders are a product of the same entropic processes. The synthesis of these ideas has profound significance for understanding processes from climatic systems to biological evolution (Ván et al., 2020). As science evolves, the emergent concept of entropy demonstrates its fundamental role in explaining not only physical processes but the axioms that govern the behavior of complex natural and technological systems, establishing it as a keystone of thermodynamic theory.

6.2. FUNDAMENTAL CONCEPTS OF THERMODYNAMICS

One of the fundamental principles of thermodynamics is the concept of entropy, which quantitatively describes the level of disorder or randomness of a system. With reference to the Second Law of Thermodynamics, entropy's implication offers basic understanding of the change in energy and spontaneity of physical phenomena. As systems evolve, they spontaneously evolve towards states of higher entropy, such that energy becomes dispersed and equilibrated over time. This built-in tendency towards randomness is consistent with the understanding that not all energy input translates to useful work, focusing on the inefficiencies involved in energy conversion processes. Consequently, the notion that closed systems will come to a point of thermodynamic equilibrium serves to underscore the general tendency toward ultimate entropy, a basis for thermodynamic theory with far-reaching consequences in many scientific fields. The dynamics of entropy and thermodynamic systems extends beyond the confines of classical mechanics, impacting more advanced phenomena like black hole thermodynamics.

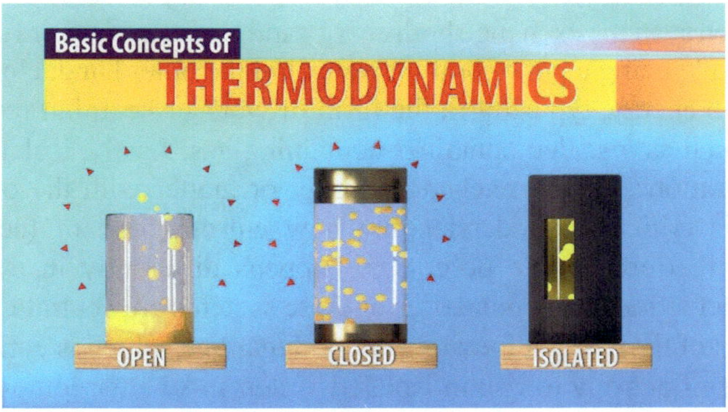

Figure 6.5. *Fundamental concepts of thermodynamics.*

Source: https://i.ytimg.com/vi/cLSYrrnjauo/hq720.jpg?sqp=-oaymwEhCK4FEI-IDSFryq4qpAxMIARUAAAAAGAElAADIQj0AgKJD&rs=AOn4CLDV_jZS4DYs-VrvpnSvK8CfktkwoGA.

Black hole thermodynamics rethinking involves geometrothermodynamics, which allows us to visualize the whole new perspective of the interaction in such extreme conditions. A Reissner-Nordström black hole, for instance, reveals that the geometry of the state space can be curved, leading out the critical points and phase transitions representing hidden thermodynamic characteristics. This new perspective overhauls conventional interpretations previously yielding conflicting outcomes. Further, the consideration of optimizing thermoelectric materials invites a critical examination of entropy generation and how it is connected with efficiency. Entropy generation is a product of irreversible processes inherent in such systems, and minimization of it is instrumental in achieving maximum efficiency, on par with the Carnot standard of efficiency.

Previous analyses depict that maximum performance entails comprehension of the interplay between thermodynamic potential and entropy generation. By way of compatibility approaches and thermodynamic analysis, scientists illustrate how under specific working conditions, entropy generation can be minimized, ultimately guiding the design of improved thermoelectric devices. Such advancements signify the strong influence of thermodynamic principles on modern technology and innovation, affirming the dominance of entropy in driving efficiencies across applications (Figure 6.5).

6.2.1. Definition of Thermodynamic Systems

Thermodynamic systems are the general structure in which thermodynamics operates. These systems come in three basic forms: open, closed, and isolated systems. Open systems exchange both energy and matter with their surroundings, a good example being a pot of boiling water. On the other hand, closed systems can exchange energy, typically in the form of work or heat, but restrict matter exchange, such as a sealed container containing gas. Lastly, isolated systems, being idealizations, do not exchange energy or matter with the outside. This distinction is critical to understanding how entropy, one of the key terms involved in thermodynamic processes, happens differently in every system type. The interactions and boundaries of these systems are central to examining the subtleties of the Second Law of Thermodynamics, which is concerned with the tendency of entropy to rise in isolated systems. A better comprehension of thermodynamic systems requires an examination of their state variables, those that characterize the systems' properties at equilibrium. Common state variables are pressure, volume, temperature, and internal energy, all of which are interrelated by equations of state specific to the materials involved. A case in point is the ideal gas law, a pressure-volume-temperature relation for ideal gases. How these variables behave under changing conditions is crucial in studying transactions of matter and energy in the context of thermodynamic processes. As investigated in many studies, such as in Kaneko et al. (1999), the interaction of these factors is crucial in the direction of thermodynamic processes in a non-reversible manner and the unavoidable rise of entropy. This reinforces the necessity of understanding thermodynamic systems, not as standalone phenomena but as part of the wider physical laws. In thermodynamic systems, entropy is a fundamental parameter that measures disorder or randomness. Entropy not only gives an insight into the second law of thermodynamics but is also very helpful for the analysis of irreversible processes. Practically, when physical systems approach equilibrium, the entropy typically increases, thus agreeing with the second laws statement that natural processes move towards more disordered states. Surprisingly, research, such as that presented in Birrell et al. (2011), explains the understanding of entropy within black hole thermodynamics, where ideas of microscopic states and gravitational interactions become related with entropy, pointing to its universal applicability. And as entropy continues to be a unifying theme for the study of thermodynamic systems, it forces us to revisit some of the traditional thinking about conservation of energy and distinction between microstates and macrostates in physical systems, this time citing interdependence of thermodynamics and quantum mechanics.

6.2.2. Types of Thermodynamic Processes (Isothermal, Adiabatic, etc.)

The examination of thermodynamic processes, in the context of entropy and the Second Law of Thermodynamics, reveals the complexity and refined dynamics of energy transformation. The primary types of thermodynamic processes are isothermal, adiabatic, isochoric, and isobaric processes, each characterized by some conditions regarding temperature, volume, and pressure. An isothermal process is characterized by a steady temperature that allows for heat transfer without an increase or decrease in the internal energy of the system. This leads to an increase in entropy, as adopted by assumptions given by the Second Law of Thermodynamics, that the total entropy in an isolated system can never decrease with time. Such processes are of significant importance in devices like heat engines, demonstrating fundamental principles of energy and work conversion underlying thermal dynamics (Bennett et al., 2006). Adiabatic processes are exactly opposite to isothermal processes, as they occur under no exchange of heat with the surrounding environment. In an adiabatic system, the change in the internal energy is solely accountable for the work done on or by the system. This establishes the inherent connection between temperature change and entropy. During the adiabatic expansion of the system, it surrenders heat in the form of work done by the surroundings, decreasing entropy at least in the first phase. But from the Second Law, it will be understood that the overall entropy of the universe, including the surroundings, still satisfies the condition that it never decreases. The processes are significant to learn, as they not only demonstrate simple thermodynamic principles but also learn about the working of actual systems like refrigerators and engines (Bennett et al., 2006). Isochoric and isobaric processes also assist in the examination of thermodynamic cycles, each of constant volume and constant pressure, respectively. During an isochoric process, volume remains constant, hence all the supplied heat gets directly fed into internal energy and, by consequence, the system's temperature. This manifests as an equivalent rise in entropy, illustrating how energy transfer acts in line with the Second Law of Thermodynamics. Isobaric processes with constant pressure experience a change in volume but with thermal equilibrium.

Both processes are crucial in defining energy efficiency in thermodynamic systems and characterizing cycles such as those of refrigerators and heat engines. Work done in such cycles turns out to be a testament to the profound relevance of thermodynamics to practical applications, constantly reinforcing the significance of entropy in energy conversion and the irreversibility of actual processes (Apertet et al., 2015) (Figure 6.6).

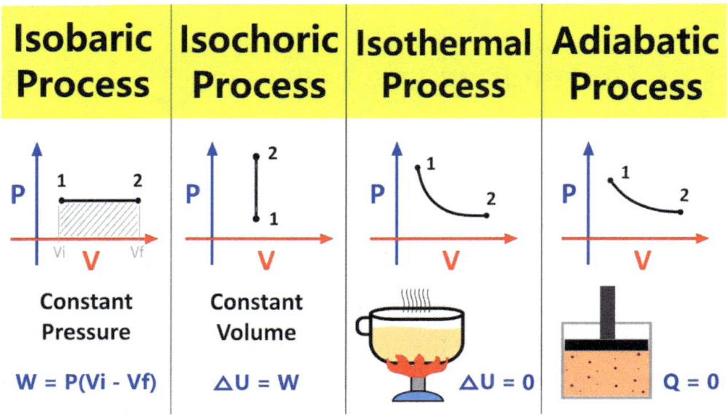

Figure 6.6. *Types of thermodynamic processes.*

Source: https://i.ytimg.com/vi/vV1fGs3JKzU/maxresdefault.jpg.

6.2.3. The Role of Energy in Thermodynamic Systems

Energy is at the center of thermodynamic systems' dynamics insofar as it directly influences their entropy and overall behavior. At the heart of such a system is the concept of energy transfer, which can be facilitated through work or heat exchange. This transfer is the basis for thermodynamics' fundamental principles, most notably the Second Law, which states that the total entropy of an isolated system will never decrease with time. The required entropy increase is matched by the spreading of energy and by the associated conversion into less work-available forms. In this context, energy is not only a quantity to be measured; it is the driving force of various thermodynamic processes, forming a simple foundation for describing the evolution of systems to thermodynamic equilibrium, and thereby illustrating the intimate connection between energy and entropy. The relationship between energy and entropy becomes especially crucial in the context of irreversible processes in thermodynamic systems. Such processes, in the sense of being in non-equilibrium conditions, naturally lead to the growth of the overall entropy of the systems due to the dispersal of energy. As the systems get disordered from ordered condition, the energy initially concentrated in specific forms gets more randomized and less useful for work output. This phenomenon is encapsulated in the Second Law of Thermodynamics, namely that energy conversions become less efficient as time progresses, resulting in waste heat and a net rise in entropy. A comprehension of this relationship provides worthwhile insight into the limitations which thermodynamic principle imposes upon energy utilization, highlighting the significance of proper handling of energy within natural and engineered systems as detailed in the literature related to thermodynamics and information. Further, the role of energy's in thermodynamic systems can possibly be strongly related to concepts within statistical mechanics, or the microstates which constitute a thermodynamic system.

Each configuration of particles corresponds to a unique set of states of energy, and the entropy of a system is really a measure of the diversity of the microstates. When the system undergoes changes in energy, the distribution of the state changes and entropy rises. Recent research has investigated this correlation more specifically, focusing on the intrinsic nature of quantum mechanics, which supports the relationships invoked in statistical mechanics in order to have a unified picture for interactions between entropy and energy. The extensiveness of these interconnections even encompasses the derivation of laws governing energy boundaries in computation and proves that concepts from thermodynamics can capture not only physical systems but also informational foundations of information processing, as stated in the example of the Landauer principle and its influence on the second law of thermodynamics (Bérut et al., 2017; Birrell et al., 2011).

6.3. THE SECOND LAW OF THERMODYNAMICS

The Second Law of Thermodynamics basically changes our perspective on energy transfer and entropy, setting forth that the overall entropy of a closed system can never decrease as time passes. According to this law, natural processes will tend towards a condition of maximum randomness or disorder, basically giving the flow of energy a direction.

For example, in any energy transfer, some energy will always be lost as a form that cannot be utilized, and this is responsible for an increase in entropy. Such inherent desire for disorder is important to identify not only for thermodynamic processes but also for the implication in other scientific fields, including chemistry and biology. In expressing the Second Law in terms of entropy, we discover also what ultimately becomes of the universe, as systems evolve toward the condition of thermodynamic equilibrium, and thereby confirm the universality of this law to all physical processes. In addition, the link between the Second Law of Thermodynamics and entropy is most critical to understanding a great number of phenomena within statistical mechanics and cosmology. The realization that entropy provides a statistical basis for the Second Law allows us to forecast and study the systems at a molecular level. In accordance with recent discussion over thermodynamic implications, the relationship between entropy and microstate organization leads to macroscopic phenomena that can be perceived, thus linking microscopic occurrences to macroscopic laws. This marriage of statistical mechanics with classical thermodynamics allows avenues for the exploration of concepts like negative temperature and what it means for the direction of time, as discussed by Klimenko et al. (2012). These types of questions not only add to theoretical understanding but challenge long-held perspectives to the forefront, placing the Second Law at center stage in contemporary explorations within physics. With the advent of an era when modified theory of gravitation is being investigated, the consequences of the Second Law are not confined to classical thermodynamics but extend to astrophysical domains. Theoretical investigations have revealed new paradigms such as scalar-vector-tensor theories of gravity, which call for a revisit of thermodynamic laws under extreme conditions like those associated with black holes. The interaction of entropy with phenomena like black hole thermodynamics shows the manner in which the Second Law deploys its application even in other areas of study, demonstrating the widespread role that entropy plays in the determination of physical system behavior. The case of black hole entropy and temperature as studied in Channuie et al. (2018) demonstrates the potential for new knowledge in the nature of gravitational phenomena and their thermal properties. As science research goes on, the Second Laws early contribution to establishing energy boundaries and entropy concepts continues to illuminate our understanding of the dynamic and structural complexities of the universe (Figure 6.7).

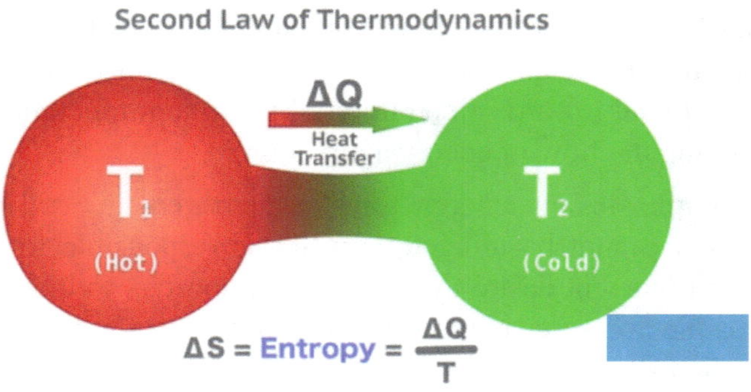

Figure 6.7. *The second law of thermodynamics.*

Source: https://www.shutterstock.com/image-vector/second-law-thermodynamics-heat-transfer-260nw-2446889003.jpg.

6.3.1. Statement of the Second Law

The Second Law of Thermodynamics articulates a fundamental principle of natural phenomena, that in a sealed system the entropy, a state of disorder, will increase over a period of time. The principle implies that energy transformations are not entirely reversible and there is always an energy that will be wasted and hence increase the entropy of the system. The use of this law is not just theoretical but also applied to areas like engineering, biology, and information theory. Understanding the function of entropies in these fields allows scientists to predict the feasibility and efficiency of processes. For instance, engineers need to consider entropy when building engines so that energy can be maximized while understanding that waste heat produces greater entropy and ideal efficiency is not possible. In the context of thermodynamic systems, the Second Law holds profound meaning, particularly concerning reversible and irreversible processes.

While reversible processes exist only theoretically and never actually occur in nature, they remain the basis against which natural processes are measured. Irreversible processes, much more common, generate increases in entropy, highlighting the inefficiencies encountered in energy conversion and transmission. For example, when a system is being heated, it will always move from the hot region towards the cold regions, supporting the Second Laws assertion that natural processes will naturally promote entropy increase. This information is relevant to the formation of technologies towards energy conservation because it alludes to the need to design systems to minimize entropy increase in case there is energy exchange, and this sets the door open for sustainable development. Recent advances in thermodynamic theory continue to make clearer the intricate relationship between entropy and probabilistic models that govern physical processes.

The MEP approach states that systems will self-organize into those states that maximize entropy production, accentuating the importance of probabilistic interactions in determining energy and matter flow. This theoretical perspective aligns with broader ideas in evolutionary theory that posit that biological systems will tend to explore the adjacent possible as well as they are able, thereby expanding their scope of function and adaptability. The Second Law, in this context, is seen as an articulation not just of a physical requirement, but of a probabilistic system acting within the macroscopic regime and within the microscopic interactions that underlie complicated biological systems (Brooks et al., 2007). Thus, an exploration of Second Laws implications can yield results linking thermodynamics, information theory, and evolutionary dynamics, enhancing our comprehension of the universes entropy-driven evolution (Figure 6.8).

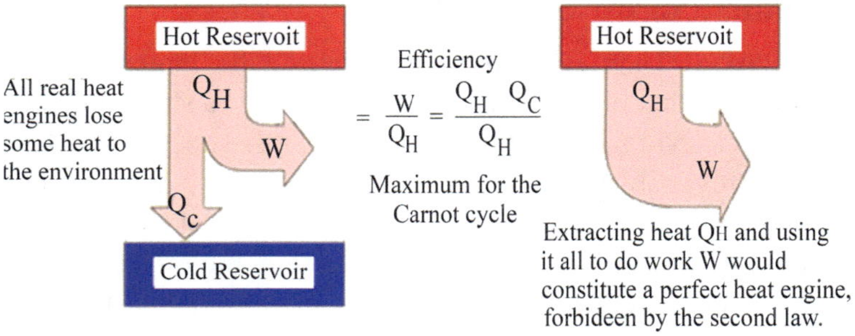

Figure 6.8. *Statement of the second law of thermodynamics.*

Source: https://lh6.googleusercontent.com/proxy/4RLeldqOw-o9FCdLQxc6w-TcAE4v7eRpBiyEktykquGBswk-mGCc_bB3PN2_ijyfHaMlVmONOECX3XK3sT-vel9X5yc108aQuhHWMKvOAmzlWg1MD8SY-.

6.3.2. Implications of the Second Law in Natural Processes

The Second Law of Thermodynamics defines that whenever energy is transferred, if neither energy is entering nor leaving the system, the potential energy always decreases, leading to an increase in entropy. The law has great importance in natural phenomena as it reinforces the fact that life and physical phenomena have a direction. A good example is the transfer of energy between trophic levels in an ecosystem, which depicts the principle quite well. Producers such as plants convert solar energy to chemical energy but only a percentage is transferred to herbivores and a smaller percentage to carnivores, resulting in a loss of usable energy at each level. Such a loss is equivalent to a gain in entropy, a measure of the energy transfer inefficiency within ecosystems that suggests an ever push towards disorder within natural systems with impacts eventually upon developmental processes and evolutionary outcomes. Additionally, the Second Law gives a description of the behavior of gases and mixing, as shown by the Gibbs paradox. In traditional thermodynamics, the mixing entropy is intuitive in such a way that it appears to be discontinuous when separating different types of gases. With the description using the quantum thermodynamic approach, the paradox is removed by the introduction of mixing ergotropy. This operational tactic shifts focus from entropy to the work available prior to and subsequent to gas mixing, accentuating the utility of quantum states in marking the difference between gases. This is an indication of the limitations of classical models and illustrates the fineness of natural processes, where entropic issues are embedded within quantum mechanics (Allahverdyan et al., 2006). Furthermore, recent theories suggest that entropy and order can emerge through probabilistic frameworks rather than solely deterministic laws, reshaping our understanding of natural phenomena. The concept of Maximum Entropy Production (MEP) reflects this idea, proposing that natural systems evolve toward configurations that maximize entropy production while maintaining spatial and temporal order. This exchange derails the general idea of entropy as merely a sign of disorder and instead suggests that it is an imperative physical drive for engaged processes. Biological organisms as they travel throughout their environments demonstrate the nuances of entropy in terms of innovation and adaptation, increasing the nearby possible while struggling with the constraints imposed by thermodynamic principles (Brooks et al., 2007). Finally, these are indicative of the intricate relationship between entropy, energy, and order and indicate the role of the Second Law in regulating physical as well as biological events.

6.3.3. Examples of the Second Law in Everyday Life

The functioning of the Second Law of Thermodynamics is demonstrated in many everyday scenarios, emphasizing that systems progress to states of higher entropy. A common illustration is the melting of ice to water. If ice in a highly ordered crystalline structure melts, it becomes a liquid state of increased disorder. Such an entropy increase is a natural transition towards thermodynamic equilibrium. In these processes, energy is transferred as heat and causes state changes in matter in accordance with the principles of the Second Law. Furthermore, if one takes into account the presence of non-conservative forces such as friction, analysis needs to incorporate thermodynamic principles for describing the transformation of energy during melting so as to explain the widespread application of the Second Law in thermal dynamics in our everyday lives (Fiolhais et al., 2013). A further interesting demonstration of the Second Law is that of the diffusion of perfume in a room. When perfume is first sprayed, the strong smell particles are highly ordered and localized. Over time, the particles move to become dispersed, moving from areas of higher concentration to areas of lower concentration, leading to a more even spread of molecules. This movement towards evenness is an illustration of the increase in entropy as the system expands. Note that this spontaneous spreading does not involve any work done from the outside; it is the increase in entropy which occurs purely due to the random movement of the molecules on their own. The Second Law decrees that this direction towards moving to equilibrium is inevitable, underlining the inherent tendency of systems to reach maximum entropy. The spreading out of disorder, as described as diffusion, is a fine example of thermodynamic concepts in action in day-to-day life (Atkin et al., 2001). Moreover, the decomposition of the composted organic material in the compost is a good example of real life of the Second Law. The compost heaps initially may be neat and comprised of well-known organic materials such as household waste and garden waste. But as this material is broken down by microbial activity, the compost itself undergoes drastically, ultimately being a mixture of decomposing, less organized material. This process illustrates the principle that ordered states, such as the original organic matter, will spontaneously tend towards greater disorder and breakdown. Energy invested in this process, although used by microorganisms to construct themselves, ends up contributing to a net gain in entropy in the environment. This explains not only how the Second Law can be observed in biological processes but also its broader significance for the understanding of energy and matter dynamics (Fiolhais et al., 2013) (Figure 6.9).

Figure 6.9. *Examples of the second law in everyday life of thermodynamics.*

Source: https://i.ytimg.com/vi/w6VNUYIUV3s/hq720.jpg?sqp=-oaymwEhCK4FEI-IDSFryq4qpAxMIARUAAAAAGAElAADIQj0AgKJD&rs=AOn4CLDZQ3x80-T6aDegeZja__CJRsGtNg.

6.4. ENTROPY IN ISOLATED SYSTEMS

Entropy is a theoretical concept that is used to account for the behavior of isolated systems, and particularly the Second Law of Thermodynamics. Entropy is really a number measure of the disorder or randomness of a system, which increases with time as the system is considered to be closed. The Second Law asserts that in an isolated system, spontaneous processes result in a net increase in entropy, emphasizing the irreversibility of certain processes. The Second Law is applicable not just to physical systems, but also to the phenomenon of times arrow—suggesting a sense of direction in which processes have a tendency to occur naturally. The notion that isolated systems reach a maximum entropy, or thermodynamic equilibrium, is central in emphasizing the tendency towards disorder and holds profound implications across broad fields of research like statistical mechanics and cosmology. Research in fluctuations of entropy within isolated systems has been of immense importance in recent research on thermodynamics. Fluctuation relations that have emerged are bringing new insights into how time-reversal symmetric microscopic dynamics produce irreversible behavior. This research bridges the gap between classical thermodynamics and more complex statistical action, providing a better explanation of why and how systems become more disordered over time. As in existing literature, fluctuation relations clarify mechanisms of dissipation and shed new light on the understanding of the arrow of time that is central to entropy study. These changes exhibit the fine interplay between microscopic and macroscopic events, illustrating that within isolated systems themselves, unorthodox behaviors take place that run contrary to classical thinking of thermodynamic equilibrium (Bochkov et al., 2011).

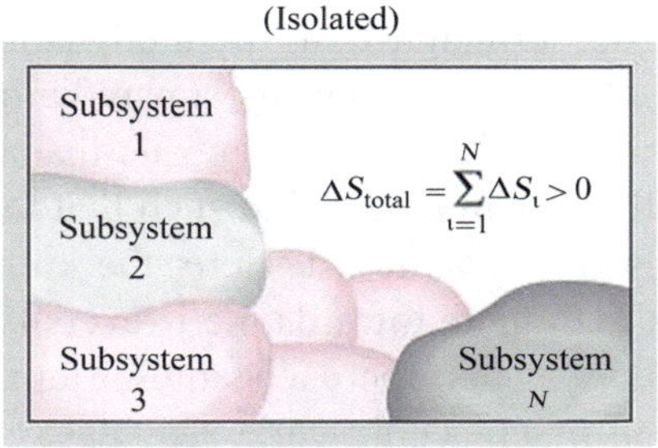

Figure 6.10. *Entropy in isolated systems.*

Source: https://encrypted-tbn0.gstatic.com/images?q=tbn:ANd9GcRy6TMvLqXgSByEBXiALxFiPU3oQDNDegCo-Q&s.

Incorporating thermodynamic laws that deal with matter and antimatter also render entropy in isolated systems more complex to grasp. By exploring CP-invariance and CPT-invariance, researchers have opened doors to challenge the underlying symmetries that govern thermodynamic principles. While the conventional formulations of thermodynamics work well in our matter-filled universe, the potential implications of these invariance principles pose intriguing questions about the nature of entropy and how it would act under extreme conditions. Investigation of these hypotheses not only enhances our knowledge about entropy in isolated systems but also inspires experimental effort to test and verify or falsify current hypotheses. Hence, understanding entropy in isolated systems calls for an appreciation of current paradigms as well as new hypotheses that can reshape the understanding of the universe and its physical laws (Klimenko et al., 2014) (Figure 6.10).

6.4.1. Definition of Isolated Systems

Isolated systems are a fundamental concept in thermodynamics, particularly when working with entropy and the Second Law. By definition, an isolated system does not exchange matter or energy with its surroundings, and therefore it is a closed system where all the internal processes occur regardless of external influences. This is one very critical attribute in thermodynamic calculations because it simplifies the study of energy conversion and changes in entropy. In isolated systems, total energy remains constant according to the first law of thermodynamics, and the intricacy of entropy interactions takes center stage. This is an experimental setup that allows scientists to study thermodynamic processes without the influence caused by external factors, thereby providing a point of reference on how entropy behaves in such systems. The significance of isolated systems extends past the mere definition to influence the comprehension of phenomena under the principles of thermodynamics. Isolated systems evolve towards a state of maximum entropy, a paradigm that encapsulates the Second Law of Thermodynamics. The law posits that in an isolated system, natural phenomena will occur in a direction such that maximum total entropy is reached until equilibrium is achieved. Here, the knowledge of the history of entropy provides us with hints towards the irreversible nature of processes in isolated systems. As systems approach equilibrium, thermal equilibrium appears as a state when macroscopic variables cease to change, guiding us to an appreciation of the correspondence between macroscopic changes in entropy and microscopic states for all forms of matter. Recent advancements in thermodynamics have broadened the definition of isolated systems to embrace a variety of situations, from those pertaining to quantum mechanics and nonequilibrium systems. The traditional parameters applied in defining an isolated system have been challenged more vigorously to keep pace with recent advancements in thermodynamics, including developments that were recently proposed. For instance, concepts such as ergotropy mixing counter-intuitive thinking by demonstrating that entropy cannot be the sole indicator of irreversible processes in quantum systems. Instead, the generalized entropy framework allows for a more general understanding and applicability to other physical systems like those that are under external fields or chemical reactions. As is reviewed in the literature, this rigorous questioning remains to redefine the boundaries and operative meaning pertaining to isolated systems, therefore promoting the overall comprehension of thermodynamics in both equilibrium and nonequilibrium conditions (Allahverdyan et al., 2006; Beretta et al., 2010).

6.4.2. Entropy Changes in Isolated Systems

The law of entropy is fundamental to the study of isolated systems, particularly in relation to the Second Law of Thermodynamics, which determines that the entropy of isolated systems will be constant or increasing. This law serves as a basis for thermodynamic processes where natural processes become irreversible under isolation conditions. When a system in isolation undergoes change, such as the mixing of gas or thermal equilibration of two differently heated bodies, the overall entropy of the system is increased.

The increase illustrates the movement of the system towards thermodynamic equilibrium, which reveals an inherent tendency for disorder to grow and thus displays the global pattern of the Second Law of Thermodynamics (Alfaris et al., 2023). Isolated systems provide a unique environment in which to learn about entropy because they are insulated from external forces that otherwise modify their characteristics. Here, entropy is not merely a characterization of energy dispersal but also the quantity of microscopic configurations that correspond to a macroscopic state. The mathematical formulation of entropy allows physicists to quantify this aspect of thermodynamics with greater accuracy. Under isolated systems, in processes of energy transformation, statistical likelihood of reaching a particular state varies, constantly driving the entropy higher or maintaining it in a constant upper bound. The behavior therefore reveals the inherent stability isolated systems achieve despite ongoing micro-scale processes, as it accentuates the prime position of entropy in disorder and energy flow systems (Beretta et al., 2010).

The consequences of entropy within isolated systems extend well into the practical realm beyond theoretical considerations; they encompass significant practical implications in a broad array of scientific domains. As a case in point, entropy's function lies at the heart of describing phase transition in materials science and predicting energy conversion process efficiency in engineering applications. While scientists investigate phenomena like quantum entanglement and nonequilibrium dynamics, the classical notions of entropy must be reinterpreted to make space for the comprehension of systems commonly beyond the realm of equilibrium (Beretta et al., 2010). Moreover, recognizing entropy's elusive nature and its relation with external interactions will pave the way for novel approaches in thermodynamics, at last enabling a greater comprehension of energy management and resource utilization within state-of-the-art technology. By describing changes in entropy in isolated systems as an expression of more general principles, we also more clearly observe the interrelation of the laws of thermodynamics and how they serve to tell us about the universe (Figure 6.11).

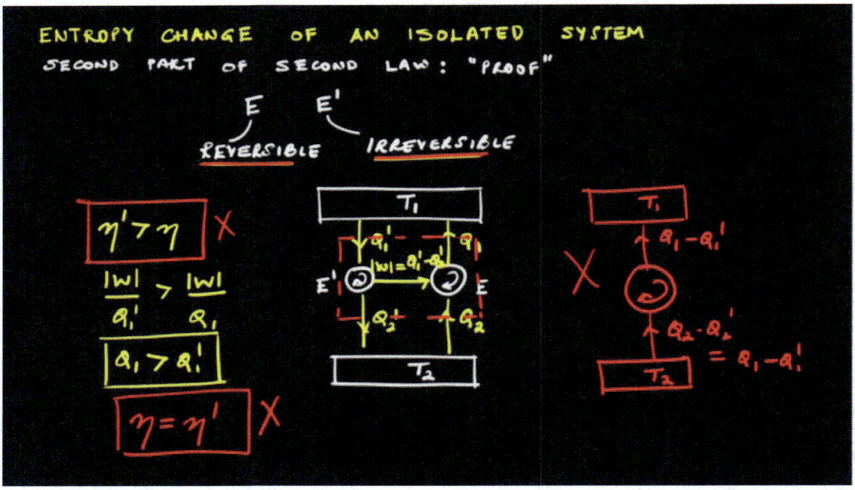

Figure 6.11. *Entropy changes in isolated systems.*

Source: https://i.ytimg.com/vi/xMh2ffQ95yY/maxresdefault.jpg.

6.4.3. The Concept of Maximum Entropy

The concept of maximum entropy is central to understanding the second law of thermodynamics, particularly in terms of the behavior of isolated systems. In short, maximum entropy is a state of maximum disorder in which a system exists with all possible microstates being equally probable. The principle states that systems naturally gravitate toward configurations that maximize entropy, and that suggests a form of natural tendency towards chaos. To this extent, the second law of thermodynamics says that the total entropy of a closed system will never decrease over time and goes on to state that processes occur spontaneously in the direction of increasing entropy. This kind of process is the foundation of all natural phenomena, either in thermal transfers or chemical reactions, and reinforces the ubiquitous role of entropy in determining the evolution of systems to equilibrium. Along with that, the entropies' and information theory interactions unveil additional features of maximum entropy as a concept to define how it aligns with a system's informational content. According to recent developments in theory, the maximum entropy production principle is in accordance with the maximization concept of informational content, with entropy being taken into account by a probabilistic inference approach (the MaxEnt principle). As systems change to states of higher entropy, information about specific configurations is lost, ultimately leading to a uniform distribution of microstates in which chaos reigns supreme. Such a perspective reaffirms that understanding entropy involves a process that treats thermodynamic regulations and information theoretically, thus bridging the gap between physical laws and their implications on information in various processes, including those that govern life itself. Moreover, in the realm of nonequilibrium thermodynamics, the maximal entropy approach provides fascinating perspectives on the dynamics of complex systems. The intersection of spatial and temporal orders sheds light on the intricacy of how physical and biological systems tend to steer processes toward maximal entropy by natural means. While the consequences of such notions would appear to suggest an emergent randomness over time, they also outline concealed patterns that regulate short-term behavior and claim that order and disorder are not separable. As biological systems, for example, strive to define an extended adjacent possible, they act in ways consistent with the generation of maximum entropy as well as to foster stability within their immediate environments. These results capture the idea that maximum entropy is more than a state of randomness but is also an enabling principle revealing the cyclical balance between order, complexity, and the inexorable tendency towards equilibrium, as a useful framework in which to view the second law of thermodynamics (Atkin et al., 2001; Brooks et al., 2007).

6.5. APPLICATIONS OF ENTROPY

Entropy is not a theory in the world of thermodynamics but a significant measure with real-world applications across an array of scientific and engineering disciplines. Entropy quantifies the level of randomness or disorder of a system such that scientists and engineers can predict how energy moves and transforms under various processes. One of the most familiar applications of entropy is in heat engines, where the way that entropy works must be comprehended in order to get the highest level of performance. Based on studies, however, students often struggle with the relationship between entropy and the Second Law of Thermodynamics, particularly the workings of heat engines, such as the Carnot cycle. This subtlety underscores the necessity of teachers emphasizing these relationships in order to foster an in-depth understanding of thermodynamic principles (Christensen et al., 2015). Another significant application of entropy is in chemical thermodynamics, particularly with respect to reaction spontaneity and equilibrium. The Second Law of Thermodynamics states that in an isolated system, spontaneous processes will tend to move toward a condition of increased entropy. As such, students and professionals in chemistry can apply the concept of entropy to determine which reactions are going to occur spontaneously. Chemists can, using the Gibbs free energy equation involving both the enthalpy and entropy, make predictions on reaction spontaneity based on conditions. Such an association reflects the suitability of entropy in connecting thermodynamic concepts with chemical behavior in a basic model of learning complex systems. The relevance of this information transcends the realm of academia to impact other sectors such as the pharmaceutical sector, where reaction optimization is crucial for efficient drug discovery (Fronsdal et al., 2010). Besides its application in chemistry and engineering, entropy also plays a vital role in environmental and ecological science. The concept of entropy can be applied to explain energy transfer in ecosystems, illustrating the process by which sun energy is transformed and allocated across the various trophic levels. The Second Law of Thermodynamics is proved by this concept in that ecological processes tend to lead to greater entropy. As human society is confronted with the urgency of climate change and resource depletion, a clear understanding of entropy is crucial. By the use of entropy laws, scientists are able to develop more sustainable approaches that minimize disorder and energy loss in ecosystems. This focus on sustainability not only elevates the relevance of entropy in ecological research but also highlights the merging of physical principles with pressing environmental issues of the contemporary world, thereby affirming the pivotal role of thermodynamics in advancing a more sustainable future (Christensen et al., 2015).

6.5.1. Entropy in Chemical Reactions

The entropy theory is at the heart of explaining chemical reactions in accordance with the second law of thermodynamics. The second law asserts that for an isolated system, the entropy would rise over time, such that natural processes would proceed towards a state of more randomness and disorder. In a chemical reaction, entropy not only describes the dissemination of energies but also the multiplicity of microscopic states accessible to generate a given macroscopic state. For instance, in a transition from reactants to products, the entropy of a chemical system could increase or decrease but the overall trend must comply with the second law. Therefore, the learning of entropy in chemical reactions provides insight into spontaneity of the processes where the reactants favor arrangements that lead to greater entropy as the reaction proceeds (Barnes et al., 2000). Chemical reactions also possess enthalpic and entropic contributions that decide spontaneity and the position of equilibrium. The Gibbs free energy, a basic thermodynamic potential, is closely tied up with these amounts through the relation $G = H - TS$, in which G is free energy, H is enthalpy, T is the temperature, and S is entropy. A decrease of free energy at constant temperature and pressure indicates a spontaneous reaction, most often facilitated by enhanced entropy. Additionally, the role of entropy in living systems proves its relevance not only in thermodynamic models but also in evolutionary biology. As reactions approach equilibrium, the increase in entropy proves to be an engine force promoting development and diversification and how energy transduction across species is compatible with thermodynamic principles (Annila et al., 2008). Aside from that, the relationship between chemical reaction energy transduction and entropy highlights an energetic aspect of life itself. All chemical reactions release or store energy, contributing to the overall process of energy change within a system. The second law of thermodynamics states that systems spontaneously tend to move into states of higher entropy, describing the motivation for evolution in biochemical processes. This process of energy transduction, which is linked to the complexity of biological networks, demands strict regulation of entropic change. Through the provision of processes for amplifying or modulating such entropic change, life maintains shape as it becomes increasingly complex. Thus, education in the concept of entropy and its application in chemical reactions is key to understanding the broader thermodynamic principles behind physical and biological processes (Barnes et al., 2000) (Figure 6.12).

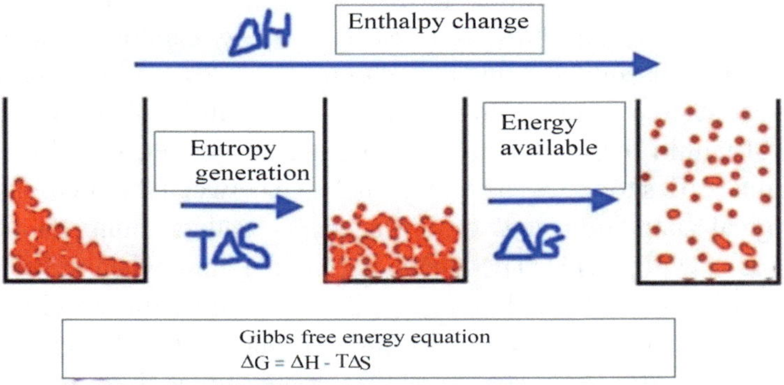

Figure 6.12. *Entropy in chemical reactions.*

Source: https://media.licdn.com/dms/image/v2/D4D12AQEHU8QsM5Hlng/article-cover_image-shrink_600_2000/article-cover_image-shrink_600_2000/0/1655975146529?e=2147483647&v=beta&t=IlvChMUfkjCWtEtHDhktAh2LO8R1ApVjViyi6D2E2Yc.

6.5.2. Entropy in Biological Systems

The entropy rule, as dictated by the Second Law of Thermodynamics, has profound implications in the case of biological systems, particularly regarding how these systems become organized and evolve in the presence of the inevitable tendency toward randomness. Biological systems operate far above thermodynamic equilibrium, a fact that allows them to create and sustain complicated structures and functions. Notably, population dynamics and biochemical reactions are characterized by equilibration mechanisms that can be formulated using mathematical constructs including evolutionary game theory and chemical reaction networks. These models demonstrate that as biological systems evolve, they equilibrate to lower free energy states in accordance with thermodynamic expectations. Hence, behavior in biological systems not only exhibits the entropy principles but also shows the inherent constraint imposed upon them by the laws of thermodynamics, bringing into the limelight the interplay of disorder, order, and energy flow in living beings (Baez et al., 2016). The use of relative entropy, or Kullback-Leibler divergence, is a significant quantitative measure for the departure of biological systems from equilibrium. Effectively, relative entropy provides a quantification of the information topographies that organisms traverse as they evolve and adapt under environmental pressures. When systems evolve, the non-decreasing nature of relative entropy signifies diminishing uncertainty and more responsive populations to their niches. It is a manifestation of the basic thesis of the Second Law, viz., that free energy cannot be maximized in the long term. Consequently, this realization elucidates the intricate dynamics of biological systems, noting how they seek to optimize their configurations for survival within thermodynamic constraints. Moreover, this drive towards equilibrium testifies to the essence of the balance between thermodynamic laws and biological adaptation, underlining a deeper understanding of life in the universe (Baez et al., 2016). There have been some newer theoretical advances that have further consolidated entropy within biology with the physical laws through such concepts as Maximum Entropy Production (MEP) and the MaxEnt principle of scientific inference. Scientists contend that these methodologies reconcile thermodynamic principles with the probabilistic nature of biological processes and hypothesize that life prefers to maximize productivity in energetic contexts. This perspective emphasizes the double character of spatial and temporal organization, implying that the evolutionary trajectory of biological systems is not arbitrary but is driven by submerged thermodynamic forces. In so far as organisms appropriate the neighboring possible, they show a strategic ordering of available resources in accordance with MEP principles. This rich interface of probability, entropy, and evolutionary biology gives a significant key to how living systems can maintain organization despite the ubiquitous tendency towards disorder, reaffirming the fundamental utility of the Second Law of Thermodynamics in the context of life.

6.5.3. Entropy in Information Theory

The concept of entropy is a cornerstone that exists in the realm of thermodynamics and information theory and serves as the bridge between physical occurrences and information processing. In thermodynamics, entropy is usually considered to be an indication of disorder or randomness within a system, encapsulating the second law of thermodynamics, which states that entropy in closed systems will grow. Entropy is a measure of uncertainty in an arbitrary variable in information theory, serving the same purpose as in thermodynamics. This dual nature points to the surprising analogy between energy and information, suggesting that the processes responsible for physical phenomena can also be used to describe the production, transmission, and annihilation of information. Landauer's principle also imposes this one by equating information erasure with an equivalent increase in entropy and thereby highlighting the computational cost in thermodynamic terms (Bérut et al., 2017). In the context of information theory, entropy is not just measured as uncertainty but elucidated in terms of the limits imposed on data processing and transmission.

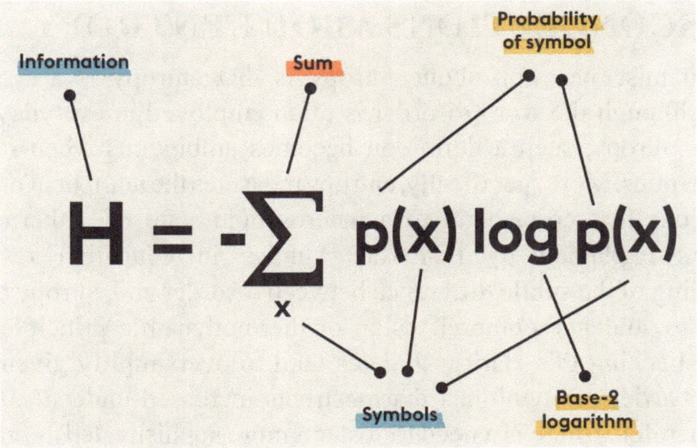

Figure 6.13. *Entropy in information theory.*

*Source: https://miro.medium.com/v2/resize:fit:1400/0*8DPGXL_lLGq8zGrc.jpg.*

As opposed to the thermodynamic entropy in classical theory, information entropy has to do with a broader domain encompassing efficiency and fidelity of communication systems. The second law of thermodynamics finds its expression in this case through limitations on computational reversibility and on the error rate. Specifically, whenever information is stored or processed, it causes an irreducible rise in total entropy in the system or the environment. This relationship is highlighted by Maxwells demon thought experiment, which, paradoxical as it sounds, illustrates the ideas of measurement and feedback control in thermal systems under the circumstances of the second law (Bérut et al., 2017). In addition, the utilization of quantum information theory has also extended the understanding of entropy, particularly within black hole thermodynamics. Here, information and entropy interact with each other in a new complexity, information degradation being synonymous to thermodynamic processes found in classical systems. The generalized second law of thermodynamics states that total entropy, including black hole and associated radiation, cannot decrease. With the inclusion of Landauer's principle in this model, it is apparent that processes like information erasure in quantum systems result in a measurable rise in entropy, thereby cementing the harmony between information theory and the principles of thermodynamics. Eventually, these findings not only widen our insight into the fundamental physical principles but also lay the groundwork for advancements in areas such as quantum computing and thermodynamic efficiency (Figure 6.13).

6.6. MISCONCEPTIONS ABOUT ENTROPY

A common misconception about entropy is that entropy is a synonym for disorder. Although the word disorder is often employed in everyday language to refer to entropy, such a definition becomes ambiguous when referring to thermodynamics. More specifically, entropy measures the number of microscopic configurations that correspond to a macroscopic state of a thermodynamic system. As explained by numerous studies in education research, an understanding of the subtle difference between disorder and entropy can largely contribute to students' comprehension of thermodynamic principles (Bass et al., 2000). Lacking this clarity, students tend to oversimplify given concepts, creating a barrier of cognitions that obstructs increased understanding of the subject. Therefore, there is a need to foster a more sophisticated appreciation of entropy in terms of statistical mechanics rather than a rough meaning of disorder. Another erroneous assumption lies in the belief that entropy can be utilized for closed systems only. While the Second Law of Thermodynamics describes that isolated systems will experience an increase in entropy, it should also be mentioned that entropy is not only present in open systems, which interact with the surroundings by exchanging matter and energy. In practical situations, the majority of actual systems are not closed; hence, understanding how entropy behaves under such conditions is essential in the analysis of thermodynamic processes under applications such as chemistry and environmental science (Geleyn et al., 2015). This misunderstanding may lead to the neglect of energy efficiency practices or ecological sustainability where the operation of entropy is inherently linked with the behavior of systems. Thus, explaining this misconception is important for students and professionals because it offers them useful information for everyday scientific practice. Besides, most people mix entropy with a lack of energy or a capacity to perform work. This idea is rooted in an incomplete understanding of energy-entropy analogy in thermodynamic processes. While it is correct that a higher entropy generally implies a spread of energy within a system, it does not imply that there is no energy but rather that the usable energy to perform work decreases. One way this applies to energy changes and entropy is in the conversion of thermal engines' energy, illustrating how they are connected (Geleyn et al., 2015). These misconceptions can produce misapprehensions regarding the effectiveness of engines and other energy systems and, ultimately, incorrect conclusions in energy management and engineering designs. Clarifying the nature of entropy as a measure of energy distribution, rather than a symptom of energy deficiency, is essential for enhancing scientific literacy and problem-solving skills in energy disciplines (Figure 6.14).

Figure 6.14. *Misconceptions about entropy.*

Source: https://www.thoughtco.com/thmb/xb6BZVK-bIakrjUE2Jj3-VIv9sY=/1500x0/filters:no_upscale():max_bytes(150000):strip_icc()/GettyImages-529148993-579ad0ba3df78c3276a7ced8.jpg.

6.6.1. Common Myths Regarding Entropy

Entropy as per the Second Law of Thermodynamics is likely to promote myths regarding what it signifies. One of them is that entropy only explains disorder. Even as the notion does suggest a drift towards a condition of increased randomness in an ensemble, it is a simplification of the idea. Entropy is precisely the number of microstates a system may have, which is a measure of the distribution of energy at microscopic scales. This refined perception indicates that entropy is not disorder in a general sense but is a measure of more variety in the microconfigurations. The vagueness of this definition typically leads to misunderstandings in scientific explanations and can bias the implementation of thermodynamic principles in physics and chemistry, needing a more specific interpretation of the term beyond randomization. Another common myth in entropy is that it is applicable solely in closed systems. This misconception overlooks the existence of entropy in open systems, systems that exchange matter and energy with their surroundings. The infamous Second Law of Thermodynamics mandates that in any spontaneous process, the total entropy of an isolated system never decreases. However, in open systems, local entropy reductions can occur, provided they are counterbalanced by greater increases in entropy beyond the local system, thus adhering to the overall direction demanded of the Second Law. This dynamic quality of entropy allows for the development of intricate structure and organization in biological and physical processes to breach the rigid interpretations emphasizing a one-way process of going to chaos without regard for the interaction of the larger system. Another myth has entropy suggesting an overall tendency towards chaos, frequently misinterpreted as a forecast of inevitable disaster in complex systems. This idea does not consider the sophisticated behavior of entropy in varied circumstances, particularly in systems governed by quantum physics or in some thermodynamic cycles. Entropy is not a mere power that brings things towards disorder, but it encompasses phenomena like producing ordered structures out of energetic processes, such as crystallization. Moreover, some black hole entropy models illustrate refined interconnections between quantum gravity and thermodynamics, further misrepresented by the reductionist myth of entropy as chaos (Peres et al., 2006; Visser et al., 2001). A satisfactory balanced understanding of entropy therefore involves an understanding of these nuances and not being swayed by reductionist myths misrepresenting its essential role in physics.

6.6.2. Clarifying the Relationship Between Entropy and Disorder

The concept of entropy and disorder forms the foundation to understand the principles of the Second Law of Thermodynamics. Entropy, the most important thermodynamic term, can be described as a measure of the number of microscopic states that correspond to the macroscopic state of a thermodynamic system. Therefore, with increasing entropy, the disorder in a system is enhanced. This is a description of the relationship that highlights the fact that systems move towards states of greater entropy, which are generally characterized by greater disorder. For instance, when ice melts to become water, the ice's crystalline structure is transformed to the more disordered structure of water molecules, thus raising the systems entropy. This unidirectional transition illustrates the tendency of natural processes to favor states of greater disorder, in accordance with the Second Laws assumption that the total entropy of a closed system will strive toward a maximum as time goes forward. The influence of the entropy-disorder relationship extends into simple thermodynamic systems to other domains, including information theory. According to one definition of entropy in this context, entropy is not just a measure of disorder but can also be described as a measure of information. In other words, as systems approach higher states of entropy, they are also losing information about their initial states of order. The concept of an ideal symmetric shape, which has a zero amount of information and maximum static entropy, is a key point of study. When systems undergo phase separation and crystallization, the processes are information minimization or symmetry maximization driven, further elucidating the coupling between information dynamics, entropy, and disorder in complex systems, as documented in research work on the Gibbs inequality and information theory principles (Atkin et al., 2001). Furthermore, the relationship between entropy and disorder is important in understanding self-organization at molecular and nanometer scales. Nonequilibrium thermodynamics breakthroughs within recent periods clarify driven systems' ability to display order as entropy increases, opposing the common belief that greater disorder means a non-organization state. Experiments in heterogeneous catalysis and molecular copolymerization demonstrate that systems may reach dynamical order through processes that are out of equilibrium. These experiments reveal that although the quantity of overall disorder is entropy, local domains can have ordered structures even under energy input. This dynamic equilibrium between increasing entropy at the macro level and the creation of localized order reflects the complexity of these systems, suggesting that the connection between entropy and disorder is contextual and dynamic (Gaspard et al., 2012).

6.6.3. Entropy and the Arrow of Time

The entropy definition is central to the thermodynamic explanation of the arrow of time. Entropy, traditionally defined as a measurement of disorder or randomness in a system, by its very nature implies directionality in time flow. With systems going in the direction of states with higher entropy, they possess an irreversible process that forcibly separates the reversible quality of microscopic physical law. This asymmetry is the basis of the second law of thermodynamics, where in closed systems, entropy would rise or stay constant, thereby deciding a temporal arrow. This is a phenomenon that accounts for our everyday experiences, where things are inclined to go from order to disorder, validating the fact that the universe is unfolding along a specific temporal trajectory (Peres et al., 2008). Moreover, the ideas of entropy have an impact that goes beyond classical mechanics into quantum mechanics, providing a more nuanced perception of irreversibility. In this quantum framework, all phenomenal information that transform to become informational in character are associated with a rise in entropy, but processes that yield lower entropy cannot be empirically tested. This description means that the second law of thermodynamics is, in a sense, a tautology about the nature of physical law—they effectively constrain the types of processes we can examine and understand. Having seen irreversible processes from this view, we see how restrictions imposed by entropy further delineate not only the direction of time, but also our understanding of physical reality itself (Batalhao et al., 2015). Finally, the relationship between entropy and the arrow of time requires a more fundamental exploration of the ground of physical theories. The investigation of nonequilibrium systems, for instance, shows entropy generation that can be measured, thus validating the preferential direction of time. Recent experimental studies, such as nuclear magnetic resonance, offer a tangible demonstration of how entropy can be directly related to processes following rapid alteration of external conditions. This empirical approach elucidates the connection between microscale behavior and macroscopic observation and establishes that the Kullback-Leibler divergence can serve as a measure to grasp the transition from a system to its time-reversed system. These advancements not only reinforce the role of entropy in determining the arrow of time but also lead researchers to explore the intricate dynamics of irreversible processes more deeply (Batalhao et al., 2015).

SUMMARY

Laws of entropy and the Second Law of Thermodynamics clarify the intrinsic nature of physical systems and what happens to them over time. Specifically, the Second Law contends that in any closed system, total entropy can never be reduced, pointing to a natural directionality of natural processes. This is the basis for understanding processes that are irreversible and helps solidify the inherent limitations on energy transformation. The applications extend beyond theoretical mechanics to practical ones, influencing our comprehension of processes ranging from chemical reactions to biological systems. Entertaining entropy as a measure of disorder in such a context greatly improves the comprehension of the universe's complicities and evolution. More recent debates have also enhanced our comprehension of the generalized second law of thermodynamics (GSLT), which accounts for the balance of energy in the universe. Already it has been shown that it is more general to describe the evolution of the universe based on the appearance of space relations and balance of energy. In thermodynamic properties analysis, both the first law and the GSLT are justified, stating that non-equilibrium systems can be accompanied by local equilibrium. These findings are contradicting the traditional interpretations, suggesting that even non-zero thermodynamic inertia systems must be conforming to the fundamental assumptions of thermodynamics but will exhibit special features. This is a vital critique that opens up wider implications of thermodynamic theory in scientific fields (Chen et al., 2018). Finally, solving complexities introduced by non-equilibrium systems further solidifies the significance of entropy in thermodynamic discourse. Validity of local equilibrium has been questioned in delayed response systems, diverging from the traditional understanding of entropy production. Negative entropy production is against the fundamental postulates of the Second Law as per current research. Redefining certain boundaries in thermodynamic law in thermodynamic inertia systems is extremely significant. Introduction of variational formulation establishes a robust backbone framework that maintains the integrity of laws while responding to complexities of such systems. Finally, these lectures deepen our insight into thermodynamic behavior and reaffirm the entropy role at the core of understanding universes evolution (Glavatskiy et al., 2015).

Entropy and the Second Law of Thermodynamics is a work that addresses fundamental principles that reveal physical processes inherent directionality. In its most basic sense, the Second Law states that in a sealed system, entropy will increase over time and be a natural tendency toward randomness. Not only does this law describe the mechanics of heat transport and energy conversion but the irreversible nature of much of thermodynamics as well. With changing systems, the increase in entropy translates into a reduction in the available energy to do work, a conclusion of far-reaching application in various fields from engineering to cosmology. Being able to comprehend these tenets allows scientists to predict behaviors and outcomes in thermodynamic systems, thus forming a foundation for innovations in energy efficiency and sustainability. An examination of entropy also reveals its function in statistical mechanics to be many-faceted, particularly in the process of phase transitions. Traditional approaches prefer to employ canonical ensembles in characterizing thermodynamic variables without paying heed to the nature of inhomogeneous systems. For instance, considering the example of the dynamics of boiling water, a traditional example in thermodynamics, the insufficiency of canonical statistics is evident. Here, phase separation of gas-liquid indicates the inadequacy of conventional models and the need for an integrated approach. Recent advances in statistical theory, for instance, Boltzmann-Planck statistics, provide a more advanced backdrop for understanding the microscopic origin of entropy, particularly of phase transitions, so that there is greater insight into this universal phenomenon and what it implies for changing systems (Gross et al., 2005). Secondly, the newly developed nonlocal thermodynamic theories demonstrate the adaptability required in contemporary science communication. These theories investigate the intricate dynamics among dynamic thermodynamic variables and entropy, providing avenues for investigations in nonlocality. By introducing modified entropy currents and current multipliers, researchers are redefining classical equations to provide more elucidating explanations on complex systems. This new focus violates traditional boundaries and extends the functioning model in thermodynamics, validating the need to consider both nonlocal and local interactions. Consequently, such investigations not only deepen our appreciation of entropy but also illuminate paths to resolving highly intricate real-world problems across various fields of study, thus reaffirming the relevance of the Second Law to advanced thermodynamic investigations (Van et al., 2003).

REFERENCES

1. Alfaris, L., Nurahman, A., Siagian, R. C., Sumarto, A., et al. (2023). *Termodinamika Lubang Hitam: Hukum Pertama Dan Kedua Serta Persamaan Entropi.* https://core.ac.uk/download/567724703.pdf (accessed on 4th October 2025).
2. Allahverdyan, A. E., Balian, R., Becker, R., Berdichevsky, V. L., Berry, M. V., Campisi, M., Carroll, S. M., Chen, J., et al. (2007). Relating the thermodynamic arrow of time to the causal arrow. http://arxiv.org/abs/0708.1175 (accessed on 4th October 2025).
3. Allahverdyan, E., Butkovskiy, A. G., Landé, A., Peres, A., Marshall, A. W., Fuchs, C. A., Janzing, D., et al. (2006). Explanation of the Gibbs paradox within the framework of quantum thermodynamics. http://arxiv.org/abs/quant-ph/0507145 (accessed on 4th October 2025).
4. Annila, A., Annila, E., et al. (2008). Why did life emerge? https://core.ac.uk/download/186331455.pdf (accessed on 4th October 2025).
5. Apertet, Y., Goupil, C., Lecoeur, P., Ouerdane, H., et al. (2015). Continuity and boundary conditions in thermodynamics: From Carnot's efficiency to efficiencies at maximum power. http://arxiv.org/abs/1411.4230 (accessed on 4th October 2025).
6. Arnaud, J., Chusseau, L., & Philippe, F. (2008). A simple model for Carnot heat engines. http://arxiv.org/abs/0812.2757 (accessed on 4th October 2025).
7. Atkin, A., Brillouin, L., Duan, Y., Elliott, M., Frank, R., Halliday, D., Hargittai, I., et al. (2001). The nature of the chemical process. 1. Symmetry evolution–Revised information theory, similarity principle and ugly symmetry. https://core.ac.uk/download/25312154.pdf (accessed on 4th October 2025).
8. Baez, J. C., & Pollard, B. S. (2016). Relative entropy in biological systems. http://arxiv.org/abs/1512.02742 (accessed on 4th October 2025).
9. Barnes, B., Benedicks, M., Bent, H., Born, M., Carnot, S., Clausius, R., et al. (2000). Inquiries into the nature of free energy and entropy in respect to biochemical thermodynamics. https://core.ac.uk/download/25283819.pdf (accessed on 4th October 2025).
10. Bass, G., Jr. (2000). *Educational Research Abstracts.* https://core.ac.uk/download/215487457.pdf (accessed on 4th October 2025).
11. Batalhao, T. B., Lutz, E., Oliveira, I. S., Paternostro, M., et al. (2015). Irreversibility and the arrow of time in a quenched quantum system. https://core.ac.uk/download/33589539.pdf (accessed on 4th October 2025).

12. Bennett, H., Kittle, C., Sun, C. P., Schroeder, D. V., Schrodinger, E., Nori, F., Callen, H., et al. (2006). Quantum thermodynamic cycles and quantum heat engines. https://core.ac.uk/download/44154918.pdf (accessed on 4th October 2025).
13. Beretta, G. P., & Zanchini, E. (2010). Rigorous and general definition of thermodynamic entropy. https://core.ac.uk/download/53606182.pdf (accessed on 4th October 2025).
14. Bérut, A., Bérut, A., Boyd, A. B., Boyd, A. B., Boyd, A. B., Barato, A. C., Barato, A. C., et al. (2017). Second law, entropy production, and reversibility in thermodynamics of information. http://arxiv.org/abs/1712.06858 (accessed on 4th October 2025).
15. Birrell, N., Hawking, S., Saida, H., Ruelle, D., Tasaki, H., Tolman, R., et al. (2011). Universal property of quantum gravity implied by uniqueness theorem of Bekenstein-Hawking entropy. http://arxiv.org/abs/1109.0842 (accessed on 4th October 2025).
16. Bochkov, G., Callen, H., Einstein, A., Messiah, A., Campisi, M., Hänggi, P., & Stratonovich, R. (2011). Fluctuation, dissipation and the arrow of time. https://core.ac.uk/download/11923064.pdf (accessed on 4th October 2025).
17. Brooks, H., Denbigh, K., Eddington, A. S., Ehrenfest, P., Einstein, A., Feynman, R. P., Gibbs, J. W., et al. (2007). *Probability as a Physical Motive*. https://core.ac.uk/download/44169300.pdf (accessed on 4th October 2025).
18. Channuie, P., & Momeni, D. (2018). On the scalar-vector-tensor gravity: Black hole, thermodynamics and geometrothermodynamics. http://arxiv.org/abs/1802.03672 (accessed on 4th October 2025).
19. Chen, Y. X., Huang, Q. H., & Tu, F. Q. (2018). Thermodynamics in the universe described by the emergence of the space and the energy balance relation. http://arxiv.org/abs/1808.10562 (accessed on 4th October 2025).
20. Christensen, W. M., Mountcastle, D. B., Smith, T. I., Thompson, J. R., et al. (2015). Identifying student difficulties with entropy, heat engines, and the Carnot cycle. https://core.ac.uk/download/214452216.pdf (accessed on 4th October 2025).
21. Curiel, E. (2014). *Classical Black Holes are Hot*. https://core.ac.uk/download/33752611.pdf (accessed on 4th October 2025).
22. DeWeese, M. R., Klymko, K., & Mandal, D. (2017). Entropy production and fluctuation theorems for active matter. https://core.ac.uk/download/655768669.pdf (accessed on 4th October 2025).

23. Fiolhais, M., & Güémez, J. (2013). Thermodynamics in rotating systems–Analysis of selected examples. http://arxiv.org/abs/1404.1786 (accessed on 4th October 2025).
24. Fronsdal, C. (2010). Heat and gravitation. III. Mixtures. http://arxiv.org/abs/1001.2579 (accessed on 4th October 2025).
25. Gaspard, P. (2012). Self-organization at the nanoscale scale in far-from-equilibrium surface reactions and copolymerizations. http://arxiv.org/abs/1203.0972 (accessed on 4th October 2025).
26. Geleyn, J. F., & Marquet, P. (2015). Formulations of moist thermodynamics for atmospheric modelling. http://arxiv.org/abs/1510.03239 (accessed on 4th October 2025).
27. Glavatskiy, K. S. (2015). Local equilibrium and the second law of thermodynamics for irreversible systems with thermodynamic inertia. https://core.ac.uk/download/43382568.pdf (accessed on 4th October 2025).
28. Goupil, C., Müller, E., Seifert, W., Snyder, G. J., et al. (2011). Thermodynamics of thermoelectric phenomena and applications. https://core.ac.uk/download/4890584.pdf (accessed on 4th October 2025).
29. Gross, D. H. E. (2005). Microcanonical thermostatic as foundation of thermodynamics: The microscopic origin of condensation and phase separations. http://arxiv.org/abs/cond-mat/0411408 (accessed on 4th October 2025).
30. Johnson, C. V., Johnston, D. A., Johnston, D. A., Brody, D. C., Weinhold, F., Hernández, G., Ruppeiner, G., et al. (2007). Geometrothermodynamics of black holes. http://arxiv.org/abs/0704.3102 (accessed on 4th October 2025).
31. Kaneko, S., Sasa, T. S., Komatsu. (1999). Thermodynamic irreversibility from high-dimensional Hamiltonian chaos. http://arxiv.org/abs/cond-mat/9911181 (accessed on 4th October 2025).
32. Klimenko, A. Y. (2012). Teaching the third law of thermodynamics. http://arxiv.org/abs/1208.4189 (accessed on 4th October 2025).
33. Klimenko, A. Y., & Maas, U. (2014). One antimatter–Two possible thermodynamics. https://core.ac.uk/download/197535379.pdf (accessed on 4th October 2025).
34. Lieb, E. H., & Yngvason, J. (2014). Entropy meters and the entropy of non-extensive systems. http://arxiv.org/abs/1403.0902 (accessed on 4th October 2025).

35. Page, D. N. (2004). Hawking radiation and black hole thermodynamics. http://arxiv.org/abs/hep-th/0409024 (accessed on 4th October 2025).
36. Peres, A., Thorndike, A., Zeilinger, B., Zwiebach, C., Rovelli, C., Rovelli, C., Rovelli, C., et al. (2006). Quantum mechanics: Myths and facts. https://core.ac.uk/download/44140221.pdf (accessed on 4th October 2025).
37. Peres, D.Z., Albert, D.Z., Albert, D.Z., Borel, E., Gyftopoulos, E.P., Reif, F., Price, H., et al. (2008). A quantum solution to the arrow-of-time dilemma. http://arxiv.org/abs/0802.0438 (accessed on 4th October 2025).
38. Reis, A. H. (2006). Constructal theory: From engineering to physics, and how flow systems develop shape and structure. https://core.ac.uk/download/62452175.pdf (accessed on 4th October 2025).
39. Uffink, J. (2001). Bluff your way in the second law of thermodynamics. https://core.ac.uk/download/11920131.pdf (accessed on 4th October 2025).
40. Van, P. (2003). Weakly nonlocal irreversible thermodynamics. http://arxiv.org/abs/cond-mat/0112214 (accessed on 4th October 2025).
41. Ván, P. (2017). Weakly nonlocal nonequilibrium thermodynamics: The Cahn-Hilliard equation. http://arxiv.org/abs/1710.04204 (accessed on 4th October 2025).
42. Ván, P. (2020). Thermodynamically consistent gradient elasticity with an internal variable. http://arxiv.org/abs/2005.13662 (accessed on 4th October 2025).
43. Visser, M. (2001). Essential and inessential features of Hawking radiation. https://core.ac.uk/download/25314477.pdf (accessed on 4th October 2025).
44. Wehrl, A. S., Holevo, C. H., Bennett, C. H., Song, D., Lubkin, E., Winstanley, E., Bardeen, J., et al. (2000). Information erasure and the generalized second law of black hole thermodynamics. https://core.ac.uk/download/49931.pdf (accessed on 4th October 2025).

CHAPTER 7

GIBBS FREE ENERGY AND EQUILIBRIUM

Contents

Introduction 241
7.1. The Concept of Gibbs Free Energy..246
7.2. Thermodynamic Principles...252
7.3. Gibbs Free Energy and Chemical Reactions..256
7.4. Equilibrium in Chemical Systems...261
7.5. The Role of Gibbs Free Energy in Equilibrium.......................................266
7.6. Applications of Gibbs Free Energy and Equilibrium................................270
Summary...274
References..277

INTRODUCTION

The Gibbs Free Energy is the core concept of thermodynamics, serving as a vital gauge of the spontaneity of phase transitions and chemical reactions. Fundamental to the idea of chemical equilibrium, the Gibbs Free Energy (G) is the amount of maximum reversible work that may be extracted by a thermodynamic system at constant pressure and temperature. Such a property not only simplifies identification of energy changes in chemical reactions but also dictates under what circumstances a reaction will spontaneously proceed to equilibrium. Thus, Gibbs Free Energy gives scientists and chemists a concrete framework for predicting directionality of reactions as well as stability in multifaceted systems, thereby amalgamating basic concepts of thermodynamics with application in scientific fields at large. Thermodynamic equilibrium is a delicate balance among reacting species aiming at attaining a state where forward and reverse reactions are of equal rates. This is mathematically represented by the equilibrium constant, which is employed to give a description of the relationship between reactant and product concentrations under specific conditions.

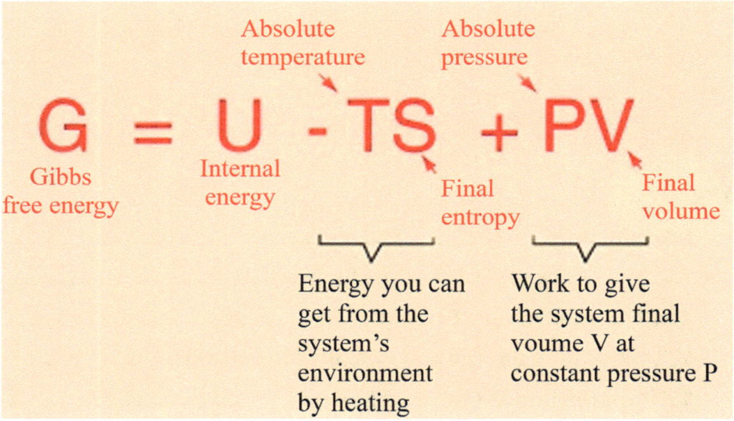

Figure 7.1. *Gibbs free energy and equilibrium.*

Source: https://equilibriumthermodynamics.weebly.com/uploads/4/7/6/6/47666193/9116227_orig.jpg.

The intrinsic relationship between Gibbs Free Energy and equilibrium identifies that every decrease in Gibbs Free Energy is a spontaneous reaction, moving towards equilibrium. As observed through fluid mixture experiments, whereby free energy minimization is used with the aim of achieving critical equations of system behavior, the crucial role played by Gibbs Free Energy in phase transitions and chemical reactions is evident, offering a science base for innovation in material processing and environmental engineering (Bruno et al., 2004). In practice, Gibbs Free Energy is a foundation for the study of many chemical reactions, such as hydrocarbon reactions. For instance, thermodynamic studies of methane reaction, such as steam reforming, show that some conditions of temperature and pressure can effectively influence product formation. In this, Gibbs Free Energy not just informs us about the feasibility of the reaction but also assists in finding the optimal operating conditions that provide us with maximum yield using minimal unwanted by-products like carbon formation (Istadi et al., 2012). Hence, it is important to understand the finer details of Gibbs Free Energy as well as the fundamentals of equilibrium for theoretical purposes and for practical use in power generation, catalysis, and chemical production processes. These findings are critical as the world makes the shift to sustainable energy options, where efficient reaction pathways increasingly important (Figure 7.1).

Definition of Gibbs Free Energy

Understanding the definition of Gibbs free energy is critical in the field of thermodynamics, particularly how it relates to chemical reactions and equilibrium. Gibbs free energy, designated by the symbol G, is the maximum possible reversible work that can be obtained from a system at constant temperature and pressure. The thermodynamic potential is also a spontaneity of measurement in chemical reactions, which affirms that a negative sign for the change in Gibbs free energy (ΔG) indicates that a reaction can occur spontaneously. In effect, the Gibbs free energy reflects the balance between entropy and enthalpy insofar as systems try to arrive at a position where G is minimized. Measuring the change in G along the path of reactions can be used to tell the reaction's outcome and hence provides the foundation for estimating chemical behavior and predicting the behavior at equilibrium in a system. The connection between Gibbs free energy and chemical equilibrium is important in that it provides a quantitative approach to examine the stability of a system. The Gibbs free energy is minimized at equilibrium, and this is a measure of stability; hence, any change in conditions will provide a change in G. The relationship between Gibbs free energy and equilibrium constant (K) can be defined by the equation $\Delta G = \Delta G° + RT \ln(Q)$, where Q is the reaction quotient. The equation indicates that at equilibrium, ΔG will be zero, i.e., there is no net alteration in the reactants' and products' concentrations. Understanding this relationship is essential for chemists and physicists because it enables the prediction of the behavior of chemical systems when subjected to varying temperature, pressure, and concentration (Heng et al., 2015). Furthermore, the Gibbs free energy concept extends its implications to processes other than simple chemical reaction, including biological reactions and environmental chemistry. The Gibbs free energy rules can account for the way in which cells can gain energy from biochemical reactions and perform work, effectively tying together principles of thermodynamics with biological processes. Specifically, the Gibbs relation is also a central component in nonequilibrium thermodynamics, wherein it connects to dissipation principles and memory effects in materials science (Ván et al., 2020). Thus, Gibbs free energy not only facilitates understanding spontaneous chemical reactions but also enables the study of energy transitions in a number of fields, further establishing itself as a vital component in the study of equilibrium in nature and man-made systems.

Importance of Gibbs Free Energy in Thermodynamics

Understanding the work of Gibbs free energy is crucial for determining the boundary of thermodynamic processes and determining the spontaneity of chemical reactions. Gibbs free energy, G, is a hypothetical thermodynamic that connects enthalpy and entropy to determine the spontaneity of the reaction at constant pressure and temperature. If ΔG is negative, then the process is spontaneous, while positive denotes it as non-spontaneous. This is why Gibbs free energy is significant not only for determining the states of equilibrium but also for allowing us to understand energy transfer in phase change and chemical reactions. This ability to relate molecular behavior to macroscopic effects makes Gibbs free energy a fundamental physical chemistry concept and materials science, allowing us to acquire insights into paths for synthesizing new materials and optimizing chemical industrial processes. Gibbs free energy is equally a significant component of researching chemical mixtures and their composition in variable conditions. Ideal mixtures of gases are researched by examining ideal gas mixtures under the Gibbs-Dalton hypothesis, observing the behavior of multi-component systems as a function of Gibbs free energy. This gives a deep insight into how various components interact with one another, becoming critical in observing reaction dynamics in real-world applications. For instance, hydrogen dissociation experiments illustrate how departures from ideality impact calculations involving Gibbs free energy and, in turn, impact reaction outcomes, especially in gravitational conditions. These types of experiments are crucial to theoretical thermodynamics development, as they allow researchers to study equilibrium as a collective property of constituents of systems while acknowledging complexity present in real-world gases and other phases (Fronsdal et al., 2010). Moreover, the significance of Gibbs free energy extends beyond ideal systems, serving as a basis for the study of surface and interfacial behavior in thermodynamics. New innovations have brought a generalized grand thermodynamic potential theory including surface work associated with solid-fluid interfaces, enhancing the research areas in phase transitions and surface interactions. These advancements not only highlight the inherent significance of Gibbs free energy in governing chemical stability and equilibrium but also lead the way towards newer material design and engineering, particularly towards addressing adsorption and catalysis challenges. By operationalizing Gibbs free energy in these contexts, scientists are able to come up with new knowledge that spans the theory and practice divide and hence make it the basis of contemporary thermodynamic theory (Rusanov et al., 2009) (Figure 7.2).

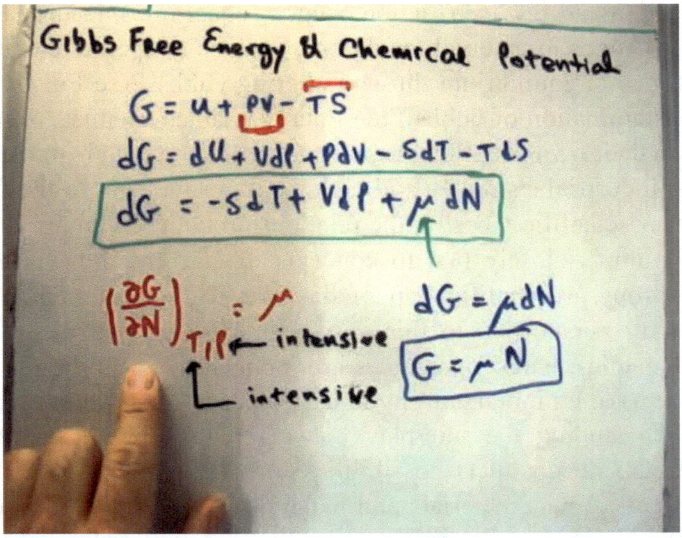

Figure 7.2. *Gibbs free energy in thermodynamics.*

Source: https://i.ytimg.com/vi/kXio3EcxX3M/sddefault.jpg.

Overview of the Relationship Between Gibbs Free Energy and Equilibrium

Centering around thermodynamics is Gibbs Free Energy (G), a crucial parameter in unraveling the conditions under which chemical equilibrium is achieved. Gibbs Free Energy provides a thermodynamic potential that helps predict the spontaneity of processes at constant temperature and pressure. The relationship between Gibbs Free Energy and equilibrium is beautifully described by the principle that a system will achieve equilibrium when the change in Gibbs Free Energy (ΔG) equals zero. In a reversible reaction, when ΔG is negative, the reaction is spontaneous in the forward direction and, where ΔG is positive, the reaction is not spontaneous. This co-dependency shows the manner in which Gibbs Free Energy not only sets the suitability of a reaction but also sets the extent to which reactants become products and ultimately helps determine equilibrium state configurations. The implications of Gibbs Free Energy extend beyond mere prediction of direction of reaction; they offer instructive insight into the energy landscape of chemical reactions. Namely, reactions involving a reduction in Gibbs Free Energy are seen to reach equilibrium at a quicker rate, fueled by kinetic forces that govern rates of reaction. Equilibrium systems have a balance of such rival rates of reactants and products, where the forward reaction and reverse reaction occur at equal rates, leading to a steady state. Also, changes in external conditions, such as temperature and pressure, lead to changes in equilibrium through altering Gibbs Free Energy. Thus, an exhaustive examination of Gibbs Free Energy enables chemists to manipulate conditions to their own benefit, enhancing the efficiency of chemical processes as varied as industrial syntheses and biological metabolism. In the framework of the broader scientific tapestry, the relationship between Gibbs Free Energy and equilibrium is closely tied to concepts of entropy and thermodynamic stability. Entropy, a quantification of disorder within a system, provides a complementary perspective to this argument. According to the second law of thermodynamics, natural processes are inclined to go in the direction that will achieve maximum total entropy in the universe. As encapsulated in recent studies, understanding the interplay between Gibbs Free Energy, entropy, and equilibrium deconstructs possibilities for new applications, such as the characterization of nanomaterials and materials under challenging conditions of stress, as addressed in Calabrese et al. (2018). Additionally, the connection to entropy principles leads to combinatorial information theory developments, demonstrating how maximizing entropy plays a key role in investigating high-probability realizations in systems of interest in statistical mechanics, as discussed by Niven et al. (2007). Gibbs Free Energy is therefore a focal point, linking thermodynamics and material sciences towards a more mature understanding of equilibrium phenomena.

7.1. THE CONCEPT OF GIBBS FREE ENERGY

The Gibbs free energy is a basic idea when it comes to the knowledge of chemical thermodynamics and equilibrium. Thermodynamic potential is the maximum reversible work that can be performed by a system under constant pressure and temperature. Thus, it is basic to deciding the direction of spontaneous processes. According to definition, a process will be spontaneous if Gibbs free energy is diminishing, i.e., mathematically as $\Delta G < 0$. Opposite to this, when $\Delta G > 0$, the system is non-thermodynamically favorable, and work from an external source would be required to make the reaction go ahead. The correlation of Gibbs free energy with equilibrium once again makes it all the more important as systems will come to evolve towards states that decrease free energy, ultimately stabilizing into a state of equilibrium where ΔG becomes zero, thereby indicating dynamic balance of the system. Moreover, Gibbs free energy is heavily related to the fundamentals of thermodynamics, especially as derived and practiced in various fields. The Gibbs relation serves as an expansion platform connecting the mechanical and thermal behavior, allowing scientists to study phase transitions and chemical reactions in an encompassing way. The thermodynamic potential can also embrace aspects of entropy and enthalpy, making its multi-functional use even greater.

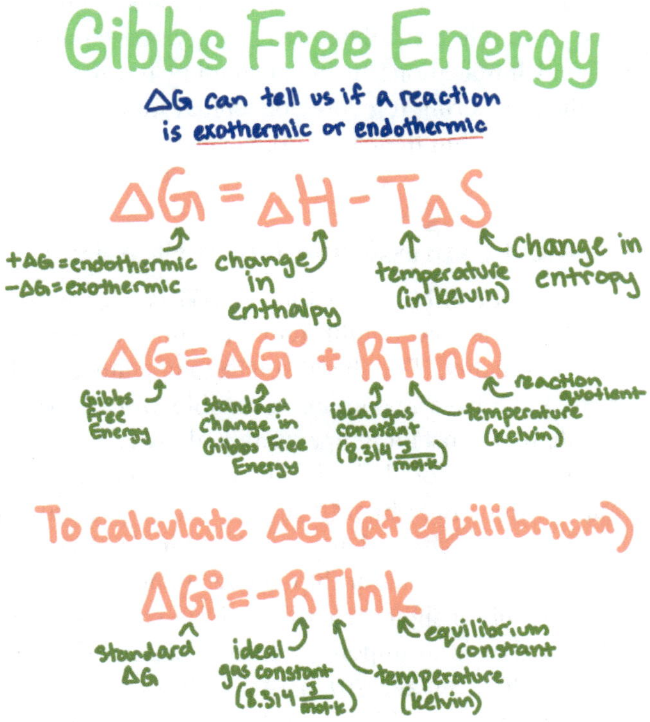

Figure 7.3. *The concept of Gibbs free energy.*

Source: https://d20khd7ddkh5ls.cloudfront.net/screenshot_2023-04-15_at_10.45.13_pm.png.

One of the most frequent applications of Gibbs free energy is in deciding if chemical reactions are possible under varying conditions, as noted by its application in derivation of constitutive relations for continuum mechanics. The uses of Gibbs free energy carry further to more advanced research, such as nonequilibrium thermodynamics and the use of nonequilibrium thermodynamics in the modeling of complicated systems, further accentuating its usefulness and relevance to the scientific community (Ván et al., 2020). With further research of the further uses of Gibbs free energy, there is a necessity to consider its use in nonequilibrium thermodynamics. Recent advances suggest that an idea of a generalized potential, analogous to Gibbs free energy, could be central in understanding stochastic processes in complex systems. This research suggests the possibility of new insights into criticality and stability in biological systems and other complex settings. By studying conditions under which systems reach equilibrium, researchers can learn a lot about the nature of different systems, ranging from chemical reactions to biological systems. Through expanding knowledge in the research on Gibbs free energy, researchers continually discover its profound implications in understanding the mechanisms behind thermodynamics, thus rendering it a core conceptual framework for upcoming studies and applications within the natural sciences (Qian et al., 2016) (Figure 7.3).

7.1.1. Mathematical Expression of Gibbs Free Energy

Mathematical equation for Gibbs Free Energy (G) is a core concept of the subject of thermodynamics and equilibrium. It's an equation $G = H - TS$, wherein H refers to enthalpy, T is the absolute temperature, and S refers to entropy. This term symbolizes maximum reversible work achievable from a thermodynamic system at constant temperature and pressure. The significance of Gibbs Free Energy is that it could give insight into the spontaneity of chemical reactions. A negative Gibbs Free Energy change ($\Delta G < 0$) signifies a reaction to be spontaneous, while a positive change ($\Delta G > 0$) indicates non-spontaneity. Not only does it highlight the relationship between enthalpy and entropy, but it also provides invaluable insights regarding phase transitions and chemical equilibria, and thus makes it an important device for scientists and engineers from all disciplines (Heng et al., 2015). Aside from being a fundamental application in thermodynamics, Gibbs Free Energy is also an essential factor in the formulation of chemical kinetics. The Arrhenius equation for reaction rates as a function of temperature can, in the context of Gibbs Free Energy, be expressed by setting equal to Gibbs Energy changes the equilibrium constant (K). The equation $K = e$^$(-\Delta G/RT)$ is an expression of such a relationship, illustrating that the stability of a chemical equilibrium is inherently connected with the Gibbs Free Energy of reactant and product. Further, with the recognition of the significance of the equilibrium constant, a value without dimensions, comes the important necessity of mathematical precision in the calculation and interpretation of Gibbs Free Energy to reaction alterations, such as temperature and pressure alterations, further relating thermodynamics to chemical conduct (Ván et al., 2020). The mathematical treatment of Gibbs Free Energy has repercussions that reach into areas beyond traditional chemistry and impacts areas such as materials science and biological processes. For instance, the Gibbs Free Energy minimization theory is employed to anticipate the development of complex structures and phases in materials in order to promote nanotechnology and metallurgy. Gibbs Free Energy governs metabolic pathways as well as biochemical processes in biological systems and provides a framework for cellular processes. As researchers continue to explore the significance of Gibbs Free Energy, novel models and approaches are developed, integrating concepts from statistical mechanics and thermodynamics to further our understanding of equilibrium dynamics. These studies emphasize the one-dimensionality of Gibbs Free Energy as a universal descriptor not only for chemical reactions but also for more general scientific inquiries in various fields (Heng et al., 2015; Ván et al., 2020) (Figure 7.4).

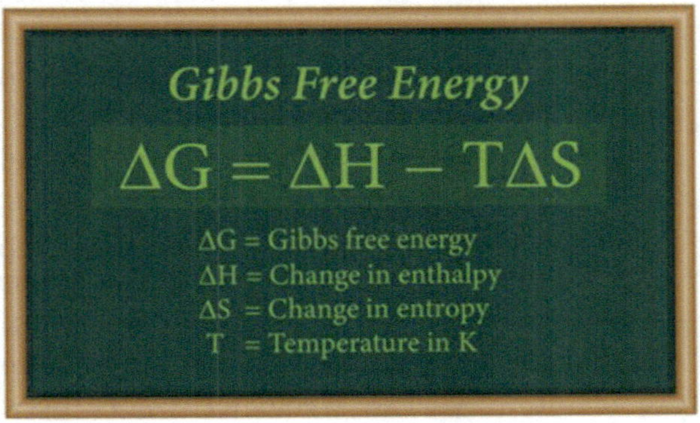

Figure 7.4. *Mathematical expression of Gibbs free energy.*

Source: https://www.shutterstock.com/image-vector/gibbs-free-energy-formula-mathematics-260nw-2321994321.jpg.

7.1.2. Components of Gibbs Free Energy: Enthalpy, Entropy, and Temperature

The relationship between enthalpy, entropy, and temperature is the backbone for the Gibbs free energy equation, which is the key in determining the spontaneity of chemical reactions. Enthalpy, the sum total of heat contained within a system, directly impacts Gibbs free energy as it also signifies the net change in energy throughout a reaction. A decrease in enthalpy, typically due to bond formation in the case of exothermic reactions, signifies that the products are more stable than the reactants. Entropic contributions, however, which quantify the disorder of a system, counteract enthalpic change. The realization of these interactions is crucial to the prediction of how the responses will move towards equilibrium, highlighting the fine balance for the thermodynamic processes of most systems. As the temperature plays the central role in the Gibbs free energy model, in this case, it plays the role of a modifier that influences the efficiency of enthalpy and entropy. Specifically, the Gibbs free energy formula incorporates temperature as a factor that multiplies entropy changes to stress the point that increasing the temperature can add weight to the influence of entropy on the spontaneity of systems. The latter is particularly relevant in non-equilibrium situations where systems are attempting to gain stability; hence, knowledge of temperatures role can prove handy in the examination of phase transitions. For instance, in phase equilibrium studies in the solid-liquid regime, temperature greatly determines the Gibbs free energy and consequently the phase composition in lipid mixtures, indicating temperatures significant role in determining the aftermath of various thermodynamic processes, (Roux et al., 2011). The interaction of enthalpy, entropy, and temperature emphasizes the significance of Gibbs free energy in attaining equilibrium between chemical systems. In industrial application, from material science to biochemical reactions, detailed understanding of these building blocks facilitates the prediction and control of reaction yields. By examining empirical relationships based on Gibbs free energy calculations, scientists can tune conditions to produce desired thermodynamic properties, resulting in optimum process efficiency and material properties. For example, the models of the temperature, enthalpy, and entropy effect on lipid phase behavior illustrate real-life scenarios where these concepts are crucial to foster knowledge in food science and biomedical applications, illustrating the universality of Gibbs free energy to attain equilibrium in different fields, (Wunderlich et al., 2009) (Figure 7.5).

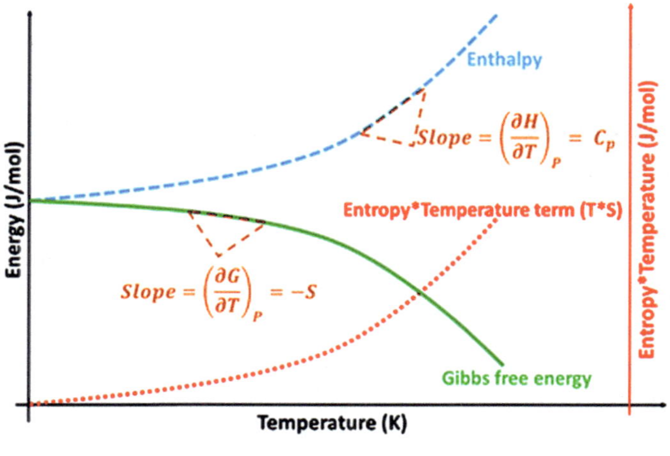

Figure 7.5. *Components of Gibbs free energy.*

Source: https://www.researchgate.net/publication/323861909/figure/fig1/AS:962138935943174@1606403206167/Diagram-showing-the-variation-of-Gibbs-free-energy-enthalpy-and-entropy-with.gif.

7.1.3. Significance of Gibbs Free Energy in Predicting Spontaneity

Understanding Gibbs free energy (G) is vital in establishing the spontaneity of a chemical reaction. Gibbs free energy is one of the thermodynamic potentials that deliver information regarding the favorableness of a process under constant pressure and temperature. An unfavorable Gibbs free energy change ($\Delta G < 0$) means that a reaction is spontaneous, whereas a positive change ($\Delta G > 0$) indicates the process is not spontaneous and will require energy. The rule enunciates the enthalpy-temperature-entropy relationship in the equation $\Delta G = \Delta H - T\Delta S$. Besides, when a reaction is in equilibrium, Gibbs free energy is at its minimum, recalling that the system is in balance in that there is no net change. Thus, the Gibbs free energy theory not only serves as a predictive process but also serves as a model for chemical equilibria.

The application of Gibbs free energy goes beyond isolated reactions to become the basis for understanding more general thermodynamic processes, such as those found in industry and the environment. For example, in biorefinery reactions to obtain sustainable fuels, Gibbs free energy is the most important criterion in determining whether the reactions are spontaneous or not. As illustrated in Jatropha seed oil conversion research, the Gibbs free energy thermodynamic analysis was found to be negative, which indicates spontaneous reaction conditions under optimized operating conditions like temperature and catalyst loading (Mary-Jane et al., 2025). This information is crucial in the design of successful pathways for biofuels and other alternate energies, mirroring the beneficial application of Gibbs free energy in real-world practice.

Through the application of these principles, scientists can optimize conditions for desired effects to result in technological and sustainability enhancements. The concept of Gibbs free energy can further be utilized in food science since it can be used to explain phenomena such as water adsorption behavior in food storage. Achira biscuits research is a case in point in indicating how Gibbs free energy coincides with thermodynamic properties like enthalpy-entropy relationships and adsorption isotherms. This discussion illustrates the relevance of thermodynamic metrics in predicting the effect of water activity on moisture content, with implications for food quality and preservation (Corbín et al., 2023). By careful analysis of such systems through the Gibbs free energy model, researchers and food industry specialists can develop ideal storage regimes that maintain product quality without waste creation. Summing up, Gibbs free energy not only represents a crucial theoretical basis for spontaneity prediction but also is used in pragmatic plans in various branches of science, making processes efficient and sustainable.

7.2. THERMODYNAMIC PRINCIPLES

Thermodynamic principles are the foundation upon which the character of systems in equilibrium can be interpreted, particularly in terms of Gibbs Free Energy. This energy status is significant since it determines chemical reaction spontaneity and phase changes. In assessing whether a process is spontaneous or not, thermodynamics provides a set of standards with focuses on energy and entropy changes. The relationship among temperature, enthalpy, and entropy can be encapsulated in the Gibbs Free Energy equation. As a key thermodynamic value, Gibbs Free Energy (G) allows us to predict the direction of chemical reactions; reactions will proceed in the direction that minimizes G, thus encouraging equilibrium conditions. Last but not least, understanding the ramifications of G for the principles of thermodynamics is important in modeling and predicting chemical and physical system processes. At the heart of thermodynamic principle is an understanding of transport phenomena, which have the potential to significantly impact systems' equilibrium state. Membrane transporters, for example, play significant roles in the regulation of the internal cell environment via the passage of substrates into and out of across membranes.

But their dynamics must obey the principles of the laws of thermodynamics to be accurately modeled. The discrepancies witnessed in most theoretical models of such systems require energetics to be taken into account. A physics-based modeling approach provides an avenue of being able to build thermodynamically consistent models, as is seen in studies of specific transporters like the cardiac Na^+/K^+ ATPase. This indicates the principle of thermodynamic laws in both chemical and biological contexts, reaffirming their role in achieving equilibrium through Gibbs Free Energy considerations. Monte Carlo simulations are an illustration of the use of thermodynamic principles in another context because they are significantly used to calculate free energy and other thermodynamic properties of complicated systems such as substitutional alloys.

Monte Carlo simulations are challenged by issues that need complex algorithms for optimization. Using high-level algorithms to close the gap between user intentions and MC codes, scientists can succeed in manipulating and determining the intended thermodynamic properties without complicating things too much. The ability to gain free energy via domains of stability efficiently applies thermodynamic theory in developing predictive models. Such computational research demonstrates the way Gibbs Free Energy serves as a unifying aspect in understanding thermodynamic behavior, further confirming its applicability in attaining chemical equilibrium among systems under study in theoretical and applied thermodynamics.

7.2.1. First Law of Thermodynamics and Energy Conservation

The First Law of Thermodynamics or conservation of energy law declares that energy cannot be destroyed or created but can be changed from one energy form to another. This fundamental principle is the cornerstone of thermodynamic theory and shows that energy is always the same in any closed system. In terms of Gibbs free energy and chemical reaction, the law specifies the need to understand how variations in energy influence the spontaneity of reactions and the state of equilibrium. Through energy tradeoff between systems and the surroundings, the constraint of heat and work relationship by the First Law helps us determine how available energy can be used to achieve work, shifting reactions toward state equilibrium. By realizing this, scientists can better realize the importance of non-spontaneous reactions, which occur upon the input of energy external to the system, emphasizing the importance of energy equilibrium in thermodynamic processes. The integration of the First Law of Thermodynamics into principles of equilibrium leads to complex relationships between energy forms and chemical potential. Gibbs free energy (G) stands out as it combines enthalpy and entropy as a function of reaction feasibility at constant pressure and temperature. When a system reaches equilibrium, the change in Gibbs free energy is zero, meaning that the energy has been evenly distributed among products and reactants. This balance accounts for how laws of energy conservation govern not only physical changes but also chemical change. It is very important to predict such changes in many fields, from materials chemistry to biochemistry, providing valuable information about the reaction routes and their importance in daily applications. Understanding these principles enhances our means of controlling conditions in a beneficial manner, steering reactions towards preferred products, for instance, in catalysis and industrial processes. Extended application of the First Law to Gibbs free energy proves that it remains valid in even more complicated systems, such as mixtures of ideal gases. According to recent publications, including those that examine models of thermodynamic behavior in mixtures, one can add the Gibbs-Dalton hypothesis to investigate energy distributions properly (Fronsdal et al., 2010). Such approaches imply that interactions between different components of gases also follow laws of energy conservation, showing the omnipresence of thermodynamics in all forms of matter. Black hole thermodynamics also sheds more light on the various facets of this law of conservation, showing how traditional thermodynamic concepts apply even in highly unfavorable conditions (Mann et al., 2018). This versatility of the First Law entrenches its role as a fundamental principle underpinning the understanding of not only historical systems but new areas of physics and chemistry as well, enhancing our comprehension of energy conservation across a broad spectrum of applications (Figure 7.6).

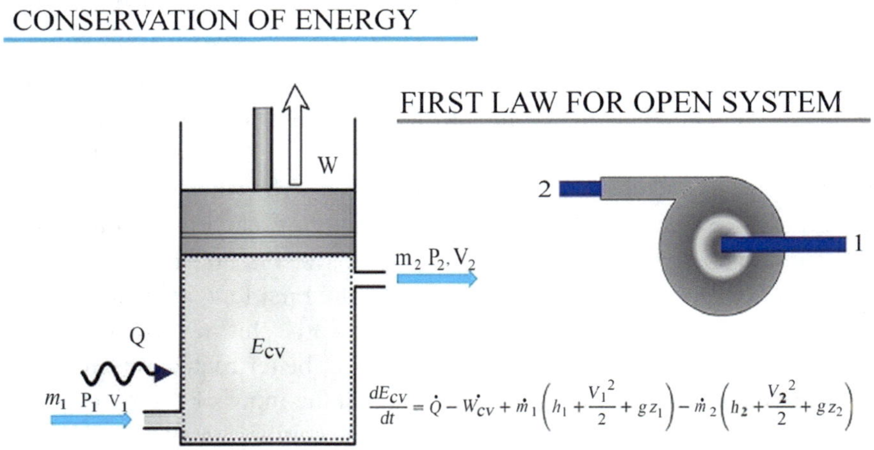

Figure 7.6. *First law of thermodynamics.*

Source: https://i.ytimg.com/vi/lbPEaaKiCww/hq720.jpg?sqp=-oaymwEhCK4FEI-IDSFryq4qpAxMIARUAAAAAGAElAADIQj0AgKJD&rs=AOn4CLDGMhchb-JR-p5igiWVrznwmbTUQ0Q.

7.2.2. Second Law of Thermodynamics and Entropy

The Second Law of Thermodynamics, one of the major principles of physical chemistry, holds that the total entropy of a closed system never decreases with time. Instead, it will grow and drive the system to a state of equilibrium. This is the law that forms the basis of understanding for spontaneous processes, those which occur automatically and will find configurations of higher entropy. The relationship between Gibbs free energy and entropy is of the utmost significance, for it is the Gibbs free energy change (ΔG) on which the spontaneity of a reaction often relies. In fact, a negative ΔG indicates a trend towards equilibrium, repeating the Second Laws influence on entropy. Thus, the Second Law not only prescribes the direction of thermodynamic processes, but it also determines the significance of entropy as a quantifier of disorder and dispersal of energy within a system, lending a conceptual framework to derive Gibbs free energy in processes towards equilibrium. As a central concept within thermodynamics, entropy directly relates to the disposition and distribution of microscopic states within a system. It informs us about the amount of energy that is not available to do work and, in addition, serves as a measure of disorder. The coordination of entropy with Gibbs free energy brings forth an even more profound understanding of equilibria in chemical reactions. The Gibbs-Dalton hypothesis states that entropy in a system of ideal gases will be accessible from the entropies of its components individually, emphasizing the additivity nature of entropy in multi-component systems (Fronsdal et al., 2010). As processes move toward equilibrium, the increase in entropy is in line with the contemporary understanding that the universe moves toward disorder, supporting the idea that chemical reactions, such as phase change, tend to proceed in the direction of increased states of entropy, increasing the overall level of energy distribution in accordance with the Second Law. Further exploration of the Second Law introduces fascinating parallels with information theory, particularly through the lens of entropy as an information quantifier. This perspective highlights the parallel between a system's disorder and the information needed to describe its microscopic states. System evolution as a response to the pursuit of equilibrium can be thought of as a process where maximization of entropy corresponds to information minimization (Atkin et al., 2001). Thus, spontaneous processes such as mixing, phase separation, and crystallization can be envisioned as driving forces towards higher entropy and symmetry states, suggesting a direct correspondence between the laws of thermodynamics and the inherent structures of nature. Therefore, the Gibbs free energy study in terms of entropy not only tells us about the processes to equilibrium but also tells us about the universes inherent disorder tendency.

7.2.3. Relationship Between Thermodynamic Principles and Gibbs Free Energy

The connection between Gibbs Free Energy and thermodynamic principles is at the core of chemical equilibrium study as well as reaction spontaneity. Gibbs Free Energy (*G*) is a privileged thermodynamic potential that accounts for the equilibrium between enthalpy (*H*) and entropy (*S*) in a system. More specifically, the equation $G = H - TS$ encapsulates how energy conversion dictates direction in chemical processes. When examining chemical reactions, Gibbs Free Energy change (ΔG) provides insight as to whether or not a reaction will go spontaneously. A negative ΔG indicates that the process is thermodynamically favorable and leads to a spontaneous movement towards equilibrium. By putting together Gibbs Free Energy and the principles of thermodynamics, researchers can better forecast and manipulate reaction conditions for achieving desired outcomes in numerous biochemical and industrial processes. Furthermore, the thermodynamic consistency of models that predict the behavior of systems, including membrane transporters, illustrates the value of abiding by Gibbs Free Energy principles. For example, the function of transporters such as the cardiac sarcoplasmic/endoplasmic Ca^{2+} ATPase (SERCA) and the Na^+/K^+ ATPase shows how energy is transmitted across membranes in a manner consonant with thermodynamic principles. These transport mechanisms rely on changes in Gibbs Free Energy to allow ions to move against their concentration gradients. Since these systems are represented in correct form, energy transfer can be discussed as a physical system constrained by thermodynamics, leading to more reliable predictions of biological behavior, a requirement from physiology to pharmacology (Crampin et al., 2018). Apart from this, the intricate relation of thermodynamics to information theory, as seen in machines such as Szilard's engine, also hints toward the importance of Gibbs Free Energy in broader contexts. In variable energy landscape systems, the thermal capacity of information can be quantified in free energy terms, which provides insight into how memory and thermal fluctuations co-operate with energy conversions. Interdependence expressed in this form extends the applicability of Gibbs Free Energy to chemical reactions into other physical systems in quantum mechanics. Thus, the intersection of thermal information capacity with non-equilibrium Helmholtz free energy highlights the manner in which Gibbs Free Energy both underpins thermodynamic processes and information dynamics, thus enhancing our knowledge of equilibrium in the many scientific fields (Gour et al., 2019) (Figure 7.7).

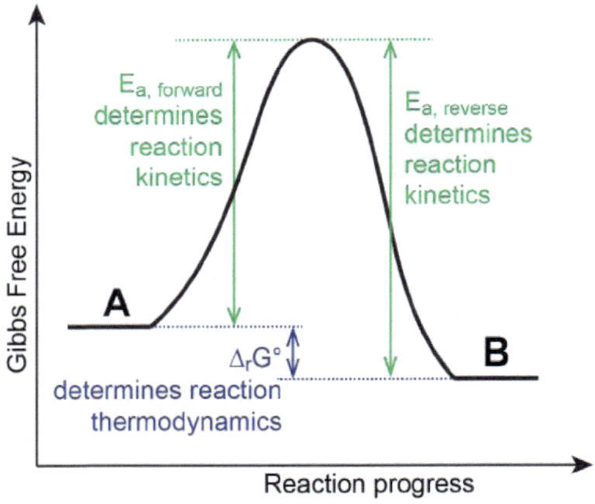

Figure 7.7. *Relationship between thermodynamic principles and Gibbs free energy.*

Source: https://chem.libretexts.org/@api/deki/files/403011/kinetic_stability.png?revision=1&size=bestfit&width=449&height=380.

7.3. GIBBS FREE ENERGY AND CHEMICAL REACTIONS

Gibbs Free Energy (G), a fundamental variable in the field of thermodynamics, plays a significant role in the determination of chemical reaction spontaneity. Gibbs Free Energy consolidates enthalpy (H) and entropy (S) into a single system so that scientists can predict if and when a process would occur under constant pressure and temperature. A negative ΔG ($\Delta G < 0$) signifies that a reaction can go on spontaneously, thereby providing the process with a thermodynamic "go-ahead." A positive ΔG indicates non-spontaneity and that energy will have to be input into the reaction if it is to proceed. Therefore, Gibbs Free Energy not only serves as a spontaneity test but also as a measure of the force driving chemical change, enabling scientists to predict the feasibility of reactions under various biochemical and industrial conditions. The role of Gibbs Free Energy is not to determine spontaneity but to help one comprehend the state of equilibrium of chemical reactions. When at equilibrium, the Gibbs Free Energy of the products is equal to that of the reactants, and hence a ΔG of zero. This equilibrium is a situation where forward and reverse reaction rates are equal, and hence a stable concentration of the reactants and products.

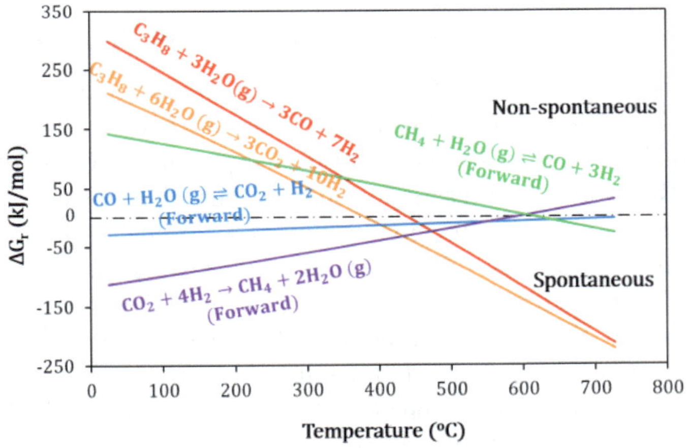

Figure 7.8. *Gibbs free energy and chemical reactions.*

Source: https://www.researchgate.net/publication/283566009/figure/fig4/AS:317044156321795@1452600625904/Reaction-Gibbs-free-energy-as-a-function-of-reaction-temperature-25-C-to-700-C-at-1.png.

Furthermore, the relationship between Gibbs Free Energy and reaction quotient (Q) is fundamental for predicting the direction of a chemical reaction. If the actual concentration of products and reactants deviates from the standard state, the Gibbs Free Energy can be used to calculate the new reaction quotient, which returns the system to equilibrium. This interaction between the concept of energies is the basis upon which an understanding of chemical dynamics and stability in various systems can be built. Moreover, Gibbs Free Energy applications are not limited to simple reaction kinetics, but extend into more complex biochemical reactions, such as membrane transport. In biological systems, the transport of ions and molecules across cell membranes required for maintaining cellular homeostasis is regulated by membrane transporters. These processes are indirectly governed by the principles of thermodynamics as the energetics of substance transport must be compliant with Gibbs Free Energy laws. Experiments highlight that thermodynamically correct representations of such transporters reflect the delicate balance of energy dissipation and maintenance of stable gradients in living organisms. These gradients being essential for processes like conduction of nerve impulse and muscle contraction, understanding the thermodynamic basis helps in unscrambling life process complexity (Crampin et al., 2018). Besides, the evolutionary character of energy transduction also validates the thermodynamic principles governing molecular behavior and stability (Annila et al., 2008) (Figure 7.8).

7.3.1. Role of Gibbs Free Energy in Reaction Spontaneity

Sophisticated understanding of Gibbs free energy is paramount while studying the spontaneity of chemical reactions. It is a thermodynamic experiment for the prediction of how likely a reaction will occur spontaneously under specified conditions. By definition, a reaction is spontaneous when it causes a decrease in Gibbs free energy ($\Delta G < 0$), an indication of the capacity of a system to do work without the provision of external energy. Conversely, energy-requiring processes exhibit positive Gibbs free energy changes ($\Delta G > 0$) because they are non-spontaneous processes. Through the knowledge that Gibbs free energy takes into account both enthalpy (ΔH) and entropy (ΔS), chemists can better explain the way energy conversions lead to the reaction feasibility, supporting the primary relationship between Gibbs free energy and reaction spontaneity in different chemical contexts. Besides, the use of Gibbs free energy is not confined to theoretical frameworks, as it affects practical scenarios such as biochemical reactions. Specifically, extensive knowledge of energy changes is essential while learning about metabolic pathways of living organisms. The spontaneity of these biochemical processes relies mainly on the changes in Gibbs free energy, which differs according to the concentrations of reactant and product in biological systems. From proteins and enzymes, the biochemical pathways have to efficiently handle energy conversions in order to keep functioning at their best. Students, as cited by Lawrie et al. (2014), have to shift from a basic understanding of Gibbs free energy to be able to understand its use in dynamic systems. This shift highlights the need to incorporate fundamental thermodynamic concepts with real-world biochemical applicability, thereby improving students' understanding of the role of Gibbs free energy in spontaneity and equilibrium. Besides, while Gibbs free energy is a fundamental concept that controls the spontaneity of a reaction, it is necessary to consider the conditions of the systems, for instance, temperature and pressure. The equilibrium constant (K) of a reaction, which is defined by Gibbs free energy, is the relative ratio of reactants and products when the system reaches the state of equilibrium. If ΔG is equal to zero, the system is in equilibrium, and the concentrations of reactants and products do not change in a net manner. Since Gibbs free energy encompasses both thermodynamic and kinetic considerations, students and researchers need to consider these variables to gain insight into the predictability of reaction spontaneity in different conditions. To this end, outcomes of research on chemical systems like $CH_3NH_3PbX_3$ highlight the mutually dependent nature of enthalpic and entropic components (Ciccioli et al., 2018). Such interdependence thereby establishes the significance of Gibbs free energy in defining the ground of chemical equilibrium and spontaneity in different areas.

7.3.2. Calculation of Gibbs Free Energy Changes in Reactions

Gibbs Free Energy calculation of chemical reaction changes provides crucial information about the spontaneity and equilibrium of reactions. Gibbs Free Energy (G) is $G = H - TS$, with H representing enthalpy, T representing the temperature, and S representing entropy, and it is the expression used for the balance between energy and disorder in a system. This relationship is the bedrock of thermodynamics, describing that reactions will proceed in a direction that reduces the systems Gibbs Free Energy. When ΔG, the change in Gibbs Free Energy, is negative, then the reaction is spontaneous; a positive ΔG shows non-spontaneity. This principle features in many fields, from biochemical pathways to industrial processes, very practically governing reaction conditions and predicting outcomes on the basis of energetic favorability. Even the determination of ΔG can be influenced to a great degree by expressing the equilibrium constant. The Gibbs Free Energy change (ΔG) at equilibrium is equal to zero, and from it the equation $\Delta G = \Delta G° + RT \ln Q$ is used to establish the relationship between ΔG and the equilibrium constant (K). $\Delta G°$ is the standard Gibbs Free Energy change, R is the gas constant, and Q is the reaction quotient. This relationship stresses that whenever reactants and product concentration changes, the Gibbs Free Energy also changes, ultimately driving the system to equilibrium. The calculation of changes in Gibbs Free Energy, therefore, not only helps to outline the thermodynamic favorability of the reaction but also elucidates how the external conditions can shift the position of equilibrium, guiding theoretical as well as practical applications in chemistry and allied fields. In real-life applications, advanced computational tools enable the calculation of Gibbs Free Energy change, enabling us to know more about complex reactions. For instance, integrative models that incorporate kinetic factors and thermodynamic potentials enable the prediction of reaction routes under varying circumstances. Particularly, uses of Gibbs free energy minimization procedures have been useful in simulating reaction behavior across numerous various environments, including advancements in current reacting flow computational code, such as SPARK. Current research asserts that these advanced calculations deliver significant improvements in the determination of reaction rates and the manner species behave in systems, thereby relating changes in Gibbs Free Energy with observed reactions (Carpenter et al., 2025). Along with this, models of metal behavior in silicate minerals highlight the importance of thermodynamic modeling to derive predictions of iron oxidation state changes under specific conditions, thus further highlighting the relevance of Gibbs Free Energy for real geochemical conditions (Morgan et al., 2015) (Figure 7.9).

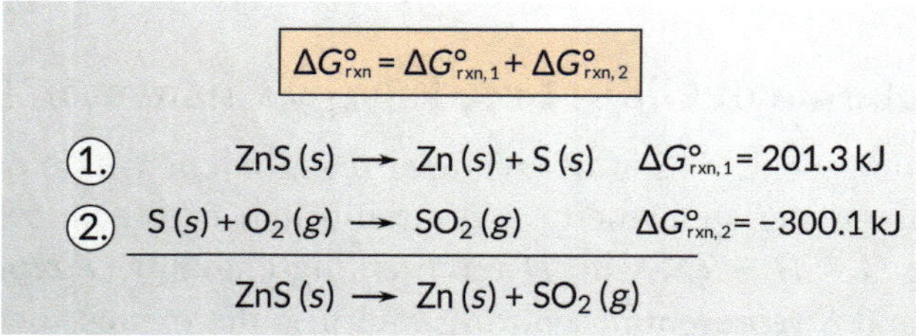

Figure 7.9. *Calculation of Gibbs free energy changes in reactions.*

Source: https://www.jove.com/files/media/science-education/science-education-thumbs/11429.jpg.

7.3.3. Examples of Reactions with Negative and Positive Gibbs Free Energy

The Gibbs Free Energy (*G*) principle is key in ascertaining the spontaneity of chemical reactions. A reaction with a negative change in Gibbs Free Energy ($\Delta G < 0$) is a reaction where the process is exergonic and will proceed spontaneously in constant pressure and temperature conditions, releasing energy to the environment. A good example of such a reaction is the glycolysis process of glucose conversion to pyruvate. This metabolic process not only provides essential energy in the form of ATP but also plays a crucial part in cellular respiration. These reactions are thermodynamically favorable as they go in the direction of maximum overall entropy of the system, resulting in the smooth operation of cellular processes and utilization of energy by living organisms. They are significant to know in fields like biochemistry, where the balance of energy is required for life to go on. Reactions with a positive Gibbs Free Energy change ($\Delta G > 0$) are also endergonic, and they require an input of energy in order to proceed. Photosynthesis is a good example of an endergonic reaction wherein carbon dioxide and water are synthesized from glucose and oxygen with the help of sunlight as the energy source. Here, the high ΔG is a reflection of the increase in chemical potential energy of glucose molecules, which is essential for the storage of energy in the plant. Spontaneity of photosynthesis does not apply and requires special conditions to occur, reflecting the delicate balance between energy consumption and conversion in biological systems. These endergonic processes are not only fundamental to biology but also to bioengineering and ecology, since manipulation of metabolic pathways can lead to sustainable energy technologies and improved agriculture practices. The relationship between negative and positive Gibbs Free Energy changes illustrates the sensitivity of metabolic networks. For instance, unfavorable thermodynamically reactions ($\Delta G > 0$) can be coupled with favorable reactions ($\Delta G < 0$) to drive biochemical reactions. Control of such reactions is necessary in homeostasis within an organism. For instance, in cellular respiration, the energy released via oxidation of NADH is utilized for ADP phosphorylation into ATP, coupling exergonic and endergonic reactions. In addition, the intricacies of these interactions can also be modeled in terms of kinetic schemes, considering thermodynamic effects and reaction elasticities derived from Gibbs free energies. This approach to modeling is crucial for understanding the dynamics of metabolic pathways and has the potential to play an important role in biochemical research and applications, as highlighted by studies that make use of such schemes to formulate thermodynamically balanced models of major metabolic transporters (Crampin et al., 2018; Liebermeister et al., 2013).

7.4. EQUILIBRIUM IN CHEMICAL SYSTEMS

Equilibrium in chemical systems is the key to comprehending how reactions happen and how they achieve a balanced state. The forward and reverse reaction rates are the same at equilibrium, leading to unchanging concentrations of products and reactants over time. This state indicates not just thermodynamic stability but also chemical interaction dynamism. The Gibbs Free Energy (G) plays a crucial role in determining the direction and magnitude of a reaction because it is both a system spontaneity measure and an equilibrium position indicator. A negative change in Gibbs Free Energy (ΔG) indicates a reaction is favorable toward product formation, while a positive value indicates reactants are favored. A knowledge of the principles is essential in order to predict reaction behavior under varying conditions, and therefore equilibrium provides the cornerstone of theoretical as well as applied chemistry. Further, the importance of solid-liquid equilibrium in chemical systems can adequately be depicted through the study of triglyceride mixtures, as documented in recent literature. Equilibrium between the solid crystalline network and liquid has a great influence on the characteristics of vegetable oils and their derivatives. Research shows that design of solid-liquid equilibrium in triacylglycerol systems can be combined with Computer-Aided Mixture and Blend Design techniques to create new mixtures with improved performance and stability. These advancements not only expand the use of renewable resources like vegetable oils but also provide critical information about the thermodynamic properties of these mixtures under varying temperatures and compositions. This integration of equilibrium-related concepts with practical applications emphasizes the significance of understanding the governing thermodynamic principles of such systems and, consequently, aiding innovation across domains such as food science and materials engineering (Roux et al., 2009). In discussing Gibbs Free Energy in relation to equilibrium, it is necessary to understand its association with entropy and thermodynamic work. Entropy, a measure of randomness or disorder, is directly proportional to the capacity of a system to perform work. The relationship between free energy and entropy is most relevant in the biological reactions, where energy transfer and conservation occurs through coupled reactions involving freely diffusible intermediates. Free energy is specifically a Second-Law energy and plays a key role to allow thermodynamic processes at finite speeds. This strategy defines how free energy and concentration dynamics compete in chemical systems and define the state of equilibrium. Further, understanding these thermodynamic interactions can inform the energy conversion and reaction routes, hence achieving an overall understanding of the rules guiding chemical equilibria both in nature and industrial processes (Barnes et al., 2000) (Figure 7.10).

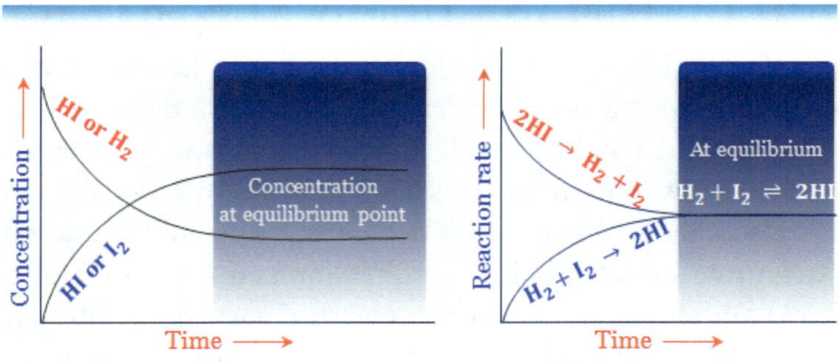

Figure 7.10. *Equilibrium in chemical systems.*

Source: https://www.priyamstudycentre.com/wp-content/uploads/2019/03/chemical-equilibrium-reaction.png.

7.4.1. Definition of Chemical Equilibrium

The law of chemical equilibrium is an intrinsic subject in the disciplines of chemistry and thermodynamics and defines a situation where the rate of the forward reaction is equal to the reverse reaction. This is a state at which the reaction rates of reactant and product are not changing over time and achieves a dynamic equilibrium in a closed system. This state of equilibrium can be quantitatively described by the equilibrium constant, which is a ratio of the product and reactant concentrations at equilibrium. Another comprehension of this equilibrium state can be derived from the concepts as outlined in thermodynamics, particularly those involving Gibbs free energy. In a chemical reaction, the system moves towards equilibrium as it becomes a free energy-minimizing state, and this points to the importance of energy considerations in determining stability and position of equilibrium in a given reaction system (Heng et al., 2015). Gibbs free energy is a key factor in understanding not only the conditions under which chemical equilibrium is attained but also the spontaneity of the reactions. When the Gibbs free energy change (ΔG) of a reaction is negative, the reaction will spontaneously proceed towards products to produce equilibrium. Conversely, a positive ΔG indicates that the reaction is not spontaneous, driving it in the reverse direction. Gibbs free energy and chemical equilibrium are related through the way reactions shift according to changes in concentration, temperature, and pressure as determined by Le Châteliers Principle. This principle outlines that if a system at equilibrium is disturbed, it will re-equilibrate in a way to oppose the perturbation, hence re-establishing that chemical equilibrium is both a static and dynamic state of matter (Adamatzky et al., 2019). Furthermore, chemical equilibrium may be characterized in terms of Gibbs free energy to describe more general application across non-standard chemical kinetics. For instance, in biological systems, the balance that is offered by chemical equilibria is crucial for metabolic pathways and for cell signaling. Thermodynamic features of biological reactions that tend to run near equilibrium enable cells to save energy resources. The concept is extended to the study of complex dissipative structures, e.g., chemical waves in reaction-diffusion systems, where equilibrium problems are intertwined with nonequilibrium thermodynamics. This integration of concepts highlights the relevance of Gibbs free energy in the prediction of the behavior of reactions and the stability of the system across various scientific disciplines (Adamatzky et al., 2019).

7.4.2. Dynamic Nature of Equilibrium

Dynamic nature of equilibrium is a fundamental principle to know in the study of chemical systems, especially when applied to Gibbs Free Energy.

Concentrations of reactants and products do not shift with time in equilibrium, yet it is not a static condition; rather, it is a dynamic balance between forward and reverse reactions. Dynamic equilibrium points out the fact that although macroscopic changes are halted, molecular interactions are still occurring. Gibbs Free Energy equation serves as an important player in this case by signaling directions, spontaneity, and feasibility of reactions. With systems at equilibrium, Gibbs Free Energy is at a minimum, a sign that indicates equilibrium within the distribution of energy over the course of the reaction. It is a principle that can be observed throughout systems because the role of free energy bears witness to the fact that even at equilibrium, kinetic activity remains the defining state of the system, a fact that comfortably lies within Ge et al. (2016). In addition, the dynamic equilibrium is not specific to chemical reactions but extends to physical reactions as well, thus bearing witness to the cross-domain applicability of Gibbs Free Energy. Membrane transporters, for example, that regulate cellular environments bear witness to how biological processes are managed according to thermodynamic principles. These transporters, who conduct homeostasis by keeping solute concentrations in equilibrium, operate according to Gibbs Free Energy gradients. Since they move things from one side of a membrane to another, it is the concurrent movement of ions and the creation of gradients that guarantee that equilibrium is a temporary position where redistribution continuously occurs instead of an endpoint. This complex dynamic underscores the importance of models that are thermodynamically valid, as seen in the research on transporters that can be unrealistically presented if modeled poorly. Careful knowledge of such systems enhances our understanding of equilibrium and demonstrates how dynamic processes shape thermodynamic reactions, as established in research such as those in Crampin et al. (2018). Generally speaking, the dynamic nature of equilibrium, based on Gibbs Free Energy theory, makes classical ideas about stability in chemical and biological systems difficult to approach. Rather than equating equilibrium as a static interval, it would be more apt to view it as a complex ballet of reactions and adjustments constantly being modified by modifications in energy. The emergence of macroscopic laws derived from microscopic interactions proves the dynamical character and suggests that equilibrium is in a state of fluctuation under forces activated by external conditions such as pressure, temperature, and concentration. As systems evolve, reaction rates evolve correspondingly, pointing to the dynamic nature of equilibrium states as systems respond to changing conditions. Understanding these dynamics not only contributes to the theory of Gibbs Free Energy but also provides us with significant insight into how systems will act under varying thermodynamic conditions, paving the way to further theoretical chemistry advancements and real-world applications in fields like biochemistry and materials science (Figure 7.11).

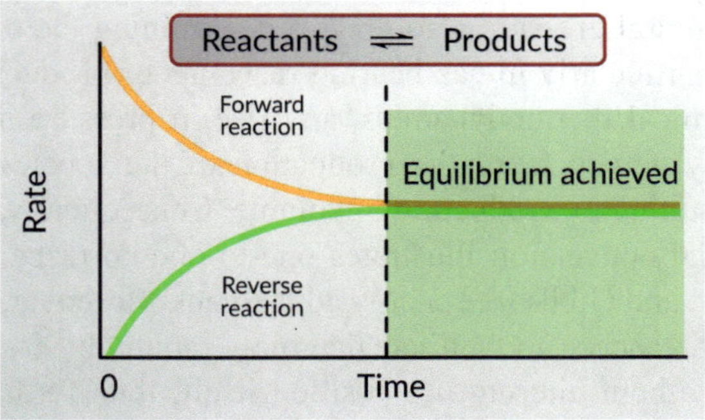

Figure 7.11. *Dynamic equilibrium.*

Source: https://www.jove.com/files/media/science-education/science-education-thumbs/11385.jpg.

7.4.3. Factors Affecting Equilibrium Position

Equilibrium position of a chemical reaction is heavily regulated by many intrinsic and extrinsic factors, which have the ability to alter the topography of Gibbs free energy. Concentration of reactants and products in the system is one of the key factors. According to Le Chatelier's Principle, when there is a change in the concentration of either reactants or products, the system adapts to counteract this change by shifting the equilibrium position. An increase in concentration of reactants tends to drive the equilibrium towards the production of products at a loss of free energy. Conversely, if product concentration is increased, the equilibrium will move to the reactant side, depicting the dynamic nature of chemical equilibria because of changes in concentration. This equilibrium is crucial to help us understand how reagents behave in a closed system and thus constitutes the basis of thermodynamic equilibrium and stability principles. Temperature is also a critical factor that affects the position of equilibrium of a reaction since it has a direct effect on the Gibbs free energy. For exothermic reactions, an increase in temperature causes the equilibrium to shift in the direction of the reactants because the system prefers to absorb the additional heat provided to it and hence promotes a reverse process. For endothermic reactions, a temperature increase will tend to favor product formation, which attests to the fact that equilibrium is temperature-sensitive. Temperature dependence of Gibbs free energy is described in the Gibbs-Helmholtz equation, where it is described how the changes in relation to temperature can lead to the shift in the state of equilibrium. These changes occur as the system tries to come into a configuration with the lowest free energy, which reflects the interplay between thermal energy and chemical potential in the dynamic framework of reaction equilibria. Extramural pressure also plays a determining factor in the position of equilibrium, particularly in gas-bearing reactants or products. According to the laws of chemical thermodynamics, any rise in pressure in these systems will cause the system to favor the production of the species that has fewer moles of gas, resulting in a reduction in volume. Consequently, this pressure to chemical potential conversion illustrates one-to-one correspondence between change in volume and Gibbs free energy adjustment. Moreover, external factors such as catalyst presence can facilitate that more rapidly by easier achievement of equilibrium without altering the position itself, thus further verifying the concept that even equilibria are displaced by the external factors but not their intrinsic outcome of minimizing free energy in a stable system. It is essential to comprehend the role that these factors play in order to predict and manage chemical processes in industrial context and theoretical studies (Antonov et al., 2009; Roux et al., 2011).

7.5. THE ROLE OF GIBBS FREE ENERGY IN EQUILIBRIUM

It is essential to comprehend the idea of Gibbs free energy in order to understand how chemical systems achieve equilibrium. Gibbs free energy (G) is a thermodynamic potential which can provide information about the spontaneity of process at constant pressure and temperature. The Gibbs free energy change (ΔG) becomes the point of reference when a reaction is approaching equilibrium. When ΔG is negative, the reaction will be spontaneous and will favor product formation. When ΔG is positive, it suggests a non-spontaneous process and reflects reactants under given conditions as favored. Thus, the relationship between Gibbs free energy and equilibrium is fundamental in that it defines in which direction a reaction will occur to attain a balanced state between reactants and products in a system. The system will have its lowest Gibbs free energy when it is in equilibrium, implying the system is in the most stable state. This can be exemplified through the example of membrane transporters, thereby hinting at the integration of thermodynamics and cellular processes. For instance, Crampin et al. (2018) describes how membrane transporters function under the constraints of thermodynamics, shaping cellular environments via the movement of substrates. The models that were created for these transport processes are thermodynamically consistent, something that validates the concept that Gibbs free energy is not just a theoretical construct but an instrumental tool as well in predicting real biological behaviors. With the use of Gibbs free energy in this regard, one can appreciate how cells can maintain themselves in homeostasis and efficiently utilize energy via transport mechanisms such that dynamic equilibrium is attained. Gibbs free energy is also essential in phase equilibria, especially in complex mixtures such as lipids. Phase behavior prediction is of key importance in lipid chemistry for application in food chemistry and pharmaceuticals. Experimental work presented in Roux et al. (2011) shows that Margules 2-suffix excess Gibbs energy model is successful in modeling the non-ideal behavior of lipid mixtures and simulating solid-liquid phase equilibria with high accuracy. By optimizing Gibbs free energy to establish phase compositions with temperatures, researchers can estimate melting profiles and other critical properties needed by end-users. The ability to predict these shows the broader applications of Gibbs free energy in experimental and industrial applications, which are the basis of its role in providing the means to understand equilibrium in intricate systems, whether biological or chemical in origin (Figure 7.12).

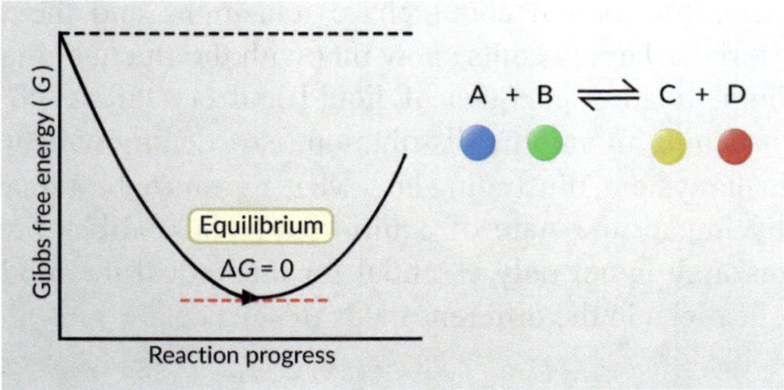

Figure 7.12. *The role of Gibbs free energy in equilibrium.*

Source: https://www.jove.com/files/media/science-education/science-education-thumbs/12160.jpg.

7.5.1. Gibbs Free Energy Change at Equilibrium

The concept of Gibbs Free Energy (G) is critical to understand the state of equilibrium of a system. When the system is in equilibrium, the Gibbs Free Energy change, ΔG, is zero, i.e., the system is in a state of minimum free energy, where the forward and reverse reactions occur at the same rate. This is crucial in thermodynamics as it implies that there is no net change within the system. The law is founded on the observation that reactions will seek to move in a direction that minimizes the free energy. In the assessment of chemical processes, identification of conditions under which ΔG is zero is crucial in determining spontaneity of reaction and equilibrium stability. For mixtures, such as ideal gases, change in Gibbs Free Energy at equilibrium is governed by various factors of thermodynamics. Action principle can provide insight into complex systems, as shown in current studies of mixtures of ideal gases where traditional methods may not be sufficient (Fronsdal et al., 2010). The application of the Gibbs-Dalton hypothesis extrapolates gas mixture behavior, leading to more comprehensive conclusions about the entropy and phase behaviors of such a system. As an example, when the concentrations of different components change, the Gibbs Free Energy will also change, indicating changes in potential energy attached to the system. This then illustrates how molecular interactions and external conditions can affect the system toward equilibrium. In addition, recent experimental studies elucidate the variation of Gibbs Free Energy near equilibrium conditions, particularly for high-density liquid phases like the supercooled phase of silicon (Gautam et al., 2012). By focusing on the excess Gibbs free energy (G^e) against different temperatures and phases, researchers unveil important information about phase transitions and the variations of entropy in materials. These results show that with the fluctuating temperatures near equilibrium, relative energies of liquid and crystalline phases change, leading to broadening in volume distribution. Broadening indicates a delicate balance within the system, illustrating how what appears to be a small fluctuation will drastically impact the state of equilibrium. Thus, Gibbs Free Energy at equilibrium research is not only essential for theoretical thermodynamics but also in its applications in the different fields of science.

7.5.2. Relationship Between Gibbs Free Energy and the Equilibrium Constant

The relationship between Gibbs Free Energy (G) and the equilibrium constant (K) is fundamental to the research on chemical reactions as well as the spontaneity of chemical reactions. When a system is at equilibrium, the change in Gibbs Free Energy (ΔG) is zero, and that would mean that the system has a minimum energy level. The equilibrium constant, K, is used to measure the ratio of product to reactant concentration at this energy level. According to the thermodynamic principles, this is represented by the equation $\Delta G° = -RT \ln(K)$, where R is the gas constant and T is the temperature in Kelvin. The above equation indicates that a negative change in Gibbs Free Energy is an indicator of a favorable reaction with a higher K which denotes a product-forming ability. Conversely, a positive ΔG denotes an unfavorable reaction and decreased K, emphasizing the thermodynamic stability of equilibrated reactants. This correlation extends into the applied uses of biochemical and material science fields. As an example, the Gibbs Free Energy can predict the direction of biochemical reactions, influencing metabolic pathways and cellular energy generation. This is of critical significance for membrane transporters, which play significant roles in cellular homeostasis. The majority of mathematical models of such transporters are likely to be unsuccessful in accommodating thermodynamic consistency, and therefore they become inaccurate in describing cellular behavior. A good framework of Gibbs Free Energy and equilibrium constants is important in the development of detailed models that adequately represent the thermodynamic constraints of these systems, hence predicting better cell regulation and energy interconversions within biological situations (Crampin et al., 2018). Equilibrium, nevertheless, is not limited to individual reactions but also to complex systems such as mixtures of ideal gases. In such a scenario, Gibbs Free Energy enables easier comprehension of thermodynamic characteristics. The Gibbs-Dalton hypothesis that can be used in such mixtures is that the total pressure of a gas mixture is equal to the sum of the partial pressures of all constituent components. When dealing with such mixtures in research, significant focus is on the change in entropy as a result of component interactions, especially under variable conditions, i.e., where gravitational forces are involved. These complicated scenarios may be addressed adequately with the application of thermodynamic modeling, leading to observations that connect classical thermodynamic theory to modern statistical method. This integrated approach can improve our knowledge concerning ideal and actual gases, further clarifying the interactions between Gibbs Free Energy and equilibrium (Fronsdal et al., 2010) (Figure 7.13).

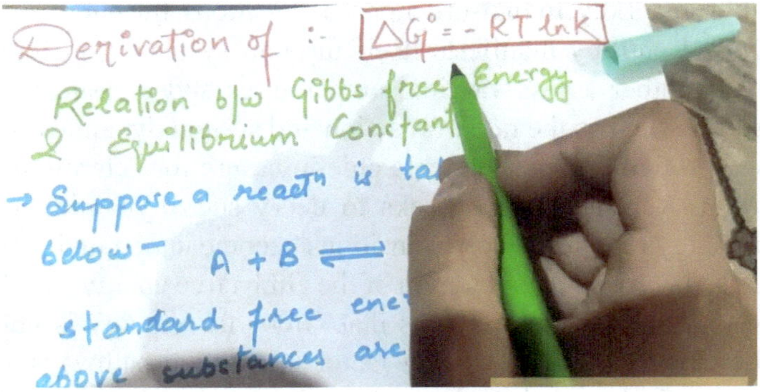

Figure 7.13. *Relationship between Gibbs free energy and the equilibrium constant.*

Source: https://i.ytimg.com/vi/h96Pi2of_Zo/maxresdefault.jpg.

7.5.3. Implications of Gibbs Free Energy on Reaction Direction

Gibbs free energy is perhaps the most important term in thermodynamics, particularly when looking at reaction direction and spontaneity of chemical reaction. Any reaction will proceed in the direction that minimizes the Gibbs free energy, causing it to move towards equilibrium and gain stability within the system. If ΔG, the Gibbs free energy change, is negative, then the free energy of the products is lower than that of the reactants, indicating the reaction to be spontaneous. If ΔG is positive, then energy must be supplied to make the reaction occur. Therefore, understanding these energy conversions in chemical systems is a necessity since it not only establishes reaction feasibility but also provides information on yield and the equilibrium concentration, therefore affecting some industrial and biological applications. The consequences of Gibbs free energy extend beyond the simplicity of a reaction's spontaneity; they also highlight the relationship between the reaction's direction and entropy, particularly in biological settings. Mechanisms in biological systems, such as membrane transport processes, illustrate how Gibbs free energy is intertwined with entropic considerations to control cell processes. For example, membrane transporters' effectiveness is governed according to thermodynamic models in which energy and entropy must be accounted for in reaction models. Awareness of this in mind, model inaccuracies are possible when these processes are modeled non-thermodynamically consistently, and therefore unrealistic outcomes follow. These contradictions accentuate the need for sophisticated models that utilize Gibbs free energy in a proper manner to predict the action of transporters in biochemistry (Crampin et al., 2018). Moreover, Gibbs free energy implications in reactions can also be observed during phase transitions and other state changes where equilibrium is key. In first-order phase transitions, the interaction of Gibbs free energy and entropy manifests itself through systems transforming from one phase to another across varying temperatures and pressures. With these changes, aspects such as the rate of reaction and system dynamics comprehend the complexity of Gibbs free energy, a chief measure for determining the state of equilibrium. Recent research seeks to decry such dynamics, particularly under high-energy regimes, whose transitions become uncertain if conventional models are used without consideration of the Gibbs free energy paradigm. Such research in phase transitions justifies that Gibbs free energy is not merely a theoretical construct but a significant factor in reaction pathways and system behavior in chemical as well as physical conditions (Brachmann et al., 2000).

7.6. APPLICATIONS OF GIBBS FREE ENERGY AND EQUILIBRIUM

Gibbs free energy is at the heart of thermodynamic system application, particularly in the prediction of the spontaneity and feasibility of chemical reactions. By the application of Gibbs free energy change (ΔG), scientists can determine whether a reaction is spontaneous under certain conditions or not. A negative ΔG means that a reaction can spontaneously proceed, thus paving the way for optimizing and designing industrial processes. Various computational methods, such as the open-source Thermochemical Equilibrium Abundances (TEA) code, utilize the Gibbs free energy minimization application in calculating molecular species abundances in various systems. These calculations allow researchers to deliver high precision in the simulation of complex chemical states, enhance predictions for procedures such as those occurring in hot-Jupiter exoplanets, and ultimately elevate our understanding of planetary atmospheres and their chemical composition (Blecic et al., 2015). Furthermore, the applicability of Gibbs free energy extends beyond theoretical prediction; it also has implications for the interpretation of physical phenomena such as phase transformations and material properties. For instance, in the case of surface science, considerations from Gibbs free energy can be used to account for liquid behavior on ordered surfaces. Experiments have shown that various geometrical configurations of pillars influence liquid drop contact angles, which give insight into adhesive properties of significance in materials science. The relationship between drop pressure and apparent contact angle shows how Gibbs free energy principles can be employed to predict and manage surface interactions like impalement transitions and force equilibrium for drops (Bartolo et al., 2009). This interaction not only adds to fundamental research but extends with relevance to everyday life in areas like coating technologies and microfluidics. Finally, utilization of Gibbs free energy in states of equilibrium indicates its fundamental character in chemical as well as physical systems. Under equilibrium conditions, the Gibbs free energy in a system reaches a minimum, i.e., the quantity of reactants and products remains constant over time. This concept is central to the understanding of chemical equilibria and how one can shift reactions in a favorable direction through Le Chatelier's principle under certain conditions. Chemists are able to maximize yield and selectivity in synthesis processes by manipulating variables such as concentration, temperature, and pressure based on calculations using Gibbs free energy. Hence, the applications of Gibbs free energy and equilibrium principles are at the core of seeking both theoretical information and applied information across different scientific disciplines and propelling technologies toward improving energy, technology, and sustainability endeavors.

7.6.1. Industrial Applications in Chemical Manufacturing

The intricate relationship between Gibbs Free Energy and equilibrium is at the core of most industrial applications within chemical manufacturing, particularly optimizing efficiency in energy consumption. Phase change heat transfer control remains significant, impacting various industrial operations such as crystallization and distillation. Extensive heat transfer is vital during such processes but results in thermodynamic inefficiencies when it is experienced at low heat flux levels. Improvements in surface fabrication technology via micro- and nanostructures have emerged in an effort to enhance mechanisms of phase change heat transfer. These engineered surfaces have the potential to play an immense part in multiphase processes, allowing for improved thermal conductivity and efficiency in petrochemical to pharmaceutical applications. This technology not only allows for current manufacturing requirements, but it also opens up routes for new chemical reactions that were not possible due to inefficient heat transfer, therefore adhering to sustainable energy practices, as discussed in Attinger et al. (2014). In chemical manufacture, the ability to control reaction dynamics by new means confers significant advantages in terms of Gibbs Free Energy. Eutectic-form reactions, involving greater than one phase, introduce genuine significance in the creation and utilization of alloy material. For instance, four-dimensional electron microscopy enables researchers to view phase reactions with magnificent spatial resolution and temporal resolution. Through direct observation of nucleation and phase growth within a controlled medium, engineers can extrapolate influential thermodynamic parameters required for uses such as alloy development and nanomaterial development. Not only does this information enhance the efficiency of manufacturing processes but it also ensures reduced material wastage and energy consumption, highlighting the invaluable contribution of Gibbs Free Energy principles in making industrial processes more efficient, as indicated by Chen et al. (2017). Furthermore, with the emergence of advanced computational models merged with real-time data analysis, there is greater insight into the thermodynamic principle of chemical manufacture. Such technology allows producers to predict and maximize under which conditions reactions achieve equilibrium and thus maximize yield while minimizing energy consumption. Strategic applications of Gibbs Free Energy measures have the power to deliver revolutionary change in process design. For example, by better manipulating temperature and pressure conditions, manufacturers can position the equilibrium point to maximize product yield. Ultimately, it is through these thermodynamic concepts that industries are able to innovate within their sustainability limits, encouraging a more responsible chemical manufacturing process. This marriage of theory and application not only benefits manufacturers but also informs the larger effort of environmental responsibility in the chemical industry.

7.6.2. Biological Significance in Metabolic Processes

Gibbs free energy's relationship with metabolic processes has a critical bearing on the understanding of biological significance at a cellular level. Metabolic processes are governed by changes in Gibbs free energy, thereby deciding whether or not reactions take place spontaneously under specific conditions. The thermodynamic concept of Gibbs free energy enables one to predict the direction of a metabolic pathway, thus resulting in the generation, storage, and utilization of energy within an organism. Furthermore, the optimization and regulation of such metabolic processes allow for dynamic modulation under shifting internal and external environmental conditions. This adaptability is essential in maintaining homeostasis because organisms need to efficiently manage energy resources to maintain growth, repair, and reproduction. By studying these processes, researchers can advance knowledge on cellular health and the mechanisms behind different metabolic disorders. In metabolic pathways, one cannot overstate the significance of control and regulation of enzymatic reactions. Metabolic pathways consist of coupled biochemical reactions that enable substrates to be interconverted to products by enzymes. Such reactions' elasticities, which quantify how responsive they are to changes in fluxes and concentrations, can be determined via Gibbs free energy. Studies show that their modellability and description are able to offer a deeper insight into metabolic control systems, altering the predictions made relative to metabolic behavior. For instance, through the construction of kinetic models with the inclusion of allosteric regulation and metabolite levels, scientists are able to create probabilistic predictions of metabolic network dynamics. This not only supports mechanistic understanding of cellular metabolism but also enables the assessment of potential effects from enzyme inhibitors, changes in gene expression, and genetic variations and hence highlights the stability of Gibbs free energy within biological systems (Liebermeister et al., 2013). In addition, the thermodynamic forces behind metabolic processes also disclose the second law of thermodynamics constraints offered by the biochemical conversion energy requirement. Analyzing chemical thermodynamic stability, it is clear that biological molecules are able to experience conversions in consonance with the energy flow dictated by Gibbs free energy. Such an understanding extends beyond cellular respiration into broader implications in our perspective of natural processes, such as the genesis of hydrocarbons from biological materials. Researchers illustrate how, under specific conditions, such as pressure and heat, biological matter becomes more stable hydrocarbons, highlighting interdependence of life and geologic processes that characterize our planet. This association underlines the significance of thermodynamic theory in biological processes, verifying the intrinsic role of Gibbs free energy and equilibrium in metabolic reactions that sustain life (Alekseev et al., 2002).

7.6.3. Environmental Implications in Reaction Dynamics

It is essential to realize the implications in reaction dynamics under the environment when looking at the principles of Gibbs free energy and equilibrium. Most chemical reactions, particularly those in supercooled liquids, show how the environmental conditions influence reaction mechanisms and product creation. For instance, in the supercooled state, the increased dynamical length scale {\\\\xi}_{\\\\alpha} is at the core of crystallization process evolution. With the decrease in temperature, {\\\\xi}_{\\\\alpha} can become larger than the crystal nucleus size, {\\\\xi}_M, allowing novel crystallization processes that lead to intricate crystalline morphology. These dynamics not only provide an explanation for the supercooled liquid stability but also make clear the fine-grained environmental factors that affect material properties, which might otherwise remain unexplored in classical reaction dynamics research (Anderson et al., 2010). It is also important to consider feedback from reaction dynamics to environmental health, particularly in atmospheric chemistry. Processes occurring at aqueous droplet and film surfaces are significant in influencing air quality and climate. Molecular dynamics simulations are helpful tools in the research of such processes on an atomic level, particularly the behavior of ions at aqueous interfaces. These interactions play significant roles in atmospheric aerosols since they decide cloud formation, radiative forcing, and climate in general. The facts that are gained from these simulations enhance our understanding of the ways that reaction dynamics can lead to environmental change and how highlighting these reactions within the general discipline of climate science and environmental sustainability is imperative (Baer et al., 2013). Finally, environmental implications are intricate in reaction dynamics and also extend to implications for ecological impact prediction and ecological impact management. The relationship between kinetic variables and thermodynamic laws, such as Gibbs free energy, reveals the behavior of reactants and pollutants in various environments. Such information acquired through research on supercooled aerosols and liquids in the environment shows unprecedented routes and chemical reactions that result in environmental harm. Because new models and theories emerge out of ongoing research, incorporating these chemistry information can provide added predictive capacities, ultimately channeled into policy formulation for limiting adverse environmental impacts. An understanding of how equilibrium adapts to environmental pressures is important in the creation of effective response to evolving ecological threats, hence bridging theoretical concepts of chemistry with good environmental stewardship.

SUMMARY

The Gibbs Free Energy and how it relates to chemical equilibrium provides profound insight into thermodynamic processes. Gibbs Free Energy is a central parameter in the decision-making process regarding the spontaneity of a reaction and the conditions under which a system will achieve equilibrium.

Through the spontaneity of a reaction, when a chemical reaction takes place, the change in Gibbs Free Energy (ΔG) informs us if the process is thermodynamically favorable. A negative ΔG indicates that the reaction is spontaneous, and a positive ΔG indicates that it will need more energy inputs. Having this general principle as a guide, chemists and physicists can predict what happens in reactions and tailor them so that some products are preferred. Thus, the concept of Gibbs Free Energy is not a theoretical abomination but a useful tool in experimental and applied chemistry. The conversation is further extended when implications of Gibbs Free Energy are considered concerning complicated systems and mixtures. As a point case, the kinetics of mixtures of ideal gases demonstrates the intricate interaction between different constituents in achieving thermodynamic equilibrium. The Gibbs-Dalton postulate makes it simple to grasp how such mixtures act, particularly under varying conditions of pressure and temperature. This approach is paradigmatic in that it projects itself based on the principles of the ideal gas laws taking into account the deviations observed in real systems. Such types of explorations, particularly those relating to gravitational influences, underscore the applicability of Gibbs Free Energy principles and their application in various environments, whether it be classical thermodynamics or modern theoretical explorations such as those relating to quantum mechanics, as intimated by the work of Antonov et al. (2009). Furthermore, the broader implications of learning Gibbs Free Energy become apparent in new computational models and methodologies. New studies reveal how the extrapolation of traditional theories can result in new equilibrium and transitional state knowledge for thermodynamic systems. Novel methodologies such as Monte Carlo simulations have been made available, revealing how thermodynamic theories continue to advance and have real-world utility. The capacity to analyze systems in which negative heat capacities are described is representative of a high level of understanding about thermodynamic fluctuations and their influences on chemical reactions and physical states. By integrating principles pertaining to Gibbs Free Energy with novel methodology, researchers are able to analyze sophisticated equilibrium conditions, providing new avenues for innovation in theoretical as well as applied thermodynamics, as observed in the analyses presented in Fronsdal et al. (2010).

The Gibbs Free Energy concepts form the heart of chemical equilibrium and the energy feasibility of reactions. Gibbs Free Energy (G) is the maximum reversible work that can be accomplished by a thermodynamic system at constant temperature and pressure. As one observes a system in equilibrium, the Gibbs Free Energy change (ΔG) is zero, as it signifies that the rates of the forward and reverse reactions are in balance. This equilibrium highlights the fact that not only does Gibbs Free Energy determine the direction in which a reaction proceeds, it is also an important determinant of spontaneity. A negative

ΔG signifies that a process is spontaneous, while a positive ΔG signifies non-spontaneity, which is vital in the prediction of chemical system behavior under various conditions. Another vital subject discussed concerning Gibbs Free Energy is its use in mixtures and their nature under non-ideal conditions. Systems composed of mixtures of ideal gases, such as in discussion under the Gibbs-Dalton hypothesis, emphasize the complexity of multi-component systems. The principle of action, established for systematic examination of thermodynamic mixtures, allows more precise exploration of interactions and formation of various states, such as the dual freedom of degrees in a binary gas mixture. Further, this approach is also inclusive of real gas analysis, which is observed from the van der Waals equation, linking theoretical models and experimental data in order to advance knowledge of phase behavior in systems (Fronsdal et al., 2010). Fluid-solid equilibrium also helps show that Gibbs Free Energy can be applied to phases of matter. The Gibbs ensemble method is a perfect example of this since it is applied to simulate solid-fluid transitions, effectively calculating phases' chemical potential and finding conditions for equilibrium. The method is particularly critical in the determination of phase coexistence, which relies on accurately describing the Helmholtz free energy in the solid phase. Through the use of these thermodynamic principles, scientists can comprehend fluid-solid coexistence, revealing what occurs at the foundation of material science. Hence, the merging of Gibbs Free Energy with phase equilibrium studies not only enriches theoretical models but also becomes a stepping stone towards more advancements in applied areas, such as the simulation of bulk materials under various conditions (Quirke et al., 2004).

REFERENCES

1. Adamatzky, A., Cover, T. M., de Groot, S. R., Fermi, E., Glansdorff, P., et al. (2019). Thermodynamics of chemical waves. https://core.ac.uk/download/286378792.pdf (accessed on 4th October 2025).
2. Alekseev, Vladimir A., Bendeliani, Nikolai A., Kenney, J. F., et al. (2002). The evolution of multicomponent systems at high pressures: VI. The thermodynamic stability of the hydrogen-carbon system: The genesis of hydrocarbons and the origin of petroleum. http://arxiv.org/abs/physics/0505003 (accessed on 4th October 2025).
3. Anderson, S. R., Berthier, L., Biroli, G., Brazovskii, S. A., Bryngelson, J. D., Capaccioli, S., Chang, S. S., et al. (2010). The ultimate fate of supercooled liquids. http://arxiv.org/abs/1006.5840 (accessed on 4th October 2025).
4. Annila, A., Annila, E., et al. (2008). Why did life emerge? https://core.ac.uk/download/186331455.pdf (accessed on 4th October 2025).
5. Antonov, V. A., Beck, C., Bohr, N., Bohr, N., Chavanis, P. H., Gross, D. H. E., Gross, D. H. E., et al. (2009). Thermodynamic fluctuation relation for temperature and energy. http://arxiv.org/abs/cond-mat/0702031 (accessed on 4th October 2025).
6. Arvind K., Angell, C. A., Bennett, C. H., Chandler, D., Limmer, D. T., Stillinger, F. H., Grochola, G., et al. (2012). Nonmonotonic dependence of the absolute entropy on temperature in supercooled Stillinger-Weber silicon. http://arxiv.org/abs/1108.3119 (accessed on 4th October 2025).
7. Atkin, A., Brillouin, L., Duan, Y., Elliott, M., Frank, R., Halliday, D., Hargittai, I., et al. (2001). The nature of the chemical process. 1. Symmetry evolution–Revised information theory, similarity principle and ugly symmetry. https://core.ac.uk/download/25312154.pdf (accessed on 4th October 2025).
8. Attinger, D., Betz, A. R., Das, A., Frankiewicz, C., et al. (2014). Surface engineering for phase change heat transfer: A review. https://core.ac.uk/download/38936179.pdf (accessed on 4th October 2025).
9. Baer, Marcel D., Levin, Yan, Mundy, Christopher J., et al. (2013). Simulation and theory of ions at atmospherically relevant aqueous liquid-air interfaces. http://arxiv.org/abs/1306.6858 (accessed on 4th October 2025).
10. Barnes, B., Barnes, M., Benedicks, M., Bent, H., Born, M., Carnot, S., Clausius, R., et al. (2000). Inquiries into the nature of free energy and entropy in respect to biochemical thermodynamics. https://core.ac.uk/

download/25283819.pdf (accessed on 4th October 2025).

11. Bartolo, D., Carbone, G., Afferrante, L., Nosonovsky, M., Nosonovsky, M., Quéré, D., & Richard, D. (2009). Microstructure superhydrorepellent surfaces: Effect of drop pressure on fakir-state stability and apparent contact angles. http://arxiv.org/abs/0911.2690 (accessed on 4th October 2025).
12. Blecic, J., Bowman, M. O., & Harrington, J. (2015). TEA: A code for calculating thermochemical equilibrium abundances. http://arxiv.org/abs/1505.06392 (accessed on 4th October 2025).
13. Bon Corbín, J., Carcel, J. A., Collazos-Escobar, G. A., Garcia-Perez, A., et al. (2023). Model-based investigation of water adsorption in achira (*Canna edulis* K.) biscuits. https://core.ac.uk/download/613661247.pdf (accessed on 4th October 2025).
14. Brachmann, J., Dumitru, A., Fraga, E. S., et al. (2000). Hydrodynamic models for heavy-ion collisions, and beyond. https://core.ac.uk/download/25299469.pdf (accessed on 4th October 2025).
15. Bruno, E., Bryk, P., Cuesta, J. A., Dietrich, S., Evans, R., Evans, R., Evans, R., et al. (2004). Integral equations for simple fluids in a general reference functional approach. http://arxiv.org/abs/cond-mat/0410185 (accessed on 4th October 2025).
16. Calabrese, S., Hueckel, T., & Porporato, A. (2018). Thermodynamic relations among isotropic material properties in conditions of plane shear stress. http://arxiv.org/abs/1811.03187 (accessed on 4th October 2025).
17. Carpenter, M. H. (2025). *A Generalized Chemistry Version of SPARK*. https://core.ac.uk/download/pdf/42830650.pdf (accessed on 4th October 2025).
18. Carrillo Le Roux, G., Gerbaud, V., & Teles dos Santos, M. (2009). Solid–liquid equilibria modelling for triacylglycerols exhibiting multiple solid phases. https://core.ac.uk/download/12041192.pdf (accessed on 4th October 2025).
19. Carrillo Le Roux, G., Gerbaud, V., & Teles dos Santos, M. (2011). Phase equilibrium and optimization tools: Application for enhanced structured lipids for foods. https://core.ac.uk/download/12042356.pdf (accessed on 4th October 2025).
20. Chen, B., Fu, X., Jagadish, C., Lysevych, M., et al. (2017). Dynamics and control of gold-encapped gallium arsenide nanowires imaged by 4D electron microscopy. https://core.ac.uk/download/84316922.pdf (accessed on 4th October 2025).

21. Ciccioli, A., & Latini, A. (2018). Thermodynamics and the intrinsic stability of lead halide perovskites $CH_3NH_3PbX_3$. https://core.ac.uk/download/159268116.pdf (accessed on 4th October 2025).
22. Crampin, E. J., Cursons, J., Gawthrop, P. J., Pan, M., et al. (2018). A thermodynamic framework for modelling membrane transporters. http://arxiv.org/abs/1806.04341 (accessed on 4th October 2025).
23. Ezechukwu, C. M. J., Igbokwe, P. K., John, M. C., John, S. U., Nwabanne, J. T., & Onyenanu, C. N. (2025). Sustainable bio-aviation fuel production from biobased oil: A novel hydrogen-free biorefinery approach. https://core.ac.uk/download/653776366.pdf (accessed on 4th October 2025).
24. Fronsdal, C. (2010). Heat and gravitation. III. Mixtures. http://arxiv.org/abs/1001.2579 (accessed on 4th October 2025).
25. Ge, H., & Qian, H. (2016). Mesoscopic kinetic basis of macroscopic chemical thermodynamics: A mathematical theory. http://arxiv.org/abs/1601.03159 (accessed on 4th October 2025).
26. Gour, G., Gu, M., Ma, J., Narasimhachar, V., et al. (2019). Quantifying memory capacity as a quantum thermodynamic resource. http://arxiv.org/abs/1806.00025 (accessed on 4th October 2025).
27. Heng, K., Lyons, J. R., & Tsai, S. M. (2015). Atmospheric chemistry for astrophysicists: A self-consistent formalism and analytical solutions for arbitrary C/O. https://core.ac.uk/download/212343442.pdf (accessed on 4th October 2025).
28. Istadi, I. (2012). Thermodynamic analysis of synthesis gas and higher hydrocarbons production from methane. https://core.ac.uk/download/11728597.pdf (accessed on 4th October 2025).
29. Lawrie, G. A., Rowland, S. L., Wolfson, A. J., Wright, A., et al. (2014). Student conceptions about energy transformations: Progression from general chemistry to biochemistry. https://core.ac.uk/download/19491859.pdf (accessed on 4th October 2025).
30. Liebermeister, W. (2013). Elasticity sampling links thermodynamics to metabolic control. http://arxiv.org/abs/1309.0267 (accessed on 4th October 2025).
31. Mann, R. B., & Mbarek, S. (2018). Reverse Hawking–Page phase transition in de Sitter black holes. http://arxiv.org/abs/1808.03349 (accessed on 4th October 2025).
32. Morgan, D., Shim, S. H., & Xu, S. (2015). Origin of Fe^{3+} in Fe-containing, Al-free mantle silicate perovskite. http://arxiv.org/abs/1607.02777 (accessed on 4th October 2025).

33. Niven, R. K. (2007). Combinatorial information theory: I. Philosophical basis of cross-entropy and entropy. http://arxiv.org/abs/cond-mat/0512017 (accessed on 4th October 2025).

34. Qian, H., & Thompson, L. F. (2016). Nonlinear stochastic dynamics of complex systems, II: Potential of entropic force in Markov systems with nonequilibrium steady state, generalized Gibbs function and criticality. http://arxiv.org/abs/1605.08071 (accessed on 4th October 2025).

35. Quirke, N., & Sweatman, M. B. (2004). Simulating fluid-solid equilibrium with the Gibbs ensemble. https://core.ac.uk/download/9023170.pdf (accessed on 4th October 2025).

36. Rusanov, I., Shchekin, A. K., Tatyanenko, D. V., Gibbs, J. W., & Kuni, F. M. (2009). Grand potential in thermodynamics of solid bodies and surfaces. http://arxiv.org/abs/0910.1543 (accessed on 4th October 2025).

37. van de Walle, A., Asta, M., Binder, K., Ceder, G., Ceder, G., de Fontaine, D., de Fontaine, D., et al. (2002). Self-driven lattice-model Monte Carlo simulations of alloy thermodynamic. https://core.ac.uk/download/4875547.pdf (accessed on 4th October 2025).

38. Ván, P. (2020). Thermodynamically consistent gradient elasticity with an internal variable. http://arxiv.org/abs/2005.13662 (accessed on 4th October 2025).

39. Wunderlich, B., Kondepudi, D., Jones, G. O., Scherer, G. W., Öttinger, H. C., Gutzow, I., Hodge, I. M., et al. (2009). Non-equilibrium thermodynamics: Structural relaxation, fictive temperature and Tool-Narayanaswamy phenomenology in glasses. http://arxiv.org/abs/0910.0026 (accessed on 4th October 2025).

CHAPTER 8

ENERGY TRANSFORMATIONS IN PHYSICAL AND CHEMICAL PROCESSES

Contents

Introduction 281
8.1. Types of Energy 286
8.2. Energy Transformations in Physical Processes..................................291
8.3. Energy Transformations in Chemical Processes................................296
8.4. The Law of Conservation of Energy 301
8.5. Real-World Examples of Energy Transformations 305
8.6. Implications of Energy Transformations...310
Summary..314
References...316

INTRODUCTION

The intricacy of energy conversions is a fundamental aspect of both chemical and physical processes. Energy, or the capacity to do work or bring about change, is transformed in various forms when substances undergo state changes from one to another. For instance, in physical processes, energy in the form of heat will bring about such state changes as melting or boiling. Conversely, in chemical reactions, changes in energy are notable in bond cleavages and formations, which are features of reactant conversion into product. Comprehensive knowledge of such changes is required, not just for theoretical study but also for application in a variety of uses, for example, material science, environmental science, and industrial processes. At the center of energy change study is the fundamental investigation of mechanisms behind the changes. Researchers use a range of approaches to study these processes, and lo and behold, the energy exchange processes are not so simple. For example, recent research has examined low-intensity electric field effects on coal samples, demonstrating that energy conversions are possible through nontraditional pathways, including destabilizing coal organic matter (COM) microstructures.

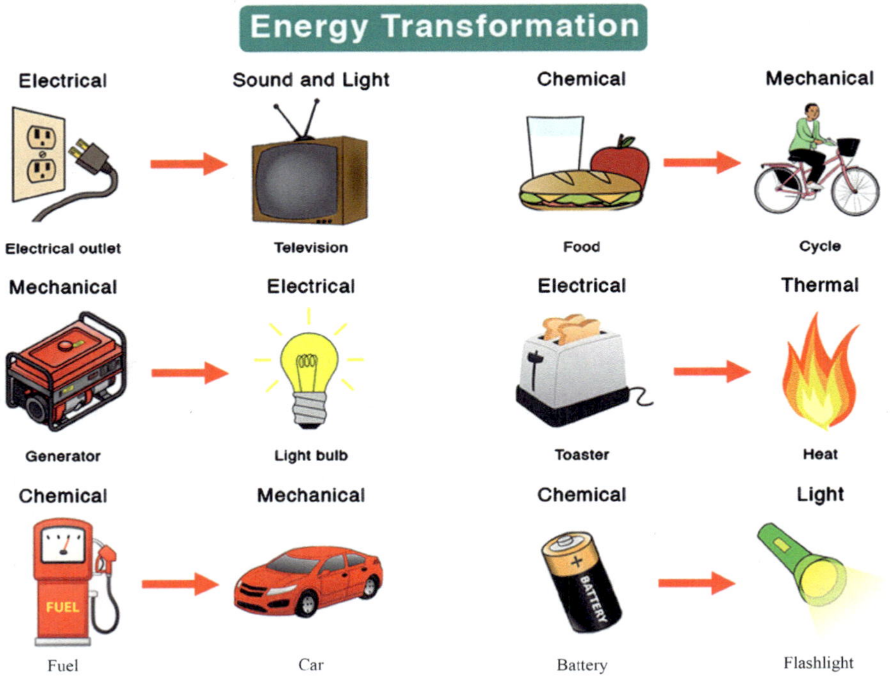

Figure 8.1. *Energy transformations.*

Source: https://www.sciencefacts.net/wp-content/uploads/2022/03/Energy-Transformation.jpg.

The findings highlight the importance of other methods for changing material characteristics, which can improve the efficiency of coal conversion technologies. Concerning the future technologies, understanding these complex relationships provides innovation doors for energy generation and use. Apart from that, the energy conversions are not just confined to common issues but extend to theoretical research as well, e.g., exploring antimatter. The thermodynamic framework shows how energy conducts itself everywhere, emphasizing principles that control both matter and antimatter interactions. While standard thermodynamics is adequate for the majority of uses, progress in the appreciation of CPT invariance suggests a closer investigation of energy properties in various contexts. As scientists explore the relative properties of CP- and CPT-invariant generalizations of thermodynamics, they highlight the importance of energy transmutations in all aspects of physical reality. It is through such interdisciplinarity that enriching our knowledge of mechanisms that drive energy and its transformation in physical and chemical situations becomes absolutely essential (Figure 8.1).

Definition of Energy Transformations

Energy transformations are the transfer of energy from one state to another and are fundamental principles of physical as well as chemical processes. In physics, energy can exist in a number of forms such as kinetic, potential, thermal, electric, and chemical energy. Under the guidance of the laws of thermodynamics, which explain the conservation and transformation of energy within a system, energy is transformed from one form to another. The energy conservation principle, or the first law of thermodynamics, is that it states energy cannot be created or destroyed but only changed. Awareness of these changes is necessary because it enables scientists and engineers to use energy effectively and guarantees processes operate effectively. Additionally, energy changes play a significant role in thermodynamic systems, where they enable phenomena such as reactions and phase transitions. For instance, during a phase change, say from a solid to a liquid state, the substance emits or absorbs energy, and this results in a particular change in temperature and state. Nucleation is one such process where a metastable phase transforms into a more stable phase, illustrating the intricacies of energy transformations. According to classical nucleation theory, the process is a delicate balance between input and output of energy as a system evolves via nucleation, growth, and coarsening of the new phase (Allen et al., 2009). Insight into such energy changes enhances materials science and kinetics and enables the creation of various technological applications. Finally, the significance of energy transformations extends to systems analysis of systems away from equilibrium. The Thermodynamic Covariance Principle (TCP) demonstrates that energy transformations must be covariant under Nonlinear Thermodynamic Coordinate Transformations (TCT). This ensures the fundamental laws of thermodynamics apply equally well in all states of a system irrespective of energy transformations occurring within it. The interplay between conservation laws and energy transformations is firmly established in mathematics by Noethers theorem between conserved charges and symmetries of physical systems (Evslin et al., 2016). From a comprehension of energy transformations across the contexts, scientists are able to better predict and control material properties and chemical reactions, ultimately improving energetic solutions for material synthesis and energy management.

Importance of Studying Energy Transformations

Energy transformation plays an essential role in comprehending the fundamental principles of physical and chemical processes. Energy transformation refers to the process where energy converts from one type into another, facilitating the fundamental phenomena of technology and nature. For instance, during chemical reactions, chemical energy stored in bonds is converted to heat and light, mirroring concepts of conservation of energy. This transformation not only powers life responses, such as photosynthesis but also determines technology advances, such as efficient engines and substitute power sources. Scientists and engineers can optimize processes, increase efficiency, and minimize environmental risks by analyzing these transformations, illustrating the broad-based significance of this field in contemporary days. Energy conversions are furthermore applied to more advanced scientific disciplines, particularly material science and quantum physics. The latest studies indicate that the mechanochemical activation of materials, as introduced in Bondarenko et al. (2018), could increase energy conversion process efficiency. Methane gas hydrate synthesis and its role in hydrocarbon synthesis is one indication of how energy conversions can develop new solutions in energy production. This mechanochemical approach uncovers the interface between chemical conversion and physical processes, creating the potential for new technologies based on efficient energy utilization. Studying such conversions allows scientists to innovate from domains of environmental stewardship to industrial processing, demonstrating the continued relevance of understanding energy conversion in current research. Also, energy transformations have mainstream and cutting-edge theories that are addressed in the study, such as those in quantum mechanics and black hole research. The theory of thermal radiation and the behavior of energy within non-stationary black holes, addressed in Singh et al. (2017), widens our theoretical frameworks and allows us better insight into energy interactions on a cosmic scale. These state-of-the-art theoretical studies not only advance our understanding of underpinning physical concepts but also inspire new technologies in all the scientific fields, from astrophysics to materials engineering. By integration of different areas of research, the importance of energy transformations is further highlighted, challenging researchers to study these ideas as we move toward a cleaner and more technologically advanced world (Figure 8.2).

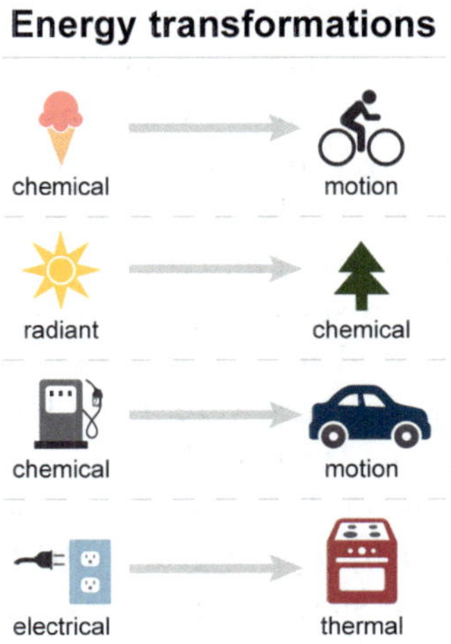

Figure 8.2. *Importance of studying energy transformations.*

Source: https://www.eia.gov/energyexplained/what-is-energy/images/Energy-Transformations.png.

Overview of Physical and Chemical Processes

Physical and chemical process is a broad field of study that explains phenomena of energy transformation. Physical processes refer to changes in the state or shape of a substance without altering its chemical composition, such as melting, freezing, and boiling. These changes are described by thermodynamic principles under which energy is conserved and converted from one state into another. For instance, when ice melts with heat, energy is passed to the molecules so they can change state from solid to liquid without altering the chemical water structure. This is a critical distinction for learning how energy travels throughout different systems, particularly in environmental and industrial situations where heat management is vital. Consequently, the knowledge of these important processes provides one with a basis for learning complex energy dynamics in scientific fields such as chemistry and physics. Chemical reactions, however, involve changes that lead to alterations in the molecular composition of compounds and thereby result in new products. The reactions tend to release or absorb energy, and indicate a more intricate interaction between matter and energy. For example, combustion reactions are very popular chemical reactions that convert reactants like hydrocarbons into products while releasing huge amounts of heat and light energy. The study of such reactions provides an understanding of the principles of energy transformation mechanisms that power anything from engines to living forms. In addition, the study of reaction kinetics and thermodynamics allows scientists to be in a position to predict how different variables can affect the efficiency and outcome of chemical reactions. Moreover, the technology advancement, such as 4D electron imaging, has enabled scientists to observe such events at atomic and molecular levels and realize dynamic behavior previously unseen, as testified by Shorokhov et al. (2009). The coming challenges of the declining situation of teaching science, particularly in the field of standards assessment, emphasize the need for a firm ground in physical and chemical processes. The fact that many states were given low grades for their science education standards points to a systemic lack of developing scientific fluency in students, posing an imminent threat to future innovation and technological development, as highlighted by Lynch et al. (2012). As energy conversions are at the forefront for addressing global issues such as climate change and managing resources, strengthening education systems to effectively teach such an understanding is imperative. By bridging the gap between theoretical knowledge and practice, education can inspire the next generation of scientists to tackle energy issues. Encouraging improved standards in science teaching eventually results in a scientifically literate population that is able to understand and influence better the intricate interactions involved in physical and chemical processes.

8.1. TYPES OF ENERGY

The various types of energy are used in the transformation process of physical and chemical systems. Kinetic energy or energy of movement is essential to understanding dynamic systems, whether a speeding car or colliding molecules in a chemical reaction. Molecular motion at the molecular level is the cause of thermal energy in substances, which in turn influences their states—liquid, solid, or gas. Besides, the transformation of kinetic energy into other forms, such as potential energy during the displacement of an object, demonstrates interdependence between energy forms in physical processes. The observation of the aforementioned changes is the foundation of the study of systems dynamics because the energy interchanges define the matter behavior under various conditions and demonstrate the essential principles of energy conservation and conversion. Other than kinetic energy, potential energy forms a key constituent in energy conversions in chemical and physical reactions. Potential energy is also known as stored energy, which typically exists within systems with forces but without movement, like in chemical bonds in molecules. When a chemical reaction occurs, the potential energy stored in such bonds is released or absorbed, and that energy gets converted to kinetic or thermal energy. Potential energy is a key factor in the reactivity and stability of chemical compounds. Besides, the concept of energy landscapes helps to describe potential energy variation in a system and furnish data on reaction paths and the regulation of reaction rates. An understanding of the interaction of kinetic and potential energy allows scientists to predict more precisely and regulate outcomes of chemical conversions. However, forms of energy like thermal, electrical, and chemical emphasize the complexity of physical and chemical conversion of energy. Thermal energy, a result of kinetic energy relating to the random movement of particles, impacts phase transitions and reaction kinetics. Electrical energy, generated from the movement of charge particles, plays a vital role in electrochemical reactions, driving reactions that convert chemical potential into utilizable energy. In addition, as underscored by recent advances, 4D electron imaging has opened new avenues of observing such energy changes with unprecedented precision to enable detailed studies of molecular interfaces and phase transitions at atomic resolutions (Shorokhov et al., 2009). By themselves, these types of energy illustrate the intricacy of energy changes, and it is thus better appreciated how energy triggers physical interactions and chemical reactions in many systems.

8.1.1. Kinetic Energy

Kinetic energy, the energy of an object due to its motion, is one of the central ideas of the general physical and chemical processes of energy conversion. The kinetic energy (*KE*) of an object according to classical mechanics is determined by the formula $KE = 1/2\ mv^2$, where m is mass and v is velocity. This is what illustrates that an object's kinetic energy rises with the square of its velocity, showing just how much dramatic energy change comes from comparatively minute velocity change. Such changes are found in everyday phenomena, whether something as mundane as a passing automobile or as a system with things as complex as gas molecule actions, where the total kinetic energy shows the temperature and leads to phase changes. This basic concept is the foundation for further study of energy changes in macroscopic as well as microscopic systems, adding to our understanding of the material world. Moreover, kinetic energy is also a crucial component in chemical reactions, particularly molecular collisions and reactions. Kinetic molecular theory states that particles move faster as temperature increases, which boosts collision rates and energy. This is particularly crucial in reaction dynamics since sufficient kinetic energy will often be required to surmount activation energy barriers for chemical reactions to occur. As an illustration, in nucleation phenomena—where metastable phases mature into more stable forms—the kinetic energy of individual molecules controls the likelihood and rate at which nuclei are generated. As discussed, the process of transformation in a phase change, such as crystallization from solids to liquids, demonstrates the influence of kinetic energy on not only the activity of substances but also on the inherent processes of energy conversion that define chemical reactions (Allen et al., 2009). The extension of kinetic energy principles to the overall system of thermodynamics and phase transitions demonstrates its significance in energy conversion. When systems move towards equilibrium, molecules' kinetic energy decreases and transforms into potential energy, seen in processes like solidification and condensation. For example, when a gas is cooled, the decrease in kinetic energy results in movement towards liquid state with the release of energy to the surroundings, thereby verifying the law of conservation of energy. In addition, complementarity between kinetic and potential energy is further elucidated through mechanisms such as molecular diffusion, where spreading molecules is an energy conversion that governs many physical and chemical systems. Through these similar dynamics, we are given invaluable insight into the intricate balance of energy conversions between matter states, ultimately building up our picture of the physical universe (Ramabrahmam et al., 2018) (Figure 8.3).

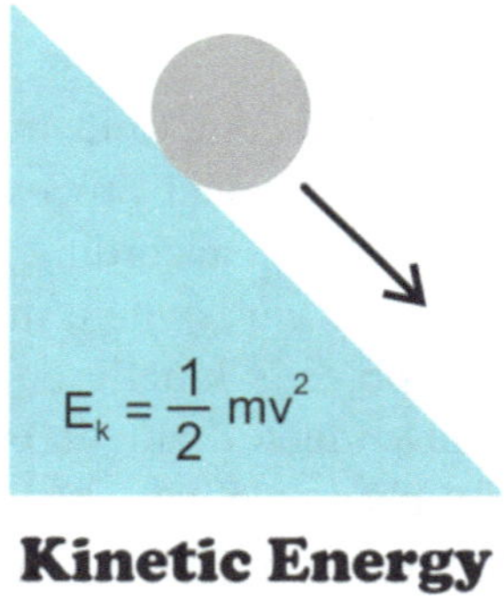

Figure 8.3. *Kinetic energy.*

Source: https://cdn.testbook.com/1753869574796-Screenshot%202025-07-30%20 at%203.29.23%E2%80%AFPM.png/1753869576.png.

8.1.2. Potential Energy

Potential energy, basically defined as the energy an object possesses due to its state or position, is at the heart of understanding energy conversion in both physical and chemical processes. Potential energy is easiest to observe in gravitational systems, where an object at some height possesses gravitational potential energy relative to a reference level. For instance, if a ball is held at a height, it can perform work as it falls due to the gravitational force acting on it. In the same way, in mechanical contexts, potential energy in chemical systems has its origin in the arrangement of atoms and molecular bonds, hence its significance in energy dynamics. Therefore, comprehending how energy is stored as potential energy and converted to kinetic energy in various processes is key to an appreciation of the laws of energy change in various branches of physics and chemistry. Potential energy is tightly linked with the stability of molecular forms in most chemical processes. Chemical potential energy is stored in the bonds of atoms; when chemical reactions destroy or form the bonds, there is a change in energy. The nucleation process most clearly illustrates such a change because it is the initial part of a phase transition where an unstable state transitions to a stable one.

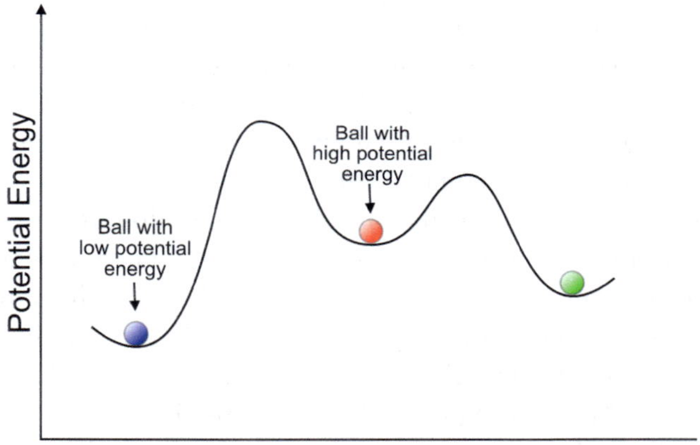

Figure 8.4. *Potential energy.*

Source: https://sop4cv.com/images/PECurve-balls.png.

This change is also one that frequently accompanies fluctuations in potential energy when outside conditions, such as pressure or temperature, cause the system's stability to change. As has been discussed in research on nucleation, knowledge of the fluctuations in potential energy with such changes can shed light on larger mechanisms within material science, such as materials solidification or liquid droplet condensation (Allen et al., 2009). Hence, potential energy is a crucial factor in determining the energetics of chemical reactions and phase transitions. Some further elaboration on potential energy in terms of energy change illustrates the law of conservation of energy. Potential energy can be converted to other forms, primarily kinetic energy, and vice versa in isolated systems. This fundamental rule underlies numerous processes, such as the conversion of gravitational potential energy to kinetic energy as an object falls. Also, mathematical frameworks that describe these conversions are significant to note system behaviors under varying conditions. With ongoing research into matters like the nature of time, some features of potential energy transformations might have consequences for our understanding of energy interactions in cosmic and quantum realms (Ramabrahmam et al., 2018). Lastly, potential energy is not static; it is an active component that enables several processes in physical and chemical contexts to take place, showcasing its pivotal position in the dance of energy transformations (Figure 8.4).

8.1.3. Thermal Energy

Thermal energy, another key player in energy transformations, plays a crucial role in numerous physical and chemical processes. It is a form of kinetic energy that accompanies the motion of atoms and molecules, which dictates the state of matter and temperature. When a material is supplied with thermal energy, the temperature is elevated, leading to phase changes such as boiling or melting. These changes are associated with large energy conversions, in that the provided thermal energy disrupts the orderly configuration of particles to allow phase changes from solid to liquid and gas. The relevance of thermal energy is apparent in ordinary occurrences such as the increase in temperature of water when it is heated on a stovetop, showing how heat can induce molecular movement to bring about a state change and thereby its pivotal role in the practice of energy changes. Apart from phase change, thermal energy is also pivotal in more complex thermal processes, including chemical reactions. For instance, heat energy is usually absorbed or released in endothermic and exothermic reactions, respectively. The interaction reveals the fundamental alterations that occur when the reactants are transformed into products. Additional examination shows how thermal energy influences reaction kinetics because higher temperatures will most likely increase reaction rates due to increased agitation of molecules. Such concepts provide the foundation in laboratory and industrial settings, where thermal energy management can optimize yields and reaction yield. Knowledge of the chemical change intricacies of thermal energy, as defined in classical nucleation theory, can identify the underlying processes controlling such essential processes (Allen et al., 2009). Moreover, the field of thermal energy covers its application in modern technologies, specifically in terms of efficiency and sustainability. Thermal energy that comes from combustion processes is a key provider of energy, but at the same time creates problems concerning emissions and conservation of energy. Research on systems such as underground coal gasification shows the importance of thermal energy management in difficult rock formations to guarantee energy extraction efficiency and minimize environmental impact (Dereviahina et al., 2018). Thermal energy management technologies involve developing materials to maximize thermal insulation or improve thermal-to-useful energy conversion, for instance, electricity. Thus, the study of thermal energy transcends traditional boundaries, revealing its fundamental importance not just under theoretical frameworks but also in practical applications that frame contemporary energy solutions (Figure 8.5).

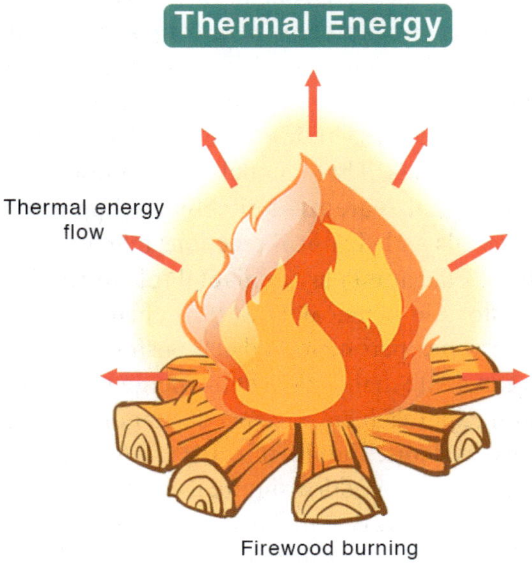

Figure 8.5. *Thermal (Heat) energy.*

Source: https://www.sciencefacts.net/wp-content/uploads/2022/02/Thermal-Energy.jpg.

8.2. ENERGY TRANSFORMATIONS IN PHYSICAL PROCESSES

Energy transformations are responsible for most physical processes, particularly those involving state or phase change. For instance, when water freezes, energy is being absorbed in the form of heat to undergo a wonderful energy state transformation. This process illustrates the processes of thermodynamics where absorption of thermal energy enables molecular motion and interaction, basically altering the state of matter. The dynamisms of energy transfer in physical processes illustrate the relationship between kinetic and potential energy. These transformations are not merely abstractions; they also have real practical implications. Identification of these processes enables scientists and engineers to create new applications in fields such as material science and environmental engineering, improving technologies that leverage efficient energy utilization and physical state control to create desired outcomes. Furthermore, the field of energy transformation encompasses more than just state change and involves more complex interactions, including chemical and physical bond formation. The quantum mechanical models, e.g., the one in Baskevych et al. (2018), showcase the intricate interdependencies between energy at potential and the character of molecular interactions on the atomic level. The stability of chemical bonds, for example, is a direct result of the energy profiles for the interaction between point charges and diatomic molecules, giving precious details on the stability and activity of various compounds. These models, in addition to strengthening our theoretical understanding, also are used directly for application, directing the synthesis of new materials and chemical reactions that are essential for various industrial applications. Education plays a pivotal role in passing on information about energy conversions in physical as well as chemical processes. Through the utilization of hands-on experiments and demonstrations, as set out in Arce et al. (2001), students can relate to the concept of energy in a more practical sense. This hands-on learning approach enhances understanding and promotes critical thought about energy systems. Future scientists and educators working with these ideas are most capable of addressing real-world issues, such as resource management and sustainability. By emphasizing the interdependence of energy conversions in various contexts, learning architectures can bring up a generation to develop solutions that more efficiently convert energy, connecting theoretical expertise and real-world application in an ever-changing technology landscape.

8.2.1. Phase Changes (e.g., Melting and Freezing)

Phase transition, particularly melting and freezing, is significant in studying energy transformations in both physical and chemical processes. Melting, in which a solid is converted to a liquid, occurs when thermal energy is added in order to allow molecules to overcome the intermolecular forces binding them in a rigid structure. It is not only a physical change; with it comes the endothermic absorption of massive amounts of energy, the latent heat of fusion. Conversely, freezing is an energy transformation when a liquid solidifies in a solid form that can be understood from the concept of molecular kinetics. When the temperature goes down, molecules are deprived of kinetic energy such that they are able to achieve a more organized form. This intricate dance of energy based on temperature changes manifests itself as measurable changes in material properties and is the foundation of most scientific investigation. Recent studies also describe the complexity of phase transitions, especially in other types of substances. The presence of amorphous solids such as the one formed by supercooled water expands the range of melting behavior even more.

Amorphous solids can be in different states, and different results are formed by melting, such as crystalline ice. Most notably, phase transitions between such amorphous phases can be caused by nonequilibrium transformations that are intrinsically different from those of equilibrium systems. One such example is melting from a denser amorphous solid at low pressures, which can occur in a sequence of well-distinguishable steps, with a phase diagram showing the coexistence of various phases of the material. The experiments for the transformations show multiplicity of energy-redistribution pathways during phase transitions (Chandler et al., 2014). In addition to the qualitative understanding of phase transformations, quantitative calculations are an essential part of modeling such complex processes. The accuracy of numerical simulations in predicting melting and solidification behavior often lies in coefficients that simulate fluid dynamics in the mushy zone during phase change. Of significance, variations in the permeability coefficient can cause variations in predicted phase change rates and solid-liquid interface morphologies. With robust computational methods being applied, for instance, the finite-volume numerical techniques, sensitivity of the predictions can be very informative of phase-change behavior. In particular, proper calibration of the simulation can lead to independent prediction concerning the permeability coefficient, and this is significant in modeling isothermal melting and solidification. This technique assists scientists in accurately recording the complexities of phase changes and energy transformation in various materials (Ebrahimi et al., 2019) (Figure 8.6).

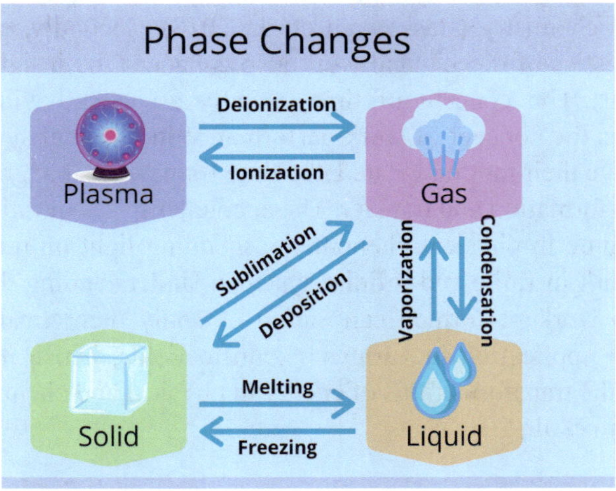

Figure 8.6. *Phase changes.*

Source: https://sciencenotes.org/wp-content/uploads/2020/10/Phase-Changes.png.

8.2.2. Mechanical Work (e.g., Moving Objects)

Mechanical work is a fundamental concept in the analysis of energy conversion, particularly in the motion of objects. Describing the product of force and the distance it travels, mechanical work represents the direct relationship between applied force and the motion of an object. In a more cosmic sense, this work is a natural part of most physical processes and phenomena and serves as the means by which energy is transformed. For example, if an object is lifted upwards, gravitational potential energy increases because of work done against gravity. The process is employed to show the conservation and conversion of energy, a phenomenon at the core of classical mechanics and a variety of chemical and physical phenomena in nature. The definition of mechanical work is extended to chemical systems, especially when considering the influence of energy changes by molecular interactions. It has been studied how different forms of mechanical work influence stability of chemical bonds and energy levels within molecules. For instance, simulation of a quantum mechanical system demonstrates that chemical bond potential energy may vary very strongly with external forces such as point charges on diatomic molecules. Identification of such interactions has the potential to give insight into stability in chemical structures and influence practical applications, such as catalyst design or synthesis of new materials. Quantum mechanical models developed based on such work specifically describe how mechanical work is involved in energy transformation of chemical systems and thereby unifies physical mechanics and molecular chemistry (Baskevych et al., 2018). Secondly, examining the time characteristics of mechanical work also enhances the insight into energy transformations. The concept of time, usually associated with mechanical motion, defines the concept of work performance in physical systems. Bodies under force have their motion located in a time framework that governs the rate of energy transformation and transfer. The encounter raises questions regarding the nature of time in physical phenomena, shedding light on the character of mechanical work in finite and infinite systems. Understanding these temporal coordinates to work performed can enrich not only theoretical analysis but also utilitarian application in various scientific fields, hence improving our knowledge of the transformations of energy in physical and chemical processes (Ramabrahmam et al., 2018).

8.2.3. Energy Transfer in Systems (e.g., Conduction and Convection)

Energy transfer processes, namely conduction and convection, are foundations in physical and chemical systems. Conduction is the method by which heat energy is passed from one medium to another via direct contact among particles where atoms or molecules collide and exchange kinetic energy. Conduction is highly sensitive to the properties of the material, e.g., its thermal conductivity, which dictates how good energy can diffuse. For example, metals, with good thermal conductivity, conduct heat readily in comparison to insulators like wood or rubber. As established in corresponding studies, particle-level conditions can also affect energy transfer sensitivity, making it important to have a thorough understanding of conduction for purposes such as material design and thermal management in engineering applications (Ebrahimi et al., 2019). Conversely, convection is another primary mechanism of energy transfer, which is brought about by the movement of fluids—gases or liquids—where warmer, less dense regions rise and colder, denser regions descend. Convection can be either natural or forced, with the former being caused by temperature-induced density changes and the latter through external means, such as mechanical fans. The effects of convection are particularly notable in heating and cooling systems, influencing the efficiency of energy transfer within various applications. In addition, fluid speed and surface area also significantly influence convective heat transfer coefficients, also showing how complex behaviors arise from basic principles.

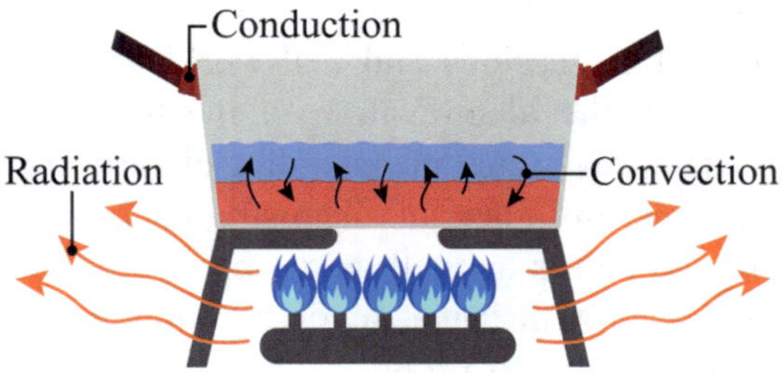

Figure 8.7. *Energy transfer in systems (e.g., conduction and convection).*

Source: https://www.shalom-education.com/wp-content/uploads/2021/03/image-19.png.

Recent studies also cover the way ambient conditions, i.e., radiation and reaction parameters, render convective systems more complicated to analyze, underlining the necessity of precision within thermal modeling (Adanhounmè et al., 2018). Furthermore, understanding the way energy is transferred through conduction and convection is crucial in thermal phase-change processes. In such transformations, energy is absorbed or released upon the change in the state of substances from solid, liquid, or gas state. For instance, during melting, thermal energy is absorbed, resulting in a state transformation of the material without changing temperature. Studies in phase-change material behavior indicate the importance of these phase transitions as functions of various parameters, including mushy zone length depending on the permeability coefficient governing motion of fluid in the phase-transition zone (Ebrahimi et al., 2019). Such a linkage not only emphasizes the importance of accurate modeling in energy transfer but also demonstrates how interdependent the processes of conduction and convection are in achieving optimum system efficiency in a range of scientific and industrial applications (Figure 8.7).

8.3. ENERGY TRANSFORMATIONS IN CHEMICAL PROCESSES

The intricate dance of energy transformations and chemical reactions is a foundation of chemistry and engineering. Chemical reactions involve the breaking and forming of atomic bonds that lead to energy state changes. These can be classified into endothermic and exothermic reactions. Exothermic reactions release energy, typically heat, and cause temperature increase in their surroundings, for example, combustion reactions. On the other hand, endothermic reactions take in energy, creating a cooling effect, illustrated during processes like photosynthesis. The comprehension of these changes is equally important to theoretical discussions and real-world applications in different spheres. For this reason, the identification of the transformation of energy during these reactions improves our understanding of the basic principles behind chemical transformations. In industrial chemistry, energy transformations are utilized to maximize processes as well as to enhance energy efficiency. The coal conversion process, for instance, involves intricate analyses where the organic content is taken apart to produce various energy-dense products. It has been proven that external factors, e.g., low-intensity electric fields, may impact the efficiency of such conversions, revealing new avenues for energy optimization (Kamchatnyi et al., 2018). Relevance of these findings includes enhancing the efficiency and safety of processing and mining of coal, which is vital in the face of the environmental implications of fossil fuel consumption. Through integration of such new techniques, chemists are able to produce more sustainable practices that enhance yields while minimizing waste and energy consumption. The molecular processes involved in energy conversion also provide insight into dynamic processes governing chemical reactions. Several techniques, including differential scanning calorimetry and thermogravimetric analysis, allow researchers to monitor changes in energy in such processes as a whole. Regularity of thermal fields in chemical interaction underlines the importance of knowing how molecules behave in different conditions. They are in line with broader speculation regarding time and energy in physical systems (Ramabrahmam et al., 2018). As researchers continue to break down the secrets of energy transformations during chemical reactions, such discoveries not only improve scientific knowledge but also aid in influencing the development of cleaner, better technologies that can assist in limiting the environmental impact of industrial processes.

8.3.1. Exothermic Reactions

Exothermic reactions are a fundamental form of chemical reactions involving energy release as heat. These reactions form a net decrease in the systems enthalpy, which accounts for their significance in all energy conversion scenarios. When chemical bonds are formed during a reaction, energy is released, and it contributes to the surrounding environments thermal energy. An example of such an occurrence is combustion reactions, in which fuels burn using oxygen to produce carbon dioxide, water, and plenty of heat. Understanding the principles behind exothermic reactions not only improves students' knowledge of chemical bonding and energy transfer but also reflects on the interdisciplinary nature of such processes in physics, chemistry, and biology (Dreyfus et al., 2013). Exothermic reactions extend into implications from biochemistry, where metabolic pathways trap energy, to industrial applications requiring efficient energy management. The applicability of exothermic reactions also goes beyond elementary chemical education; they are at the core of real applications, such as in mechanochemical reactions of materials engineering. Mechanochemistry demonstrates how exothermic reactions can provide novel routes to synthesis under specific conditions of the environment, resulting in novel materials with desired properties. This transformation occurs under non-equilibrium conditions, where the natural solid reactivity is amplified by the creation of surface defects and interphases that facilitate reactions producing nanoparticles with certain properties (Achimovičová et al., 2013). Being able to carry out such reactions in high efficiency not only represents energy conversions but is also in line with the growing emphasis on green strategies within chemistry and materials science. Therefore, knowledge of the processes in exothermic reactions can be used in teaching as well as in applications. Secondly, thermodynamic concepts involved in exothermic reactions can be used in other areas of science to relate ideas in physical chemistry to bioprocesses. The energy release from the processes is significant in sustaining life since it corresponds directly to metabolic processes responsible for cellular activities. An understanding of energy transformation through exothermic reactions gives students a complete overview of chemical processes and their applications in different fields, from environmental science to industry (Dreyfus et al., 2013). Moreover, learning about such reactions in classrooms can improve critical thinking and problem-solving abilities because of their complex interaction with other chemical processes. Hence, the study of exothermic reactions contributes not only to chemistry knowledge but also to interdisciplinary connections that foster a better appreciation of the conversion of energy in physical and chemical processes in the majority of fields of study (Figure 8.8).

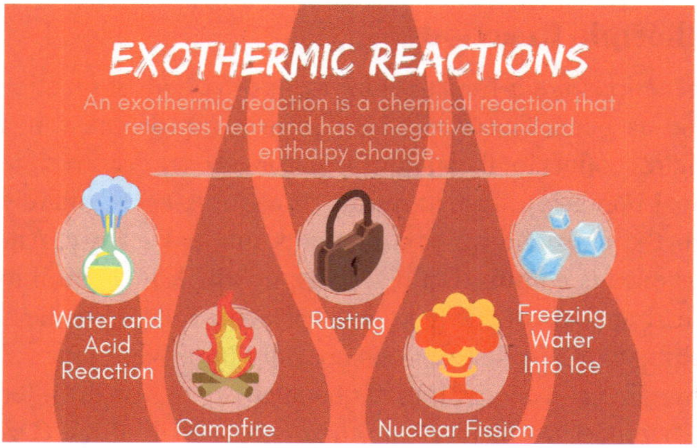

Figure 8.8. *Exothermic reactions.*

Source: https://sciencenotes.org/wp-content/uploads/2021/01/exothermic-reactions.jpg.

8.3.2. Endothermic Reactions

Endothermic reactions are a major feature of energy conversion in physical and chemical transformations, whereby heat is absorbed to cause the reaction to proceed. Such reactions need energy input, usually in the form of heat, to assist in bond-breaking in reactants. As a result, endothermic reactions have a tendency of lowering the temperature of the surrounding environment, thereby contrasting with exothermic reactions, which release energy. This intrinsic characteristic of endothermic reactions goes a long way in highlighting their crucial use in culinary science, biochemical reactions, and industrial reactions. For instance, the dissolution of certain salts into water might be endothermic, producing a cold solution, which can beautifully illustrate the process of energy transformation under a given scenario. Endothermic reactions are not merely significant in understanding basic thermodynamic principles, but also for analyzing intricate systems such as those outlined in advanced chemical processes. For example, the reaction of gold atoms with ethylene and acetylene highlights endothermic reactions, where certain gold adducts' synthesis, for example, $Au(C_2H_4)$, requires energy input to facilitate the bond-forming and rearrangement process. This intricate interaction between reactants and the energy input necessary illustrates the overall themes of energy conversion in chemical reactions. The outcome of the reaction dynamics illuminates under what circumstances particular products are formed, revealing the significance of energetics in dictating reaction pathway and reactant and product stability. The information concerning endothermic reactions hence is also capable of unveiling knowledge regarding energy efficiency during chemical synthesis and material formation.

Endothermic reactions also encompass more than individual chemical systems, extending implications to environmental processes as well as biological systems. Photosynthesis in the environment is a typical illustration of an endothermic reaction in which plants utilize sunlight to convert carbon dioxide and water to glucose and oxygen. The important biological process demonstrates how energy conversion sustains life on Earth based on energy input from the sun to trigger chemical change. Understanding such reactions not only contributes to our knowledge of bioenergetics but also introduces us to the importance of energy transformations in ecological processes. As we learn more about the relationship between chemical reactions and the transformation of energy, it can be realized that endothermic reactions are unavoidable for the sustenance of life and for other processes, emphasizing their importance under the umbrella of energy transformations in physical and chemical processes (Figure 8.9).

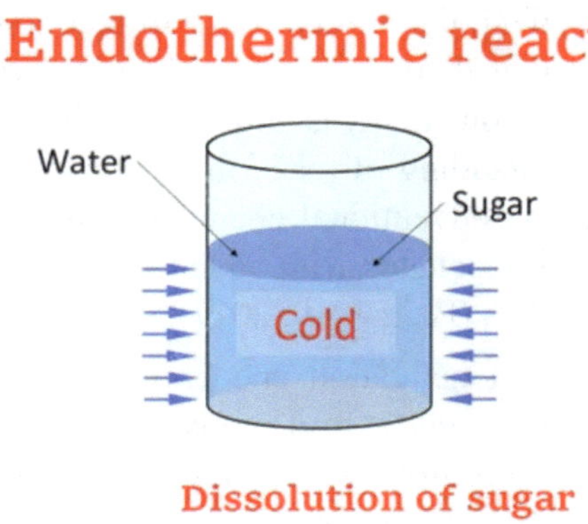

Figure 8.9. *Endothermic reactions.*

Source: https://blogger.googleusercontent.com/img/b/R29vZ2xl/AVvXsEi8_12fPJED7g6x9f9DrTjdnQhsEZFnbB4u9kWUV4nGv9Xb9Q4BkSzvsvCoM_x70PDyeZ9eW57KNxSfJ56VX28XJ-gJXr9TkLyg4EjcqQStTBYls4MGxCKa-WTJEFOllR1gg1rAMrbm6VG9RdLIzrmnqMR7snj9kqjE1Bq6TOtM1pBmZhBkw-3pcUX48w/s640/endothermic%20reaction.webp.

8.3.3. Activation Energy and Reaction Rates

Activation energy is a key concept in the rates of chemical reactions since it is the main obstacle that has to be overcome for reactants to be transformed into products. The energy barrier stands at the heart of the transition state theory where reactants must reach an energy state or activated complex before proceeding with the reaction. The significance of activation energy is captured in the Arrhenius equation, which quantitatively describes the rate constant of reactions in terms of temperature and activation energy. The equation illustrates how varying temperature can profoundly impact reaction rates; as temperature rises, more molecules possess sufficient energy to overcome the activation energy, with an increased likelihood of successful collisions. Hence, the concept of activation energy not only explains the kinetics of chemical reactions but also has broader implications in the form of energy conversion in physical and chemical processes.

Combining the principles of statistical mechanics, the theory of activation energy may further be extended using the relativistic effects paradigm for reaction rates. A few new theories have proposed that rates of chemical reactions might be influenced by relative motion among observers, merging special relativity ideas into kinetic theory. For instance, as described in recent research, when a fast-moving observer travels, available kinetic energy to reactants is altered. By this relativistic effect, the observed rate of reaction would therefore be decreased. This implies that there is a reconsideration with respect to the conditions under which activation energy is understood. Such an understanding not only enhances our understanding of reaction kinetics but also engages an elaborate interaction between conventional chemical perceptions and modern theory-based schemes, which resonates with advances in quantum chemically based reaction dynamics models (Baig et al., 2017).

Besides, the application of activation energy is not confined to typical chemical reactions but its use cuts across many fields including statistical mechanics and biochemistry. In complex systems such as biochemical pathways, the activation energy controls the rates at which enzymes catalyze reactions and thus cellular functions. Such interaction introduces the statistical nature of particle interaction and stochastic processes underlying such reactions. Evidence of particle currents displaying universal statistical behavior across networks confirms such an assertion to the effect that seemingly disparate activation energies shape the pattern of such systems significantly (Chen et al., 2010). In conclusion, studies on activation energy and reaction rates not only provide a basis for chemical kinetics but also represent a more universal understanding of energy conversion that feels intuitive to numerous scientific disciplines.

8.4. THE LAW OF CONSERVATION OF ENERGY

The Law of Conservation of Energy asserts that energy cannot be generated nor destroyed but can be transformed from one form of energy to another. This is a general principle in physical and chemical processes that guides our understanding of energy transformation in various systems. Conversion of energy is a phenomenon encountered in physical processes such as mechanical work and heat transfer. For instance, as kinetic energy is created from potential energy when an object drops, the total of the energy remains constant in the process. The event is proof of the continuous process of transformation of different forms of energy based on the law of conservation, hence opening avenues for study of more complex systems that appear to exhibit complex energy change.

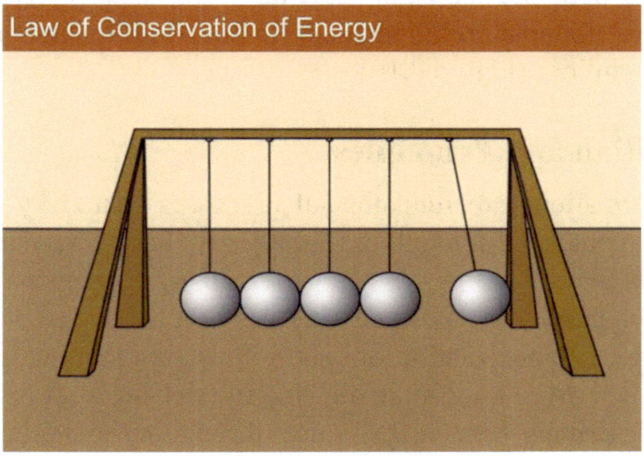

Figure 8.10. *The law of conservation of energy.*

Source: https://i.ytimg.com/vi/Lh5HyI7b-KE/maxresdefault.jpg?sqp=-oaymwEmCIAKENAF8quKqQMa8AEB-AH-CYAC0AWKAgwIABABGH8gWCgsMA8=&rs=AOn4CLAFpdflWdQ8DuLVyttW7IEJpiNncw.

In chemical reactions, the Law of Conservation of Energy is a cornerstone on which reaction dynamics may be comprehended. In chemical reaction, reactants are transformed into products accompanied by change in energy in the form of light and heat. Such alterations are indicated within the framework of exothermic and endothermic reactions wherein the system energy fails to alter though energy is either released or absorbed. The concept of conservation is best explained by examining biochemical pathways that regulate cellular metabolism. Here, some pathways have the ability to regulate the flow and transformation of energy, demonstrating the broad application of the law of conservation in portraying energy exchange in intricate biological systems, as defined under metabolic network analyses (Cercignani et al., 2014). New thermodynamics concepts broaden the scope of the Law of Conservation of Energy to non-equilibrium conditions. In such conditions, which are typical in driven systems, there are intricate dynamics generating unexpected energy flow. Statistical mechanics serves to demystify these complexities by characterization of stationary non-equilibrium states with consistent energy flows despite external conditions (Sole et al., 2015). By analyzing the properties of such systems, researchers can examine emergent behaviors as a result of energy conversions within. Such an understanding of non-linear processes adds to our knowledge of energy conservation under various conditions and contradicts the classical notion of equilibrium, emphasizing the applicability of laws even under dynamic and complex conditions. The research not only adds to our theoretical foundation but also circles back into functional applications in scientific disciplines (Figure 8.10).

8.4.1. Definition and Principles

Energy transformations are fundamental processes shared by both physical and chemical contexts, defined essentially by the conversion of energy from one to another. This process is governed by a number of underlying principles developed by physics and chemistry. For instance, the first law of thermodynamics, or conservation of energy law, says that energy can neither be created nor destroyed but can be transferred or changed. It is a fundamental principle for describing how energy flows through systems, in a mechanical engine or a cell or in a chemical reaction. In every case, energy transformations enable different functions, from heat produced in burning to chemical potential energy that is stored in molecular bonds. A fundamental comprehension of these transformations is essential to intellectual research and practical applications across most scientific fields. The second law of thermodynamics introduces a second fundamental principle governing energy transformations, i.e., the principle of entropy. Entropy, which is usually termed as an indicator of disorder, mandates that energy conversion is never perfectly efficient, with some energy always lost as waste heat in operations. The law emphasizes the systems' tendency to seek a state of higher entropy over time, which influences the viability of energy transformation efficiency. For example, when a motor car engine converts fuel's chemical energy to kinetic energy, it also produces heat energy radiated into the environment and contributes to overall entropy. Such laws can shed light on system inefficiencies and push innovations toward energy conservation and sustainability, which lead to effective energy management policies directed toward environmental stewardship. Further, energy conversion laws govern specific applications and theoretical frameworks that encompass physical and chemical processes. As demonstrated in studies of multi-component membranes, energy conversions can possess complex dynamics that rely on structural and contextual characteristics of their systems. The distinctive character of lipid-protein membranes, for instance, challenges traditional wisdom about energy conservation and dissipation, indicating elaborate relationships among thermodynamic properties and membrane function. The above findings explain how the energy transformations can vary dramatically from situation to situation and underscore the significance of close understanding of closure properties in chemical compounds. Lastly, the identification of determinants of structure affecting the efficiency of energy transformation enhances our comprehension of natural as well as engineered systems, presenting a wider framework for further research and technology development based on energy utilization (Backer et al., 2005; Earley et al., 2016).

8.4.2. Applications in Physical Processes

The intricate ballet of energy transformations has a central role to play in many physical processes that occur in nature and have far-reaching applications. One such major area of application is in materials science, where advanced imaging techniques such as 4D electron imaging enable scientists to visualize how the physics of materials unfolds at atomic and nanoscale resolutions. This capacity not only fortifies the idea of structural dynamics but also brings about understanding into the material behaviors under differing physical stimuli. Also, the ability to observe processes in real time allows scientists to fine-tune material properties at the development stage, hence making it convenient to translate innovations in nanotechnology and create intricate materials engineered for specific purposes, hence solidifying the central role of energy conversions in these physical processes (Shorokhov et al., 2009). In mechanochemistry, energy transformations take place distinctly in the form of phase transitions and atom interactions induced by mechanical stress rather than thermal energy. The process shows the efficiency of triggering chemical reactions in nonequilibrium conditions with immense benefits compared to traditional methods. For instance, mechanochemical processes have been applied in materials engineering where reactivity of the solid phase can be enhanced through the formation of reaction-promoting defects in the solid state. The outcome of such approaches goes far beyond mere synthesis; they ensure the production of nanomaterials with desired properties at low energy expenditure and waste generation, demonstrating how innovative energy conversions can reimagine conventional manufacturing processes (Achimovičová et al., 2013). Further, the study of physical processes related to energy conversions has significant bearing in various industrial fields, primarily in heterogeneous catalysis and extractive metallurgy. Through the utilization of the new properties of mechanochemistry-derived nanomaterials, industries can generate higher yields with increased efficiency. The control over solid-state energies selectively allows it to be able to design catalysts that will perform optimally at mild conditions, bypassing the energy requirement used in traditional catalytic reactions. Therefore, these technologies of energy transformations not only reflect academic relevance but also have the potential to make great strides towards sustainability and financial viability for the industries that rely on such innovations (Achimovičová et al., 2013) (Figure 8.11).

Figure 8.11. *Energy transformations applications in physical processes.*

Source: https://i.ytimg.com/vi/O4Rxb-3jIqQ/maxresdefault.jpg.

8.4.3. Applications in Chemical Processes

With respect to chemical reactions, the application of energy transformation is one of the major concepts towards helping various types of reactions, especially during phase transition. For instance, classical nucleation theory is a simple model which explains the way that metastable phases transform into more stable states. This theory characterizes the initial stages of phase changes, for example, nucleation, growth, and coarsening, that are very important in processes such as condensation of droplets from supersaturated vapor and solidification of materials. The combination of temperature and pressure work as factors of primary importance that may trigger destabilization of a system and ultimately decide the direction of the transformation. By recognizing these energy transfers and what they signify, chemists can more effectively manage conditions to achieve intended results in chemical synthesis, thereby optimizing efficiency and yield in production (Allen et al., 2009). Besides, the recent advances in 4D electron imaging technology have considerably enhanced our insight into the energy transformations during chemical reactions. By providing access to atomic and nanoscale structural dynamics, this technique offers unprecedented opportunities for the investigation of chemical reactions and phase transitions in real time. Time, introduced as the fourth dimension, allows researchers to image dynamic changes occurring during molecular interface interactions and within nano(micro)mechanical systems. With this direct observation capability, improved knowledge of how energy is transferred and transformed in various chemical processes becomes possible. By learning more about these interactions, scientists can create better techniques and come up with new ways to address issues within materials science and chemical engineering (Shorokhov et al., 2009). Finally, the use of principles of energy conversion in chemical process operations extends beyond normal methods to encompass new technologies and sustainability projects. With the drive for cleaner chemical processes because of environmental concerns, energy consumption optimization is needed. Techniques such as photochemical reactions, in which the energy of sunlight drives chemical reactions, show the possibilities of energy-efficient substitutes in industrial processes. By considering energy conversions in these processes, chemists can optimize reaction pathways, save energy, and reduce waste. The continuous research on energy applications in chemical processes not only helps in achieving improved efficiency but also serves global sustainability considerations, promoting innovations required for a sustainable future in chemical production practices.

8.5. REAL-WORLD EXAMPLES OF ENERGY TRANSFORMATIONS

Energy transformations are important in many practical applications, showcasing the fundamental concepts of physics and chemistry. One prime instance is the process of converting solar energy into chemical energy in terms of photosynthesis in plants. In this very complex procedure, chlorophyll absorbs sunlight and facilitates the process of converting water and carbon dioxide into glucose and oxygen. It is this process of energy conversion that is not only essential in keeping the plants alive but also the entire food chain. The significance of this transformation affirms the interdependence of natural systems in that it is the foundation of energy transfer in ecosystems—transferring energy from one state to another, hence reflecting the intricate interactions among physical and chemical processes in nature (Ramabrahmam et al., 2018). A case in point is also the transformation of chemical energy in batteries into electrical energy. In daily technology items, batteries operate through electrochemical reactions that discharge stored energy. For instance, lithium-ion batteries from mobile phones and electric vehicles convert the chemical energy trapped in lithium compounds into electrical energy as ions move through an electrolyte. Not only do these transformations power gadgets, but also contribute significantly to the creation of green technologies, such as electric vehicles that reduce carbon emissions. A tour of such energy transformation forms—efficiency and limitations—emphasizes the growing need to create more advanced energy storage technologies in harmony with environmental stewardship (Arce et al., 2001). Moreover, the combustion process is a straightforward but highly significant illustration of energy transformation from chemical to thermal energy. When the fossil fuels, such as oil, coal, or natural gas, are burned, the chemical energy stored in such fuels is released as heat. This energy is harnessed in power plants to generate electricity or in the case of vehicles to provide propulsion. While this energy conversion is efficient, it has been a source of worry in terms of greenhouse gas emissions and global warming, prompting research into less damaging alternatives. Understanding these energy transformations and their influence on us drives innovations in renewable energy technologies to minimize environmental impacts, hence further driving a necessity to make our way through the complex dance of energy conversion, utility, and sustainability in our daily lives (Ramabrahmam et al., 2018).

8.5.1. Photosynthesis in Plants

Photosynthesis is a primitive process whereby plants, algae, and certain bacteria capture light energy to trap chemical energy, primarily in the form of glucose. This intricate process not only sustains plant life but is also the basis of energy transfer in the ecosystem. The process occurs in chloroplasts, as chlorophyll absorbs sunlight and initiates a series of chemical reactions. During these reactions, atmospheric carbon dioxide and water from the soil are transmuted into glucose and oxygen. This conversion depicts the mechanism of converting energy from sunlight into chemical energy, which shows how light fuels biochemical reactions that are vital to survival. Understanding photosynthesis provides knowledge about energy flow in ecosystems and emphasizes the role of plants as carbon sinks, making the topic an integral part of energy transformation in physical and chemical processes. Photosynthesis efficiency is influenced by a variety of factors in the environment, including light intensity, temperature, and carbon dioxide concentration. All these can control the rate of photosynthesis, illustrating the subtlety of ecological systems. Conditions of optimal accuracy permit plants to maximize capture and conversion of energy, while suboptimal conditions may disrupt this process, leading to reduced biomass production. For instance, high temperatures can first boost the rate of photosynthesis, while high temperatures can cause thermal stress, damaging chlorophyll molecules and plant health. These processes demonstrate that energy transformations during photosynthesis are not just a question of the inherent biochemical processes but also of the immediate environmental context. This aspect of photosynthesis emphasizes the importance of studying these processes in the context of environmental transformation and their relevance to ecosystem function. Furthermore, the ramifications of photosynthesis go beyond individual plants and extend to influencing global biochemical cycles and climate. Through the capacity to capture carbon dioxide and release oxygen, photosynthesis is instrumental to maintaining atmospheric equilibrium as well as perpetuating life on our planet. The integration of photosynthesis with other processes within the ecosystem maintains soil health, nutrient regimes, and water balance, thereby facilitating ecosystem resilience. Additionally, finding out about photosynthetic processes can lead to developments in plant cultivation and biofuel generation, enhancing food security and climate change management. The significance of photosynthesis thus summarizes the fundamental mechanism of energy transformation as well as its general role in sustainability and environmental well-being. By looking at these key relationships, one can appreciate the manner in which the processes of photosynthesis reverberate through regional and global ecological processes in the framework of energy conversion in physical and chemical processes (Figure 8.12).

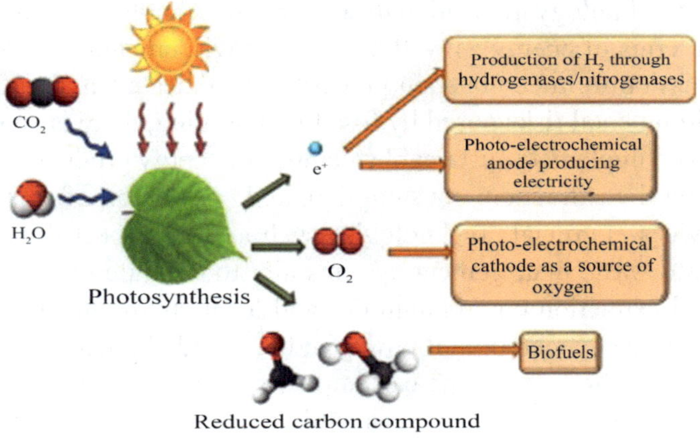

Figure 8.12. *Photosynthesis in plants of energy transformations.*

Source: https://ars.els-cdn.com/content/image/1-s2.0-S1389556714000409-fx1.jpg.

8.5.2. Combustion of Fossil Fuels

Burning fossil fuels is a fundamental process in energy conversion, during which chemical energy in hydrocarbons is released as heat and light in the oxidation process. Energy conversion is the foundation of contemporary industrial economies, driving electricity generation, transportation, and heating. Burning fossil fuels such as coal, oil, and natural gas yields carbon dioxide and water as significant byproducts, responsible for most greenhouse gas emissions. This vital transformation not only maintains human development but also generates environmental problems, as an indication of the twinning of the benefits and drawbacks of energy generation. The examination of previous trends reveals how fossil fuel energy is the foundation of economic development in the United Kingdom, demonstrating its role in relation to exergy provision and conversion into useful work, thus illustrating the intricate combination of energy availability and economic development (Warr et al., 2025). In addition, efficiency in energy conversion when fossil fuels are burned is technology, equipment quality, and method of operation dependent. It has been established that only a portion of the energy produced when combusted is efficiently converted into work or heat, and the remainder is lost as waste heat. For example, in domestic applications of heating, the conversion efficiency might be much less than that achieved in modern technology for power generation. Since energy services are obtained from these fossil fuels, they possess various qualities based on the application and temperature gradients involved in their use. This imbalance indicates the necessity for technology innovation in conversion technologies that can be more efficient in terms of energy as well as reduce environmental problems caused by high emission from combustion processes (Gudmestad et al., 2015). Based on the environmental risks posed by fossil fuel burning, there is an urgent need to develop and implement sustainable methods of energy production. The focus on expanding the extraction, consumption, and conversion of fossil fuels into energy services is crucial. Technologies such as low-impact drilling systems and improved waste management systems aim to mitigate the adverse effects of fossil fuel extraction on communities and local environments. In addition, practices such as cogeneration of power and heat and alternative methods such as the utilization of heat pumps can improve energy efficiency significantly, hence reducing the greenhouse gas emissions associated with fossil fuel combustion. In the construction of a sustainable energy future, it is crucial to adopt practices that harmonize economic growth with the preservation of the environment, enhancing the overall efficiency of energy transformations while reducing their impact on climate change (Gudmestad et al., 2015).

8.5.3. Battery Operation in Electronic Devices

Operation of batteries in electronic devices involves intricate energy conversion inherent to their functioning. Key to this functioning is electrochemical reactions whereby chemical energy is transformed into electrical energy to power components. Upon discharge of the battery, chemical reactions occur in the anode and cathode, releasing electrons that move in an external circuit, transferring energy to the device. Its efficiency is a function of various factors including the battery's chemistry, temperature, and design variations. Moreover, innovation has in recent times sought to engineer these processes for better performance, particularly in portable devices where high energy efficiency must be achieved without sacrificing functionality. These are factors that make battery usage a critical area of research under the broad category of physical and chemical process energy transformations. As the demand for portable and power-constrained electronic devices grows, saving power is increasingly important.

Batteries for these devices must balance energy delivery with requirements of efficiency and endurance. In this respect, strategies such as cutting down capacitance and avoiding unnecessary functions become factors of longer battery life and performance (Havinga et al., 1997). Handheld device architecture customarily features low-power design approaches, which can aptly reduce energy for usage. The implementation of effective communication protocols and error control techniques can guarantee that energy consumed in data transmission is optimized. This systems approach to the utilization of batteries improves customer experience and sustainability by maximizing device life and reducing electronic trash.

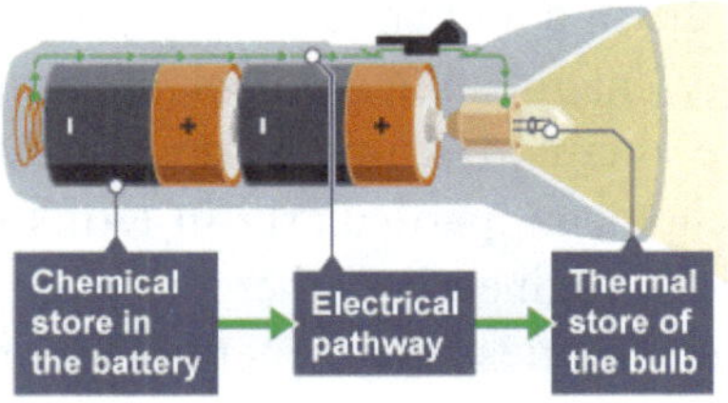

Figure 8.13. *Battery operation in electronic devices of energy transformations.*

Source: https://bam.files.bbci.co.uk/bam/live/content/zw3q8mn/small.

Emerging technologies, such as wireless sensors and mobile devices, represent a significant revolution in the use of batteries. These devices require unobtrusive energy conversions to support evolving user needs and ambient conditions in their environment (Dalmau et al., 2008). Adding sensors to these devices enhances their complexity and energy demands. Thus, effective battery management systems are essential in the sense that they enable such devices to perform best while facilitating communication among disparate elements. By using a common component model, designers make the relationship between software and hardware uncomplicated in order to enhance overall performance.

Such advancements highlight the importance of understanding battery operation and energy conversion as a basis for enhancing the ability and versatility of contemporary electronic devices (Figure 8.13).

8.6. IMPLICATIONS OF ENERGY TRANSFORMATIONS

Energy transformations are central to physical and chemical processes and have important impacts on most scientific and practical applications. The implications of the transformations go beyond the transfer of energy; they influence the efficiency and feasibility of energy systems as well. For instance, in physical systems, as potential energy is being utilized to become kinetic energy, the efficiency of energy utilization takes center stage. When energy is transferred, it can be susceptible to loss in heat dissipation, therefore becoming issues for applications as diverse as mechanical engines and biological systems. These inefficiencies require novel means of capturing and utilizing energy in a more efficient manner, emphasizing the need for advanced technology with the capability of minimizing losses and maximizing output. In chemistries, conversions of energy relate closely to the principles of thermodynamics, dictating the reaction kinetics and equilibria. Energy conversions in such a context often mean exothermic and endothermic processes with the liberation or absorption of energy manifesting appreciable impacts on reaction rate and product yield. The thermodynamic system provides a framework for understanding such phenomena, particularly in terms of energy conservation and entropy. In chemical engineering, there is a need to optimize energy conversions in industrial activities where energy input determines profitability and environmental issues. The growth of ideas in thermodynamics, including research on CP- and CPT-invariant thermodynamics, once again indicates the intricacy of understanding change in energy at micro- and macro-scopic levels, guiding our vision of nature (Klimenko et al., 2014). Moreover, consequences of energy changes are not only scientific but also ethical and societal. The shift towards sustainable energy is driven by the need to transform the consumption and production of energy. Energy conversions must be considered in consideration of the sustainability of resources, impact on the environment, and social justice in access to energy. Improvements in renewable technologies leverage the understanding of energy conversions to transform solar, wind, and hydro energy into usable and acceptable forms with fewer ecological effects. This holistic energy transformation approach calls for coordination among policymakers, scientists, and communities to develop a whole perception that connects science and societal need. Through pursuing this interdisciplinarity path, the effects of energy transformations can contribute towards building a sustainable future by prioritizing the use of ethical reasoning in scientific innovation.

8.6.1. Environmental Impact

Environmental impact due to energy transformations is a critical concern, especially with the fast rate of technology advancement. With increasing extraction of mineral resources and the creation of new sources of energy, anthropogenic pressure on the biosphere significantly increases. Such development requires an active and intensive control system to track anthropogenic changes in the natural environment. Freshwater resources are one of the sensitive areas of concern since surface land waters are extremely vulnerable to human impacts. The hydrobiological service, which surveys and controls the surface water levels of pollution, is a good instance of an organized response within environmental surveillance in general. Through applying systematic observation, society can better defend against the acceleration of existing environmental problems and thereby harmonize energy conversion into sustainable application (Abakumov et al., 1980). As energy transformation is involved in climatic and ecological system modification, regions like the Arctic are experiencing radical transformations that underline the importance of scientific collaboration globally. The Arctic in Rapid Transition (ART) program emphasizes a holistic comprehension of the ways in which the transformation impacts the physical and biogeochemical components of the Arctic Oceans. Central to this endeavor are questions about the relationships between previous sea ice switches and energy flow, elemental cycling, and biological productivity changes. Through examining these past trends in conjunction with current and projected trends, scientists can determine important feedback mechanisms that can be utilized for making future projections and environmental response. The integration of process research and present monitoring programs makes it possible to address the intricate issues regarding energy conversions and ecological health in an integrative manner, upholding the relevance of astute scientific inquiry in preventing unfavorable environmental impacts (Forest et al., 2010). Most importantly, the acknowledgment of direct ecological responses to energy conversions is essential in designing measures to prevent environmental deterioration. Since terrestrial and aquatic ecosystems are increasingly affected by energy consumption, effects on biogeochemical cycling, biodiversity, and productivity are strong. Emphasizing the effects of temperature change, stratification, and pH change illustrates the delicate balance in ecological systems where small changes cause large developments in ecosystem function. As organisms and ecosystems evolve to tolerate these changes, their resilience and adaptability are key to informing conservation planning. Thus, a thorough study of the relationship between energy conversions and their environmental implications is crucial for formulating sound environmental policies and advocating sustainable practices that help conserve our natural resources for future generations.

8.6.2. Energy Efficiency and Sustainability

The relationship between energy efficiency and sustainability lies at the heart of modern environmental science, particularly in regards to energy conversions within physical processes as well as chemical processes. Energy efficiency is the utilization of less energy to achieve the same task and prevent wastage. This is utilized in various fields like manufacturing, transport, and building design where the utilization of high-tech technology can save significant amounts of energy. The shift toward efficient energy systems is not only financially advantageous; it is also crucial in order to reduce greenhouse gas emissions and minimize the environmental footprint of human activity. As the world worries more about climate change, growing energy efficiency becomes a strategic necessity for fostering sustainability, aligning economic development with environmental protection. Sustainable practices and renewable energy technologies are increasingly becoming renowned for their ability to minimize adverse environmental impacts while propelling economic growth. This has led to innovation of new methods for recycling and recovering byproducts in energy-intensive technologies. For instance, biorefineries nutrient-recovery-first can use waste biomass to produce worthwhile products like biofuels and fertilizers. Such a strategy not only creates a portfolio of premium-value products but also addresses the real issue of nutrient depletion in agriculture, enabling sustainable practice. Through the integration of physical, chemical, and biological transformations, biorefineries can themselves be examples of sustainability by enabling principles of circular economy and reducing reliance on nonrenewable resources. By so doing, such systems help seal anthropogenic nutrient cycles, validating the argument that energy conversions have to necessarily value sustainability in order to function effectively in a highly dynamic world (Carey et al., 2016). The Anthropocene epoch is a most significant transition in planetary processes, especially in the context of energy consumption and environmental impact (Alberti et al., 2017). During this period, human activity has increasingly affected Earth's energy systems and its biosphere, ushering in challenges that require a re-evaluation of our approach to energy usage. The evolution towards an agency-dominated biosphere illustrates how civilizations are gradually shifting from traditional energy sourcing to innovative technologies that enhance energy extraction efficiency. This shift provides both evolutionary forces and new avenues for ecosystem process that can effectively seize energy while maintaining ecologies in balance. With the unfolding of the Anthropocene, the need for energy sustainability practice and policy intensifies in urgency, towards a critical nexus between energy conversions, human activity, and the sustainable future of our world.

8.6.3. Technological Advancements and Innovations

The physical and chemical processes energy transformation scene has been drastically shaped by technology innovation and breakthroughs of the last decades. New technology has optimized extraction, conversion, and utilization of various forms of energy with a significant margin of increased efficiency and minimized wastage. For instance, nanotechnology and materials science research has made it possible to develop more efficient solar photovoltaic cells that can harness the energy contained in sunlight to electricity with record efficiency. These developments not only maximize the conversion of energy but also make possible the integration of renewable energies in existing infrastructures, thereby promoting sustainability. This combination of technology and energy systems is indicative of the ongoing revolution in how energy is captured and utilized, with technology playing its pivotal role in the transformation of energy. Besides, the advent of computer-based tools has transformed research practice in the physical sciences profoundly, as concluded from studies examining information practices across numerous scientific disciplines. The ubiquity of large datasets and sophisticated computing enables researchers to investigate and model energy conversions more accurately than previously conceivable. This shift to data-based research reflects an abandonment of traditional experimental paradigms in favor of more integrated and cooperative forms of research, facilitating further comprehensive knowledge about energy dynamics. Scientists actively utilize advanced software to simulate chemical reactions, facilitating increased interdisciplinary cooperation and making breakthroughs in energy systems possible. These advancements lead to a key trend where technology not only enables energy conversion but also transforms the paradigm of scientific research in the physical sciences (Bulger et al., 2011). With regard to addressing energy challenges, innovation in energy storage technologies has become increasingly crucial. Technologies like new-generation battery systems, mainly lithium-ion and future solid-state batteries, have been in the spotlight for their ability to store energy more efficiently and release it sustainably. These technologies are especially important in the shift towards renewable energy sources like solar and wind, which are naturally intermittent. With technological advancements in storage solutions, they support stabilization of energy supply and increase the reliability of renewable energy systems and contribute significantly towards the world's shift to sustainable energy practice. Furthermore, the advancement of research in this sector reflects escalating consciousness of the contribution of energy storage solutions towards ensuring maximum energy transformations, and thus presents the congruence of technological innovation and effective energy management policies (Ashford et al., 2002).

SUMMARY

The investigation of energy transformations within chemical and physical reactions reveals the intricate nuances underlying these phenomena. The following essay has illustrated how various mechanisms of energy transitions function, emphasizing the significance of fundamental thermodynamic principles that govern natural as well as man-made transformations. In particular, the research indicates toward the critical role of chemical and physical transformations as they relate to the states of energy as well as stability of the system. From the observation of coal specimens under different conditions, as discussed in the destruction of coal organic matter, we are aware of the manner energy interactions evolve under different influences, however minor such influences might be, like weak electric fields (Kamchatnyi et al., 2018). These realizations trigger the doorway to a deeper comprehension of the energy process, one that is paramount in constructing efficiency in energy conversion technology. Moreover, the inclusion of the classical nucleation theory enhances the depth of our knowledge even more regarding energy conversions in the phase transition setting. Nucleation is an important term used to capture the demonstration of how energy is stored and released during material changes. The nucleation process itself takes on a special significance when destabilized systems ripen into stable structures, hence demonstrating the functions of energy gradients and outside perturbations as catalyzers of these conversions (Allen et al., 2009). Through nucleation, we can observe that energy conversions are not merely mechanical or thermal but also comprise a variety of stimuli that may lead to considerable physical conversions. Understanding these processes provides researchers and practitioners with the means to manipulate material properties for new functions. Lastly, observing the breadth and significance of energy transformations in physical and chemical processes instills an overall appreciation of natural phenomena and technological advancements. With ongoing studies on the interactions of matter and forms of energy, the lessons learned about their transformation will bring more sustainable processes and improved methodologies to numerous industries. With an understanding of the interconnection between energy, stability, and transformation, as established by the findings on coal and nucleation theory, we are best able to meet the challenges of resource management and environmental sustainability. Along the way, society is progressing to the point where it will use energy in a form that is both efficient and environmentally friendly, paving the way for developments ahead.

Understanding changes in energy in physical and chemical changes is important in understanding how energy is conveyed and transformed through reactions as well as phase changes. In physical changes, such as boiling or melting, energy is typically absorbed or released when the system undergoes a change from one state to another. For example, in ice melting, energy is supplied to break the hydrogen bonds that keep the water molecules locked in a solid form, transforming solid ice to liquid water. This is not altering the chemical composition of the water but does demonstrate the critical role of thermal energy in facilitating the phase changes. The understanding of these processes provides a basic basis to investigate more intricate chemical changes, in which energy changes are coupled with changes in molecular constitution. Chemical processes, however, entail the conversion of reactants to products, with an accompanying definite exchange of energy. Energy is released into the surroundings, typically in the form of heat, in exothermic reactions, while endothermic reactions require energy input to react. It is exhibited by various chemical reactions, such as combustion, where fuel molecules react with oxygen to produce carbon dioxide and water with the release of huge amounts of energy. Its application in power production and car engines indicates how vital understanding these changes is. It is at the core of research fields like thermodynamics and chemistry, which ensure developments in technology and efficiency in energy usage. In addition, the incorporation of sophisticated computational techniques has transformed the way energy changes are understood. For instance, new free energy estimations from atomistic Hamiltonians provide a better understanding of biomolecular mechanisms and materials science. By applying the Free Energy Perturbation method beyond its original scope, researchers can calculate more precise sampling of molecular conformations and therefore make energy transformation analysis more accurate. Such innovation is much better than traditional methods, as observed with simulations that involve better efficiency in sampling requirements. In addition, implications of the study spill over into laboratory science, influencing safety assessments with regard to nuclear waste disposal and the interactions of microbial processes in subsurface systems, thus indicating the broad applicability of energy conversion studies in current scientific discussion (Grubmüller et al., 2019; Humphreys et al., 2010).

REFERENCES

1. Abakumov, V. A., Gasilina, N. K., & Izrael, Y. A. (1980). Hydrobiological service for observations and control of surface waters in the USSR. https://core.ac.uk/download/11020701.pdf (accessed on 4th October 2025).
2. Achimovičová, M., et al. (2013). Hallmarks of mechanochemistry: From nanoparticles to technology. https://core.ac.uk/download/186617483.pdf (accessed on 4th October 2025).
3. Adanhounmè, A., Adomian, G., Ahmadi, G., Annapurna, P., Anwar, M., Ariman, T., Aski, M., et al. (2018). Lie symmetry analysis and numerical solutions for thermo-solute chemically reacting radiative micropolar flow from an inclined porous surface. https://core.ac.uk/download/199214473.pdf (accessed on 4th October 2025).
4. Ahmed, M., Altinay, G., Kostko, O., Metz, B., et al. (2019). Probing reactivity of gold atoms with acetylene and ethylene with VUV photoionization mass spectrometry and ab initio studies. https://core.ac.uk/download/294625774.pdf (accessed on 4th October 2025).
5. Alberti, M., Frank, A., & Kleidon, A. (2017). Earth as a hybrid planet – The Anthropocene in an evolutionary Astro biological context. http://arxiv.org/abs/1708.08121 (accessed on 4th October 2025).
6. Allen, M. P., Becker, R., Bellon, L., Berthier, L., Binder, K., Binder, P. M., Binder, D., et al. (2009). Modeling of nucleation processes. https://core.ac.uk/download/52695338.pdf (accessed on 4th October 2025).
7. Arce, J., & Betancourt, R. (2001). The design, implementation, and assessment of a new capstone course aimed at science education majors. https://core.ac.uk/download/215487647.pdf (accessed on 4th October 2025).
8. Backer, J. M., Bagatolli, L. A., Bedeaux, D., Capovilla, R., Casimir, H. B. G., Defay, R., Evans, E. A., et al. (2005). A general theory of non-equilibrium dynamics of lipid-protein fluid membranes. http://arxiv.org/abs/cond-mat/0409264 (accessed on 4th October 2025).
9. Baig, M. W. (2017). On the theory of time dilation in chemical kinetics. http://arxiv.org/abs/1203.2510 (accessed on 4th October 2025).
10. Baskevych, O. S., Bilan, N. V., Sobolev, V. V., Stefanovich, et al. (2018). Electrical charges as catalysts of chemical reactions on a solid surface. https://core.ac.uk/download/288815744.pdf (accessed on 4th October 2025).
11. Bondarenko, V., Sai, K., & Svietkina, O. (2018). Effect of Mechan activated chemical additives on the process of gas hydrate formation. https://core.ac.uk/download/168412206.pdf (accessed on 4th October 2025).

12. Bulger, M., Institute of Physics, Institute of Physics Publishing, Kyriakidou-Zacharoudiou, A., Meyer, E., et al. (2011). Collaborative yet independent: Information practices in the physical sciences. https://core.ac.uk/download/14306211.pdf (accessed on 4th October 2025).
13. Carey, D. E., Mayer, B., McNamara, P. J., Yang, Y., et al. (2016). Recovery of agricultural nutrients from biorefineries. https://core.ac.uk/download/213074743.pdf (accessed on 4th October 2025).
14. Cercignani, C., De Donder, T., Hill, T. L., Nicolis, G., Prigogine, I., et al. (2014). Irreversible thermodynamics of open chemical networks I: Emergent cycles and broken conservation laws. https://core.ac.uk/download/31208058.pdf (accessed on 4th October 2025).
15. Chandler, D., & Limmer, D. T. (2014). Theory of amorphous ices. http://arxiv.org/abs/1306.4728 (accessed on 4th October 2025).
16. Chen, H., Doi, M., Fujisawa, T., Kelly, F., Levitov, L. S., Linke, H., Chertkov, M., et al. (2010). Geometric universality of currents. http://arxiv.org/abs/1008.2749 (accessed on 4th October 2025).
17. Dalmau, M., Louberry, C., & Roose, P. (2008). Heterogeneous component interactions: Sensors integration into multimedia applications. http://arxiv.org/abs/0812.2989 (accessed on 4th October 2025).
18. De Sole, A., Arnol'd, V. I., Bertini, L., Brémaud, P., et al. (2015). Macroscopic fluctuation theory. https://core.ac.uk/download/54249161.pdf (accessed on 4th October 2025).
19. Dereviahina, N., Inkin, O., Sotskov, V., Tishkov, Y., et al. (2018). Integrated analysis of geofiltrational parameters in the context of underground coal gasification relying upon calculations and modeling. https://core.ac.uk/download/168414004.pdf (accessed on 4th October 2025).
20. Dreyfus, B. W., Geller, B. D., Gouvea, J., Redish, E. F., et al. (2013). Chemical energy in an introductory physics course for the life sciences. http://arxiv.org/abs/1308.3667 (accessed on 4th October 2025).
21. Earley, E. Joseph. (2016). How properties hold together in substances. https://core.ac.uk/download/156785500.pdf (accessed on 4th October 2025).
22. Ebrahimi, A., Kleijn, C. R., & Richardson, I. M. (2019). Sensitivity of numerical predictions to the permeability coefficient in simulations of melting and solidification using the enthalpy-porosity method. http://arxiv.org/abs/2004.14168 (accessed on 4th October 2025).
23. Evslin, J., Sonnino, A., Sonnino, G., Steinbrecher, G., et al. (2016). Symmetry group and group representations associated to the thermodynamic

covariance principle (TCP). http://arxiv.org/abs/1512.04810 (accessed on 4th October 2025).

24. Forest, A., Forwick, M., Frey, K., Mathis, M., et al. (2010). Arctic in Rapid Transition (ART): Science plan. https://core.ac.uk/download/11900464.pdf (accessed on 4th October 2025).
25. Grubmüller, H., & Reinhardt, M. (2019). Determining free energy differences through variational morphing. http://arxiv.org/abs/1906.12124 (accessed on 4th October 2025).
26. Gudmestad, O. T., & Traa, K. (2015). Sustainable use and production of energy in the 21st century. https://elar.urfu.ru/bitstream/10995/57416/1/ijepm_2016_v1_1_02.pdf (accessed on 4th October 2025).
27. Havinga, P. J. M., & Smit, G. J. M. (1997). Low power system design techniques for mobile computers. https://core.ac.uk/download/pdf/11454227.pdf (accessed on 4th October 2025).
28. Humphreys, P., Metcalfe, R., & West, J. M. (2010). Microbial effects on repository performance. https://core.ac.uk/download/55755.pdf (accessed on 4th October 2025).
29. Kamchatnyi, O., Molchanov, O., Rudakov, D., Soboliev, S., et al. (2018). Destabilization of the hard coal microstructure by a weak electric field. https://core.ac.uk/download/168413992.pdf (accessed on 4th October 2025).
30. Klimenko, A. Y., & Maas, U. (2014). One antimatter–Two possible thermodynamics. https://core.ac.uk/download/197535379.pdf (accessed on 4th October 2025).
31. Lynch, J., Lerner, L. S., Schwartz, M., Gross, P. R., Schwartz, R., & Goodenough, U. (2012). The state of state science standards 2012. https://core.ac.uk/download/71359935.pdf (accessed on 4th October 2025).
32. Nicholas A. (2002). Pathways to Sustainability: Evolution or Revolution? https://core.ac.uk/download/4379293.pdf (accessed on 4th October 2025).
33. Nováčková, K. (2009). Stability of salts and complexes of humic acids. https://core.ac.uk/download/30278457.pdf (accessed on 4th October 2025).
34. Ramabrahmam, V. (2018). The physics of timelessness. https://core.ac.uk/download/186330638.pdf (accessed on 4th October 2025).
35. Shorokhov, D., & Zewail, A. H. (2009). New light on molecular and materials complexity: 4D electron imaging. https://core.ac.uk/download/4883901.pdf (accessed on 4th October 2025).

36. Singh, T. I. (2017). Quantum radiation properties of general nonstationary black hole. http://arxiv.org/abs/1702.07346 (accessed on 4th October 2025).
37. Warr, B., Schandl, H., & Ayres, R. U. (2025). Long term trends in resource exergy consumption and useful work supplies in the UK, 1900–2000. https://core.ac.uk/download/pdf/7123125.pdf (accessed on 4th October 2025).

CHAPTER 9

THERMOCHEMISTRY OF SOLUTIONS AND MIXTURES

Contents

Introduction 321
9.1. Fundamental Concepts of Thermochemistry 326
9.2. Types of Solutions 331
9.3. Thermodynamic Properties of Solutions 336
9.4. Mixtures and Their Thermochemical Behavior 341
9.5. Calorimetry in Thermochemistry 346
9.6. Real-World Applications of Thermochemistry 351
Summary 355
References 358

INTRODUCTION

The study of thermochemistry is at the heart of understanding energetic processes in the dissolution of substances in solutions and the formation of mixtures. Thermochemistry, by definition, deals with changes in heat energy associated with chemical reactions and physical changes. The field provides fundamental knowledge, which is critical in many areas of science and industry, including pharmaceuticals, materials science, and environmental chemistry.

By examining heat differences, such as heat of mixing and heat of solution, researchers can predict how chemicals will react under different conditions, which is vital for developing efficient processes. This knowledge of heat transfers is used to develop solutions in real-world applications, demonstrating the relevance of thermochemistry in grasping solutions and mixtures with depth. When it comes to the study of the thermochemical properties of solutions, there is a need to utilize temperature-pressure relationship and molecular nature of the solvent and solute. The laws of thermodynamics behind such interactions provide a foundation of understanding how things will respond when they are mixed together, including concepts such as solubility and reaction spontaneity. Principles such as entropy and enthalpy are crucial in such interactions, governing the energy changes that accompany the creation or breaking of molecular structures.

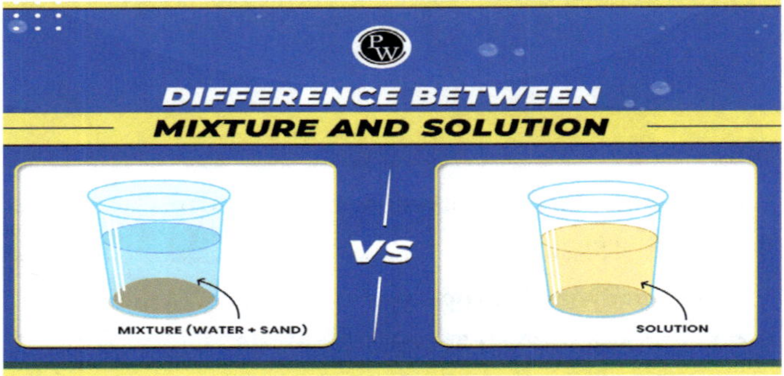

Figure 9.1. *Thermochemistry of solutions and mixtures.*

Source: https://static.pw.live/5eb393ee95fab7468a79d189/GLOBAL_CMS_BLOGS/d3f2e5bb-0299-4fb3-9f2e-7c2eb6b6600d.jpg.

By employing methodologies that quantify specific free energies and enthalpies for various chemical species, researchers can more accurately quantify these interactions and the related thermochemical processes (Horton et al., 2025). This particular understanding contributes to the enhancement of chemical mixture-based fields, particularly in instances where reactions must be tracked via complex thermochemical examination. Beyond the fundamental foundations of thermochemistry, the practical implications of research into it extend as far as the strategy of heat transfer techniques and modeling chemical systems. For example, the thermochemical effects of other planet atmospheres on the transfer of heat exhibit real-world functional applications of these principles beyond laboratory settings (Gilbert et al., 2025). Not only is it significant to aerospace engineering to understand how variations in atmospheric composition influence energetic transfers but also for overall use in climate modeling and atmospheric research. Lastly, the synergy of thermochemical knowledge with practical processes allows us to manipulate and exploit solution and mixture properties, leading to innovations in addressing modern scientific challenges. Thus, thermochemistry is a foundation discipline that exemplifies correspondence between theory and application across various disciplines (Figure 9.1).

Definition of Thermochemistry

The field of thermochemistry is fundamentally concerned with the study of heat change in chemical reactions and phase changes. Thermochemistry serves as a bridge between chemistry and thermodynamics, involving the quantitative aspect of the transformation of energy. The central idea concerns the laws of heat transfer, particularly of energy, in chemical processes. By analyzing the enthalpy changes of such reactions, researchers are able to gain knowledge about compound stability, reaction pathways, and system efficiencies. In the study of solutions and mixtures, thermochemistry aids in understanding how interactions between solutes and solvents affect thermal properties, which is extremely important in applications ranging from industrial processes to environmental chemistry. The implications of these energy changes are profound, from spontaneity of reactions to conditions at equilibrium in a range of chemical systems. The practicality of thermochemistry has applications across a spectrum of fields, including materials science, biochemistry, and engineering. As an example, the thermochemical characteristics of reactions with mixtures can dictate how materials react under changing thermal conditions. This information is crucial to planning such systems as combustion engines or heat exchangers, where temperature regulation and heat transfer are important considerations. A consideration of note is the significance of enthalpy in determining the feasibility of reaction in solutions. Investigating the thermochemical interactions that surround reaction enthalpies has the potential to reveal the energetic favorability of alternate routes, and hence enable scientists to engineer to an intended result. Thermochemical processes are therefore not only a source of deepening our knowledge of chemical principles but are also useful in designing practical entities in a number of scientific fields, and hence they represent an unavoidable area of research. Further, innovations in the analysis of thermochemical processes enlighten us about intricate phenomena like combustion and detonation reactions. Through understanding of the thermodynamic cycles governing such reactions, taken from stationary combustion waves, scientists can extrapolate pertinent parameters that enhance system efficiency. For instance, comparison between various modes of combustion allows scientists to ascertain optimal configurations purely based on energy and entropy considerations. Deflagration-detonation distinction under conditions of constant stagnation states are an explicit description of the manner in which thermochemistry explains the fundamentals governing performance in propulsion systems. This discussion not only establishes the applicability of thermochemical concepts to actual processes but also emphasizes the integrated nature of chemical reactivity and energetic problems in the design of efficient systems, thus broadening the scope of thermochemical research to actual situations.

Importance of Studying Solutions and Mixtures

The study of solutions and mixtures is of critical importance in the study of thermochemical processes because it provides information on how different substances behave at the molecular level. Solutions, as the solutes that are dissolved within the solvents, possess certain unique properties that are rather different from those of pure chemicals. These differences would lead to variations in the thermodynamic properties, such as boiling points, melting points, and vapor pressures. They need to be understood for a broad variety of applications related to the creation of chemical mixtures in industrial processes and new material creations. In addition, understanding how different components react with each other in a mixture allows chemists to predict behavior under any state of affairs, leading to effective and safer reactions. Solutions and mixtures hence need to be learned because they are vital in advancing both theoretical knowledge and applied application in chemistry, particularly in thermochemistry where energy change during phase transitions and chemical reaction is of significant importance. In thermochemistry, the thermodynamics of mixtures and solutions are of extremely significant importance in ensuring energy balance is maintained in chemical processes. From the study of such systems, mechanisms governing energy exchange can be uncovered, with relevance for fundamental research and engineering applications. A case in point is how chemical reactions in mixtures can absorb or produce heat, affecting reaction rate and equilibrium state. These thermodynamic alterations can be explained in terms of various theoretical models, pointing toward the significance of understanding the constituent parts of solutions and mixtures. Additionally, the investigation into how temperature modifies solubility and reaction rates provides valuable insight into environmental processes and industrial processes. Having the ability to control these variables further enhances our power to manipulate reaction and tailor conditions to desired outcomes, illustrating the significance that study of mixtures and solutions must hold for thermochemistry. Beyond pure science, the practical significance of the study of solutions and mixtures extends into many arenas such as pharmaceuticals, food technology, and environmental science. In pharmacy formulation, understanding the behavior of mixtures of compounds in solutions benefits in formulating effective drugs with optimal bioavailability and drug performance. Similarly, in food science, mixtures of ingredients come to define product texture and shelf life, illustrating the principles of thermochemistry. To a larger extent, addressing environmental problems, such as removal of pollutants and resource usage, is rooted in the understanding of chemical mixtures and their interactions. The investigation of the thermochemical behavior of solutions and mixtures therefore guides sustainable practices and innovations. Generally, the significance of these systems being investigated goes beyond theoretical knowledge, significantly influencing our capacity to address real-world challenges while building our comprehension of the phenomena of thermochemistry (Figure 9.2).

Figure 9.2. *Studying solutions and mixtures.*

Source: https://www.teachengineering.org/content/images/PTImages/ptHeaders/PTMixtures.png.

Overview of Key Concepts in Thermochemistry

Thermochemistry is a cohesive idea under the study of chemical mixtures and solutions, providing data on the energy changes that occur in chemical reactions and even physical processes such as dissolution and mixing. Among the ideas that are integral to thermochemistry is the idea of enthalpy, or system total heat content, which is used to quantify the exchange of heat in a reaction. Upon their formation, the enthalpy of solvation, which accounts for the energy variations accompanying solute-solvent interactions, is involved. This implies both exothermic reactions, where heat is released, and endothermic reactions, where heat is absorbed. Understanding such energy relationships is central to predicting how a solution will behave under varying circumstances, since they directly control the conditions of equilibrium, spontaneity of reactions occurring in mixtures, and temperature dependence of reactions in mixtures. In addition to enthalpy, entropy is another principle that is vital in the field of thermochemistry, and particularly with solutions and mixtures. Entropy quantifies the level of disorder or randomness within a system and plays a critical role in the determination of the viability of chemical reactions and phase change. The Second Law of Thermodynamics dictates that during natural processes, the total entropy of the system and the environment will rise, tending to force spontaneous reactions to go forward. In mixed systems, solvent-solute interaction leads to changes in both entropy and enthalpy, influencing the overall Gibbs free energy change of the process. Taking control of these thermodynamic factors can aid in various applications, ranging from solubility prediction to reaction condition design for maximum yield and efficiency in chemical processes. The behavior of thermochemical principles is also described further in the form of dynamic models that simulate the intricate interactions of mixtures and solutions. For example, advanced kinetic models like the kinetic multi-layer model for aerosol gas-particle interactions give further insight into reaction kinetics and phase behaviors. By considering all mass transport steps, such models can assess changes in particle size and chemical compositions resulting from various thermodynamic conditions. Such models generally fill the gaps in multiphase chemistry highly relevant to atmospheric processes and chemical reactions in solutions. Furthermore, results from these studies often bear relevance to other areas, such as environmental science and materials science, thus attesting to the interdisciplinarity of both pure and applied thermochemistry (Gregory et al., 2015; Koop et al., 2011).

9.1. FUNDAMENTAL CONCEPTS OF THERMOCHEMISTRY

Basic principles in thermochemistry form an essential understanding of mixture and solution behavior. Thermochemistry is the examination of energy changes that are associated with chemical reactions, particularly the heat transferred in the course of such reactions. The discipline uses the principles of thermodynamics to explain interactions at the molecular level and defines how energy is transformed and preserved. Making and disintegrating mixtures are typically endothermic or exothermic reactions and can be examined relative to enthalpy change. Specifically, when a solute dissolves in a solvent, the resulting energy changes—releasing or absorbing heat or not—greatly influence the attribute of solutions. Identifying such thermochemical principles, therefore, not only accounts for the character of solutions but also opens doors to practical applications in a wide range of scientific fields. In exploring the thermochemistry of some mixtures, one of the major areas of research involves the chemistry of interactions between hydrides and hydroxides, particularly in solid-state reactions. Studies have confirmed that sodium hydride and sodium hydroxide show distinctive properties since they support reversible hydrogen storage, an important feature for energy uses. The study of the Na-O-H system provides immense thermodynamic behaviors during dehydrogenation and hydrogenation routes. Experiments based on thermogravimetric-differential thermal analysis combined with mass spectrometry emphasize decisive events, including hydrogen desorption from certain temperature ranges. Moreover, the presence of complex phases in such mixtures indicates advanced relationships among different hydrogen forms, pointing to the fine interactions of thermochemical systems (Gregory et al., 2015). Further, the bridging of combustion reactions with thermochemistry indicates the impact of energy transformations on practical applications. Combustion can be explained through thermochemistry since it involves complex reactions that produce abundant heat. With theorists seeking to bridge the gap between theory of combustion and practical applications, it is imperative that they understand the underlying thermodynamic principles thereof. Modern studies focus on developing relationships between temperature gradients and composition of the mixture, particularly in controlled conditions such as laminar flame fronts. Though some progress has been made, the field still grapples with the transition of conceptual thermochemical concepts to more complex, multi-dimensional combustion situations. More study of these areas can result in better combustion methods, hence confirming the key significance of thermochemistry in establishing engineering sciences (Marble et al., 1956).

9.1.1. Enthalpy and Its Significance

Enthalpy is a basic idea for understanding thermochemical solution and mixture properties. Enthalpy, or the total of a system's internal energy plus the product of its pressure and volume, provides valuable information regarding how energy is modified in physical and chemical processes and reactions. It is particularly valuable in processes such as dissolution, where solvents interact with solutes and energy is either absorbed or released. These interactions are not only theoretical; they have practical applications in manufacturing operations, where reaction rates and product stability are reliant on mixing enthalpy. Through observing changes in enthalpy, chemists can even predict the way that different solvents will interact with solutes, and thereby optimize conditions for desired reactions. This understanding ensures energy is managed effectively, which is important in both university settings and industrial processes. Furthermore, the significance of enthalpy is applied to calorimetry, the process of quantifying heat change during chemical and physical reactions. With knowledge of enthalpy changes, chemists are able to estimate the amount of heat absorbed or evolved from reactions, which enables them to determine the feasibility of reactions as well as energy efficiency. For instance, enthalpy decreases in exothermic reactions but increases in endothermic reactions, illustrating how these types of changes can be applied to practical issues such as energy production. Furthermore, enthalpy serves to account for the thermodynamic principles governing solution behavior such as solvation effects and phase transitions. Marble et al. (1956) observes that the understanding of the combustion process behavior and energy transfer is a step towards the creation of advanced engineering applications based on principles of thermochemistry.

Figure 9.3. *What is enthalpy?*

Source: https://chemistrytalk.org/wp-content/uploads/2023/03/640px-Rising_temperature.jpg.

This highlights the very critical intersection point between theoretical calculation of enthalpy and practical technological innovations. Furthermore, the impact of enthalpy in the field of mixtures cannot be emphasized enough, particularly for the behavior and stability of complex systems. In solutions, enthalpy changes account for interactions between solute and solvent molecules, determining the overall thermodynamic nature of the mixture. This becomes significant when studying systems with instability, e.g., with certain gas mixtures under alternative conditions. As an example, the nature of unstable gaseous detonations or hypervelocity interactions can be characterized by the entrainment of enthalpy flow through coupled reactions and jetting phenomena. Austin et al. (2009) suggests that the transition mechanisms of these systems are dominated by the enthalpic nature of substances. Thus, enthalpy analysis does not simply provide elementary information about thermochemistry but is also a basis for applied advances in material science and engineering application. Through close examination of enthalpy, researchers are better able to control the complexities of thermochemical systems, which consequently offers improved solutions to industrial concerns (Figure 9.3).

9.1.2. Heat Capacity and Its Role in Solutions

The concept of heat capacity is a significant factor in the thermochemistry of solutions because it basically represents a substance's ability to absorb heat without undergoing an impressive temperature change. The heat capacity, which is also known, i.e., the amount of heat energy utilized in raising the temperature of a unit mass of the material by 1°C, is vastly different from one material to another. It not only varies with the solute-solvent interaction but also with their concentrations in solutions. Heat capacity needs to be known when one is dealing with changes in temperature while mixing or dissolving, particularly where the behavior of solutions depends on changes in temperature, such as phase transitions or rates of chemical reactions, demonstrating its importance in thermodynamic mixture research. Heat capacity in solutions reflects the subtle nature of molecular interaction, particularly in cases involving hydrogen bonding or van der Waals forces. For instance, mixtures of liquids' heat capacities usually show non-ideal behavior due to the new interactions at a molecular level that are established. Such interactions may lead to greater heat capacities than their pure analogs, as the solution itself gains more energy when heated. The determination of heat capacity in such solutions allows scientists to predict thermal solution behavior, an essential condition in many industries, for example, pharmaceuticals, food technology, and materials science. Thus, knowledge of heat capacity of solutions supports the process design and optimization on the basis of thermal property. In addition, the temperature dependence of heat capacity is also very important in understanding the thermodynamic stability of solutions and mixtures. With varying temperature, the heat capacities can also vary, impacting the solutions' behavior under different conditions. For example, comparing heat capacities at various temperatures can help understand the phase behavior variation of solutions, such as why certain mixtures will separate on heating or cooling. According to the new study, the relationship of heat capacity and other thermodynamic properties in some systems, for example, sodium compounds, demonstrates complex dynamics of phase interaction and thermodynamic cycles (Shepherd et al., 2006). These observations allow one to better forecast how heat will transfer through the solution, eventually determining the viability and effectiveness of the reaction, thereby the significance of heat capacity to general discussion of thermochemistry of mixtures and solutions (Gregory et al., 2015) (Figure 9.4).

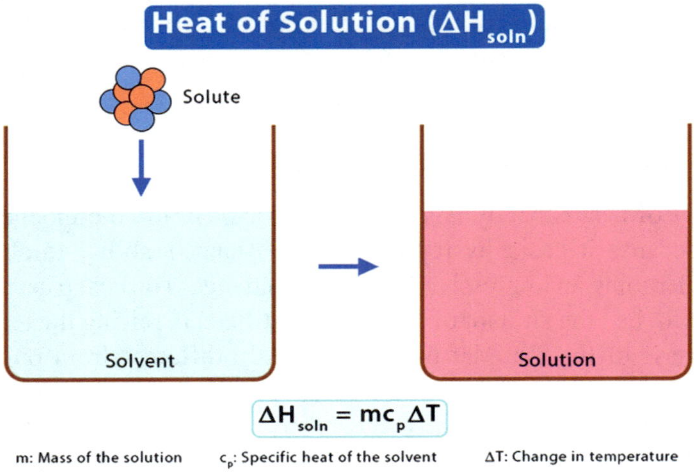

Figure 9.4. *Heat capacity and its role in solutions.*

Source: https://www.chemistrylearner.com/wp-content/uploads/2023/03/Heat-of-Solution.jpg.

9.1.3. The First Law of Thermodynamics

The first law of thermodynamics or the law of conservation of energy says that energy cannot be created or destroyed in a closed system. Instead, energy is transformed from one type to another but under the rule that the total energy within the system never increases or decreases. This fundamental idea has important implications within the realm of thermochemistry, particularly when evaluating how energy varies in chemical reactions occurring in solutions and mixtures. Through comprehension of the interconnected relationship between work, heat transfer, and internal energy, chemists are able to anticipate how various variables affect a systems energy equilibrium. Therefore, the first law of thermodynamics enhances our knowledge of energy distributions and transformations that occur in various states of matter, thereby predicting application in fields ranging from materials science to biochemistry. Within the field of thermochemistry, the first law serves as the foundation for many processes, including those within chemically reacting systems. In studying gas-particle flows, say, energy interactions of phases illustrate the conservation and conversion of energy under changing conditions. The use of principles of thermodynamics in such chemically reactive systems illustrates gas dynamics interaction with thermochemical calculation. More specifically, while initial calculations of flow can be calculated independently of energy transactions, understanding in the end involves recognizing that energy conservation is paramount, either as a heat transfer or momentum exchange due to particle collisions. Such understandings are central in exploring thermochemical solutions and mixtures' behaviors to allow scientists to condition for optimal desired chemical yields, eventually maximizing product yield and reaction efficiencies (Anderson et al., 2025). The first law of thermodynamics is also a critical explanatory framework for thermal behavior in solutions and mixtures, particularly under conditions of extremes, i.e., hypersonic flows. The thermal dynamics of energy dissipation in the processes of heat transfer reveal the complex interactions between the molecular components of a system. With the application of the first law in heat transfer scenarios involving gases and planetary atmospheres in other worlds, researchers can analyze the storage or loss of energy in thermal interactions. The analysis provides information on laminar flow behavior and total thermal efficiency in these atmospheres. Understanding the implications of energy conservation not only facilitates the forecasting of the system's behavior but also makes it possible to design new materials and chemical processes that are essential for applications ranging from aerospace to environmental science (Gilbert et al., 2025).

9.2. TYPES OF SOLUTIONS

Classification of solutions is primarily based on the phases of the solvent and solute in the mixture. It is possible to divide a solution into solid, liquid, or gas based on whether the solvent and solute are in the same phase or not. For instance, an alloy is a common solid solution in which metals like iron and carbon combine to create steel. On the other hand, when a solid dissolves into a liquid, such as table salt in water, the solution is a liquid solution. Gaseous solutions, such as air, are mixtures of more than one gas mixed well together. All types of solutions have unique thermochemical properties that decide how energy is expended or acquired upon dissolution, which is an essential part of thermochemical studies of mixtures and their interactions during chemical reactions. The nature of solute-solvent interactions is key to determining the thermodynamic properties of any solutions. The degree of solute molecule interaction with solvent molecules can be key to determining the stability and characteristics of the solution as a whole. An example of this would be polarity, an important consideration in solute solubility in water solutions in that polar solutes will tend to dissolve readily in polar solvents like water due to strong intermolecular forces. Conversely, non-polar solutes have poor interactions with polar solvents and consequently lower solubility. These interactions are not hypothetical; they find expression in real-world operations, from manufacturing drugs to creating industrially applicable mixtures. Thermodynamics of heat exchange in these interactions is determined and demonstrates to us how enthalpy changes can unmask energy transactions involved in dissolving various compounds in solution (Carlo et al., 2025). In cases such as chemically reacting flows through rocket nozzles and exhaust plumes, the categorization of solutions is further complicated. Thermochemistry of the reactions underscores the importance of flawlessly coupled flow solutions that model the effects of particle drag and heat transfer between phases. It is these interactions that lead to a complicated system in which the principles of thermochemistry—chemical equilibrium, kinetic effects, and non-equilibrium phenomena—apply (Anderson et al., 2025). It is this difficulty in solving the governing equations for such systems that indicates the need for a firm grasp of solution types to confidently predict their behavior. While researchers explore new pairs and solutions, their commitment to the integration of thermodynamic principles results in information both theoretical models and actual applications across a range of disciplines (Figure 9.5).

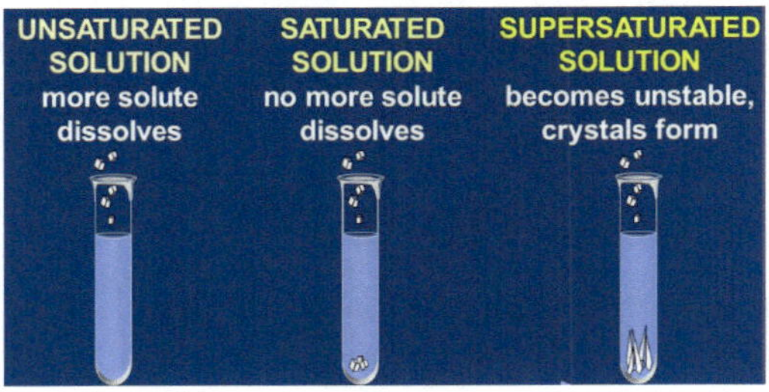

Figure 9.5. *Types of solutions in chemistry.*

Source: https://files.askiitians.com/cdn1/images/2016113-193754498-2789-g12.png.

9.2.1. Homogeneous vs. Heterogeneous Solutions

Homogeneous and heterogeneous solutions are integral to the study of solution thermochemistry and mixtures. Homogeneous solutions are homogeneous in composition in all directions and are a single phase regardless of the scale at which they are being considered. The reliability of their thermal properties renders them unavoidable across a broad spectrum of applications in application, ranging from the synthesis of chemical dilution to reactions. When mixed, the solute particles are dispersed evenly among solvent molecules, which contributes to a stable thermodynamic structure that facilitates simple calculations of enthalpy change and energy transfer. Thus, the reproducibility of homogeneous mixtures facilitates the representation of thermochemical processes in a manner that scientists can focus on variables affecting chemical reactions without the interference created by phase separation. On the other hand, heterogeneous solutions exhibit separate phases or components that do not create a homogeneous composition, which results in a mixture whose properties vary from one region to another. This property makes thermal behavior more difficult, as the different phases may have different thermodynamic properties, making it difficult to calculate energy in the course of reactions. Heterogeneous solutions may include emulsions, suspensions, and colloids, with each having different interactions that influence its thermal stability and energy transfers. For example, the presence of solid particles in a liquid will alter the specific heat capacity and thermal conductivity of the system, necessitating more advanced models to characterize their thermochemical behavior. The understanding of these complexities allows scientists to develop better predictive models, thus enhancing the efficiency and safety of chemical processes utilizing heterogeneous solutions. The consequences of homogeneous and heterogeneous states are profound in use in polymer science and atmospheric chemistry. In polymer science, models such as the Scott-Magat model for a polymer highlight the heterogeneity of polymer structures, e.g., polydispersity in molecular weight for effective modeling of behaviors like precipitation (Mohammadi et al., 2008). These approaches demonstrate the intricacies in heterogeneous mixtures in which the different molecular components interact differently, challenging to basic thermochemical assessment. Similarly, in atmospheric chemistry, understanding why heterogeneous reactions differ so markedly from homogeneous reactions helps refine models of atmospheric reactions to be more accurate, critical to predicting the impacts of climate and pollutant behavior (Allen et al., 2025). As researchers develop better thermodynamic models for homogeneous and heterogeneous mixtures, that knowledge will become progressively critical to the science and to practical applications across many industries.

9.2.2. Solvent-Solute Interactions

Solvent-solute interactions are central to the explanation of the thermochemical characteristics of mixtures and solutions. They decide the dissolution of solutes in solvents, essentially determining the thermodynamic characteristics of solutions. Whether a solvent is polar or nonpolar provides a broad level of influence over solvation dynamics. Polar solvents, for instance, such as water, stabilize ionic solutes through ion-dipole interactions, providing a favorable environment for dissolution. Conversely, nonpolar solvents won't dissolve polar solutes, and therefore there will be bad solubility. This waltz of the solute and solvent molecules is not a physical phenomenon but an array of energetics which can be measured by the use of thermodynamic principles. Knowledge of these interactions forms the basis for further exploring the extended meaning within the field of thermochemistry and application to processes such as energy transfer and phase transitions. Thermodynamics of the interaction between solvent and solute are further explained through more complex theoretical models. One of the hopeful methods is the integral equation theory of molecular liquids based on statistical mechanics. This method begins with the characterization of a molecular interaction potential force field that serves as a foundation for solvation free energy understanding. Diagrammatic analysis allows one to derive integral equations that are indicative of correlation functions between molecules in solution. Perhaps the greatest aspect of this theory is the way it simplifies complicated many-body interactions into manageable forms, increasing one's understanding of how solutes behave in various solvent conditions.

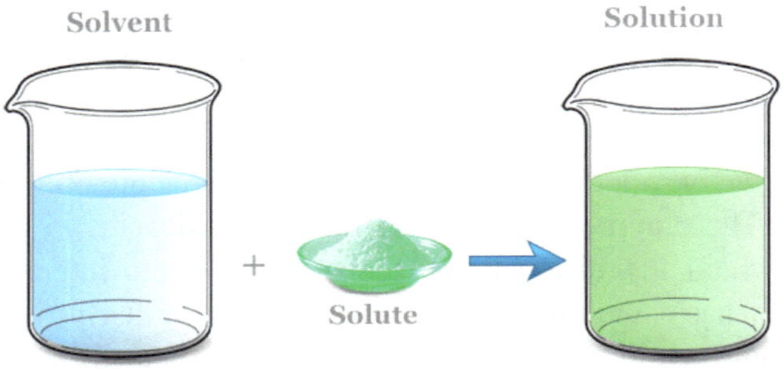

Figure 9.6. *Solvent-solute interactions.*

Source: https://www.priyamstudycentre.com/wp-content/uploads/2022/06/solution-solvent-solute-definition-types-and-examples-of-chemical-solutionsin-chemistry.png.

This analytical ability enables calculation of solvation structures and thermodynamic properties, offering insights equivalent to molecular simulations but successfully managing larger spatial and temporal scopes. This degree of rigor in theory is required for prediction and quantification of the behavior of complex mixtures. Moreover, practical applications of understanding solvent-solute interactions extend far beyond into numerous fields of application, including biochemistry and materials science. Accurate prediction of solvation structures can guide drug design, where solute (active pharmaceutical ingredient) interaction with biological solvents (e.g., bodily fluids) is central to efficacy. By utilizing methodology like the three-dimensional reference interaction site model (3D-RISM), researchers can gain a holistic view of how solvent interaction sites influence solute behavior. This model, in combination with the Kovalenko-Hirata closure approximation, makes correlation functions mappable, hence graphically representing the way solvation affects thermodynamic properties. The capacity to examine such interactions with precision not only enhances fundamental thermochemistry knowledge but also contributes to applications within scientific disciplines, ultimately closing theory and practical significance (Figure 9.6).

9.2.3. Factors Affecting Solubility

Solubility of a substance in a solvent is a complex phenomenon governed by numerous thermodynamic and environmental variables. Temperature is one of the key regulating factors of solubility; as a rule, solubility of solids in liquids increases when temperature rises. This relies on higher temperatures carrying sufficient energy to counteract intermolecular forces between the solid so that it dissolves. When gases are involved, though, increasing temperature tends to lead to lower solubility as greater kinetic energy allows space for the gas molecules to escape the solvent. Thus, the interaction between solubility and temperature is of utmost significance, not only in influencing the formation of solutions but also their stability and characteristics under varying conditions, highlighting the importance of the thermodynamic approach to account for such interactions. Another significant parameter in affecting solubility is pressure, particularly that of gases. According to Henrys Law, the solubility of a gas in a liquid depends upon and is directly proportional to the partial pressure of the gas above the liquid. This kind of relationship is central to understanding the dissolution of gases in solutions, especially in industrial applications like carbonation. The compression of gas molecules into liquid state is enhanced by increased pressure. But the pressure effects on the solubility of solids and liquids are significantly reduced, and the unique role of gases in solution is emphasized. These pressure effects are of significance in a wide range of fields, from chemical engineering to environmental science, because they offer promise of enhanced prediction and control over the solubility in phenomena of application and natural occurrences, like the properties of atmospheric gases. The nature of the solute and solvent also plays a significant role in determining solubility. The polar solvents such as water are inclined towards the dissolution of ionic and polar molecules due to the desired interactions between solute ions and solvent molecules. This is summarized by the slogan like dissolves like, which says that the similar type of interactions will increase solubility. Further, other solutes may influence solubility through other mechanisms like common ion effects or solute-solute interactions that enhance or hinder dissolution. For instance, addition of a common ion to saturated solutions can lead to decreased solubility of a salt, which is a consideration of relevance in equilibria and chemical reactions. As shown in studies of solid solutions and phase equilibria, understanding the interactions between these is essential for designing effective solutions in chemical processing and ecological protection, evidence of the multidirectionality of solubility (Hong et al., 2019; Lewis et al., 2025) (Figure 9.7).

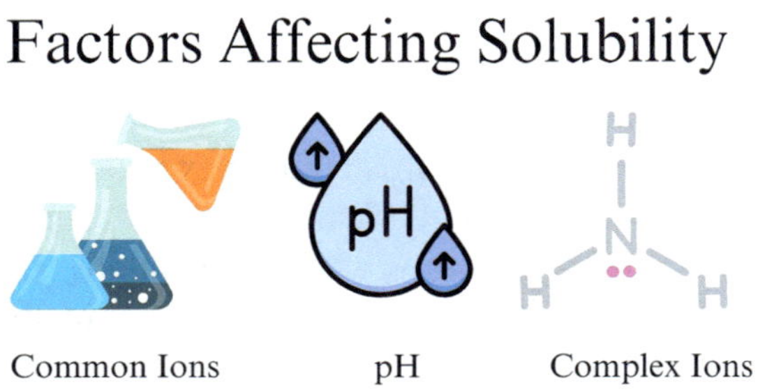

Figure 9.7. *Factors affecting solubility.*

Source: https://i.ytimg.com/vi/HWDkqEL2G1U/maxresdefault.jpg.

9.3. THERMODYNAMIC PROPERTIES OF SOLUTIONS

The thermodynamic properties of the mixtures form the backbone in describing the behavior of mixtures in chemical and physical conditions. Solutions, during their formation, undergo change in their inherent thermodynamic properties, including enthalpy, entropy, and Gibbs free energy. Enthalpic changes are particularly significant because they indicate the heat of absorption or release in dissolution and express solute-solvent molecule interactions. Besides, entropy is the most important because the mixture of substances has a tendency to bring an increase in disorder in the system. This relationship exists in the equation of Gibbs free energy, which provides a forecast for the spontaneity of reactions. In investigating these properties, researchers can infer information about compounds' stability and solubility in solutions, helping advance theoretical and experimental thermochemistry as witnessed in phenomena under investigation with respect to Gilbert et al. (2025). Investigation of the thermodynamic solution properties also involves investigating ideal and non-ideal behaviors. Ideal solutions follow Raoult's law to the letter, wherein vapor pressure is directly proportional to mole fraction. Real solutions approach non-ideal situations due to intermolecular forces and solute-solvent molecule interactions. These non-idealities are represented by activity coefficients, which measure the effect of concentration on thermodynamic properties. Awareness of these deviations is crucial to applications in biochemical engineering and material science, where accurate thermodynamic data are needed in systems design. In particular, non-ideality in biological systems is highlighted by the thermodynamic characteristics of biochemical compounds in water solution, as documented in Carlo et al. (2025). Furthermore, these thermodynamic properties must be investigated for various industrial applications, especially in drug formulation as well as separation techniques. Thermal stability, solubility, and the overall system enthalpy are all of principal importance that determine solute behavior within a solvent matrix. For instance, active pharmaceutical ingredient dissolution behavior can significantly influence their bioavailability, necessitating thorough knowledge of their solution thermodynamic interactions. Moreover, the optimization of conditions like temperature and pressure in thermochemical processes can enhance product efficiency and yield. Last but not least, the thermodynamic investigation of solutions not only contributes to improved formulations but also opens paths for breakthrough strategies in green practices, emphasizing the relevance and utility of this field in our more complicated technology-dominated world.

9.3.1. Changes in Enthalpy During Dissolution

The dissolution process of solids in solvents is intrinsically related to enthalpy changes, which is the heat evolved or absorbed during interaction. During the dissolution of a solute, there is a sequence of molecular interactions where intermolecular forces between particles of the solute are reduced. This is an energy-consuming process with a positive change in enthalpy (ΔH). However, when solute molecules become separated in the solvent, new solute-solvent interactions take place that are able to release energy, leading to a negative ΔH. The net change in enthalpy that one observes upon dissolution thus depends upon the interplay of these antagonistic processes. It is critical to realize such changes in not only thermochemistry but also material science, where data on solvation energetics determines the design of drugs and other chemical solutions. Secondly, the influence of temperature on altering enthalpy change during dissolving cannot be overstated. An increase in temperature tends to enhance the solubility of most solutes, with more enthalpically favorable interactions occurring as kinetic energy in the system increases. As more frequently and with greater energy molecules collide, the likelihood of successful solvation increases, thereby reducing the enthalpic barrier to surmounting intermolecular forces. For certain systems, such as those concerned with very polar solvents and ionic solutes, the endothermic or exothermic nature of dissolution manifests very visibly, influencing the total temperature trend of the system. These phenomena are identified by empirical observations, the argument being that varying solvent compositions or solute structures can lead to extremely varying thermodynamic outcomes for dissolution (Jaroslav Šesták et al., 2012). In practical applications, enthalpy change during dissolution becomes crucial in industrial processes requiring precise thermodynamic control. For example, the identification of the melt viscosity of the glass in glass formulation can significantly influence the product quality. Research indicates that enthalpy changes upon dissolution of oxides in silicate melts determine their viscosity profiles (Dingwell et al., 1988). Because the energetic landscape changes depending on the nature of the solute, these thermal dynamics must be accounted for by producers in a bid to optimize melting processes. Therefore, thermal solution analysis and the corresponding changes in enthalpy are of utmost significance, yielding critical information that determines efficiency and performance in chemical production and formulation, further underlining the significance of thermochemical concepts in real-life applications (Figure 9.8).

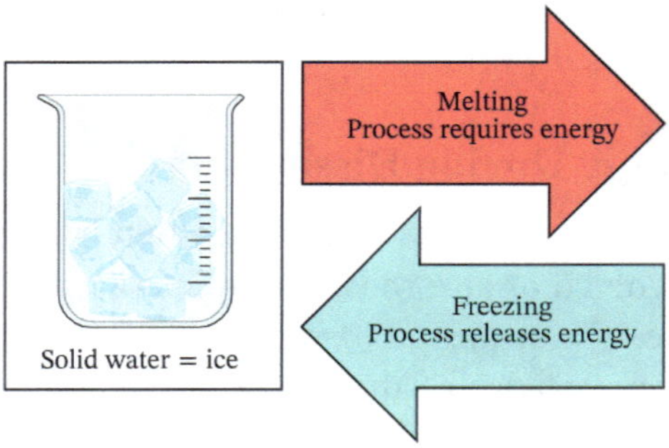

Figure 9.8. *Changes in enthalpy during dissolution.*

Source: https://images.nagwa.com/figures/explainers/240126790659/2.svg.

9.3.2. Gibbs Free Energy and Spontaneity

The concept of Gibbs free energy is of prime significance when calculating the spontaneity of chemical reactions, particularly in the context of thermochemistry. Gibbs free energy (G) is the maximum work which can be reversibly done by a thermodynamic system at constant temperature and pressure. Change in Gibbs free energy (ΔG) determines whether a process will occur spontaneously or not. If ΔG is negative, the process will be thermodynamically spontaneous, but an increasing ΔG indicates that the process will be non-spontaneous. This relationship highlights the intrinsic connection between the fluctuations in energy and the directionality of chemical reactions. Further, the significance of Gibbs free energy extends to more complex reactions beyond simple reactions, affecting other biochemical reactions and the stability of mixtures and solutions overall. Thus, understanding this thermodynamic parameter is fundamental in forecasting the behavior and outcome of most chemical systems in the laboratory as well as in nature. Gibbs free energy when applied to the realm of solutions and mixtures reveals additional facets in terms of chemical equilibria and the spontaneity of reaction. Because solutes are engaged within a solvent, their respective Gibbs free energy changes can have a profound effect on solubility and stability within molecular arrangements. For instance, the presence of solute can stabilize or destabilize certain states within a mixture and thus influence reaction pathways. Also, the spontaneity of mixing processes can be accounted for by Gibbs free energy, in which the structure of systems evolves toward the state of minimum energy. The enthalpy and entropy factors are also most critical here; a process that induces an increase in randomness (positive entropy change) may lead to the reactions despite unfavorable enthalpic contributions. Therefore, Gibbs free energy consideration is critical for appreciating the delicate balance among the thermodynamic parameters in solutions and mixtures, as expressed through their respective phase behaviors. Additionally, the application of Gibbs free energy is not simply theoretical in character; it has deeper implications in prebiotic chemistry and astrobiology as well. Life-origin studies have confirmed the thermodynamic feasibility of certain molecular pathways, especially those that incorporated hydrogen cyanide (HCN), which are believed to have taken place within the prebiotic environment. Thermodynamics of the processes are then capable of providing hints about conditions under which life's building blocks could have been synthesized, providing insight into possible pathways in a variety of extraterrestrial environments. For example, theoretical models suggest that HCN-derived polymers can be stable under some conditions, thereby supporting the hypothesis of simple organic molecule-emergent properties of life's. The interplay between Gibbs free energy and spontaneity in reaction in such a scenario is not only meaningful in regard to the fundamental character of chemical reactions but also related to their possible place in larger astrobiological stories (Middelburg et al., 2025; Sandstr\\\\uf6m et al., 2022).

9.3.3. Effect of Temperature on Solubility

Temperature is one of the determining factors of solubility of a substance in various solvents, which comes under study of thermochemical processes. With an increase in temperature, the kinetic energy of the molecules also increases and hence the interaction between particles of the solute and solvent. This generally leads to higher solubility, particularly for solid solutes in liquid. For example, all such salts as sodium chloride dissolve more in water at high temperatures. But this is not a hard and fast rule. There are certain salts, such as calcium sulfate, that are less soluble in higher temperatures, indicating that the solubility-temperature relationship could be complex and will vary with the chemical nature of the solute. Understanding these dynamics is crucial, particularly in fields such as biochemistry and environmental science, where the variation of temperature can greatly affect the behavior and reactivity of solutes within solutions (Carlo et al., 2025). The thermodynamic principles governing solubility are variation in enthalpy and entropy upon the variation of temperature. At high temperatures, the increased molecular motion is likely to result in an endothermic process of dissolution involving the uptake of heat, initiating a greater solubility of the solute. Entropy in this case is crucial since the disorder resulting from solvation prefers dissolution at high temperatures.

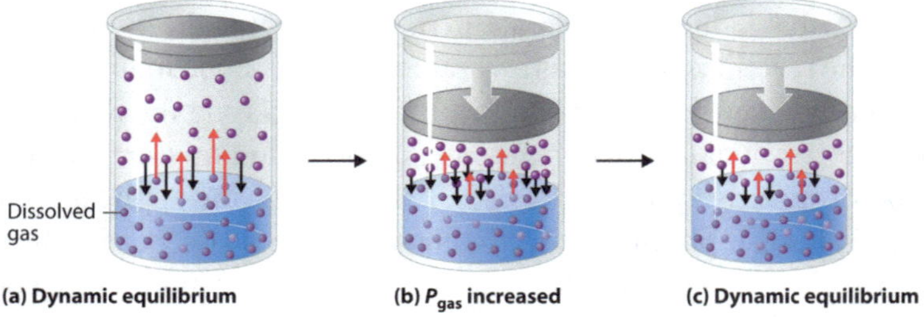

Figure 9.9. *Effect of temperature on solubility.*

Source: https://saylordotorg.github.io/text_general-chemistry-principles-patterns-and-applications-v1.0/section_17/e21684c864731d8f2e572dfcbf18b54f.jpg.

This can be seen in the dissolution of most organic molecules, where raising temperature yields a more disordered molecular shape and thus higher solubility. Dissolution exothermic reactions can be shown to have lesser solubility at higher temperatures, which can be observed in some gas solubility phenomena. These thermodynamic principles are essential to learn to predict solubility behaviors, especially in complex mixtures wherein there are multiple solutes involved (Hong et al., 2019). In industrial applications, temperature effects on solubility are a foundation for various industries like pharmaceuticals, chemical production, and food processing. Pharmaceutical formulations may often need temperature-controlled conditions to optimize drug solubility and bioavailability. For chemical production, accurate temperature control can lead to improved yields and selectivity in reactions with processes dependent on solubility. Temperature regulation is another critical function in food processing; ingredients are often dissolved or blended at specific temperatures to create optimal texture and flavor profiles. Further, environmental concerns such as climate change illustrate the relevance of understanding solubility fluctuations, as warming climates can alter solubility for toxins in aquatic systems, with potential impacts on ecological health. Hence, knowledge at a detailed level about the effects of temperature on solubility is imperative in many areas of science and industry (Carlo et al., 2025) (Figure 9.9).

9.4. MIXTURES AND THEIR THERMOCHEMICAL BEHAVIOR

Investigation of mixtures and their thermochemical behavior is vital in understanding how substances respond energetically in a solution. Mixtures, whether homogenous or heterogeneous, possess unique properties that usually differ from those of the individual components. In a thermochemical sense, the thermodynamic activity of mixtures is determined by changes in enthalpy and entropy in dissolving reactions, and this leads to a net free energy change. The elementary thermodynamic laws determine the viability and magnitude of reactions within mixtures, with the condition of equal energy exchange determining reaction spontaneity. For example, when a solid solute is dissolved in a solvent, solute and solvent molecule interactions have a significant impact on changing thermal characteristics, modifying heat absorption or release during the process. Thus, it is necessary to understand not just individual components but also their interactivity at atomic and molecular levels in order to analyze mixtures. Compounding the complexity of thermochemical analysis of mixtures is the principle of specific heat capacities, which can also vary due to intermolecular forces present in different compositions. The specific heat capacity of a mixture depends not only on the proportion of its components but also on their respective thermodynamic properties. With this, the measurement and computation of the specific heat can be valuable data regarding thermal behavior under different conditions. For example, mixtures often exhibit phenomena such as heat of solution or heat of mixing, which are quantifiable and can be utilized to distinguish between endothermic and exothermic behavior. Through this knowledge of the heat interactions of mixtures, scientists are capable of applying thermodynamic principles to define practical applications, such as chemical product manufacturing, or predicting solubility limits in advanced solutions. Such studies are of pivotal importance for the establishment of such fields as material science to biochemical engineering, in which control of thermal behavior is essential. Thermodynamic properties also form a basis of determining the stability and reactivity of mixtures, especially biochemical ones. The majority of biological reactions occur in aqueous media, where solute interaction significantly affects reaction kinetics and mechanisms. The computational-model-based principles of thermodynamics can provide insight into such behaviors. As a specific example, the application of the partition function can enable one to estimate particular free energy and enthalpy via various species in a mixture. This method is best applied in the analysis of biochemical compounds, as in research on the thermodynamic properties of aqueous solutions (Horton et al., 2025) and the measurement of energy change upon biochemical reactions in solution (Carlo et al., 2025). With further insight into these behaviors, researchers can improve the experimental design and industrial processes associated with biochemical interaction in mixtures (Figure 9.10).

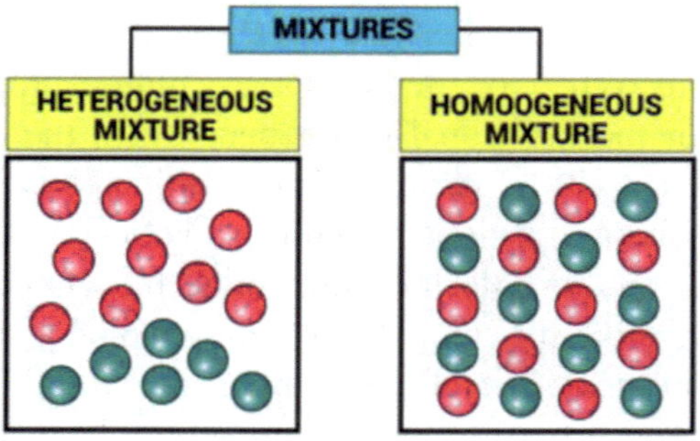

Figure 9.10. *Mixtures and their thermochemical behavior.*

Source: https://external-content.duckduckgo.com/iu/?u=https%3A%2F%2Ftse1.mm.bing.net%2Fth%2Fid%2FOIP.qw-2wpYmLzGaSTiV6EsObwHaDi%3Fpid%3DApi&f=1&ipt=6296b641d56aa36ee4633073c074b0d03dc41f0d5c12c67a-bee96bcfe044d433.

9.4.1. Definition and Examples of Mixtures

In thermochemistry, mixtures acquire a definition that is of utmost importance since it determines the interaction and behavior of substances in solutions. A mixture is defined as a material system that comprises two or more substances that have their own chemical identities and therefore distinguish them from pure substances. Mixtures can further be classified as homogeneous and heterogeneous. Homogeneous mixtures, like saltwater, exist in uniform composition and properties throughout, while heterogeneous mixtures, such as salad, are composed of various physical phases. This distinction is important because it dictates thermodynamic calculations and the energy change undergone during phase transitions and interaction between mixture components, implicating their thermochemical behavior in beneficial applications such as combustion processes (Marble et al., 1956). The utility of mixtures is demonstrated in numerous applications ranging from cooking to industrial production. Air, for instance, is a homogeneous gaseous mixture of nitrogen, oxygen, carbon dioxide, and trace components with specific thermochemical properties necessary for the sustenance of combustion. Understanding air as a mixture allows precise estimation of combustion characteristics such as flame temperature and emissions. Unlike homogeneous mixtures, heterogeneous mixtures can perform complex interactions, particularly if one is to consider the thermodynamic properties of suspensions or emulsions. Such diversity of characteristics raises important questions regarding energy efficiency, stability, and reaction rate, thereby affecting their real-world use in engineering and laboratory experiments, where phase behavior and component interaction need to be well understood (Kovalenko et al., 2015). Additionally, mixtures are also investigated to obtain information regarding fundamental thermochemical quantities, especially phenomena of solvation. The interaction of the solvents with the solutes yields a mixture that results in various solvation dynamics, which can be examined using statistical mechanics models. Techniques, such as the three-dimensional reference interaction site model, are used for charting the solvent-solute interactions, giving a sophisticated understanding of the solvation structure and its thermodynamics. This information develops a clearer picture of the way mixtures behave in various conditions, informing theoretical investigation and practical utilization in chemistry and materials science. A general view of these principles not only explains the nature of mixtures but also emphasizes how significant they are in advancing thermochemical studies and technology advancement in various disciplines (Kovalenko et al., 2015).

9.4.2. Ideal vs. Non-Ideal Mixtures

The distinction between ideal and non-ideal mixtures forms a foundation in the world of thermochemistry, particularly for mixtures and solutions. Under an ideal mixture, the components follow Raoult's Law such that the concentration and vapor pressure of each component vary directly. This is due to the fact that it is understood that intermolecular forces between different molecules are close to pure substance ones. However, in real-world scenarios, ideality is never attained due to variations in molecular size, polarity, and intermolecular forces between the molecules in non-ideal mixtures. Deviations are very important since they can affect significantly properties such as boiling point elevation, freezing point depression, and vapor pressure, with an implication of the thermodynamic behavior of solutions and mixtures. Non-ideal mixtures are complex and require rigorous examination of their thermodynamic behavior. For instance, the phenomenon of strong intermolecular interactions such as hydrogen bonding or ionic interaction has a tendency to cause positive or negative deviations from Raoult's Law, and these can drastically alter the systems behavior. Precisely, non-ideal mixtures can cause energetic or entropic alterations that affect solution stability and formation. Thermodynamic study of these mixtures is performed by calculating excess properties, which characterize the deviation from real and ideal mixture behavior. These calculated outcomes are imperative to predicting the behavior of mixtures under conditions of change and provide insight into reaction mechanisms and interactions involved. As in complicated gaseous reaction research, awareness of these mixtures thermodynamic properties unearths the most important characteristics of non-ideal systems (Austin et al., 2009). The implications of recognizing ideal and non-ideal mixtures take theoretical applications to practical industrial applications such as chemical processes and biochemical systems. The thermodynamic properties of biochemical compounds dissolved in water illustrate that the assumption of ideal behavior may not always hold true since the solute-solvent interactions are generally more important (Carlo et al., 2025).

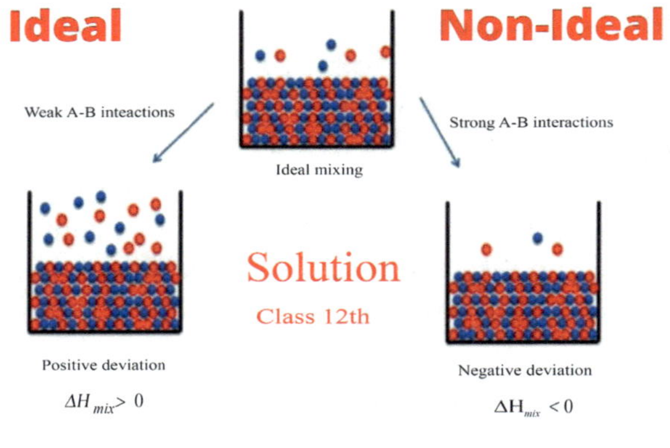

Figure 9.11. *Ideal solution vs non ideal solution.*

Source: https://external-content.duckduckgo.com/iu/?u=https%3A%2F%2Fi.ytimg.com%2Fvi%2FiJ68_Ya1chM%2Fmaxresdefault.jpg&f=1&nofb=1&ipt=7238fecc0ad0e68ce45ab123835b505eff831c46f036c1b33c4aa8245b9ec5b6.

For example, understanding of the non-ideal characteristics of mixtures is necessary in the area of enzymatic reactions in biochemistry to know how reaction rates and equilibriums are affected. This valuable appreciation leads to more efficient and predictable chemical process design. In addition, environmental chemistry is benefited by the examination of non-ideal mixtures in examining and reclaiming pollutants in diverse aqueous matrices. In the long term, an in-depth understanding of ideal and non-ideal systems is also required to solve real-world issues in scientific research and in industrial applications (Figure 9.11).

9.4.3. Raoult's Law and Its Applications

Raoult's Law, one of the cornerstone laws of chemistry, describes that the vapor pressure of a solvent in solution is directly proportional to the mole fraction of the solvent. Specifically, when a non-volatile solute is introduced to a solvent, the solvent's vapor pressure decreases in proportion to the concentration of the solute. This relationship emphasizes the colligative properties of solutions, which are properties that are dependent on the number of particles of the solute rather than their nature. Information regarding Raoult's Law is essential in the determination of solution behavior in various conditions, particularly when applying thermochemistry, as energy and heat transfer modification occur. The information assists the chemist in modifying conditions for optimal effect in industry or in the lab, effectively utilizing solvent-solute interactions to achieve maximum efficiency and effectiveness in chemical reactions. Applications of Raoult's Law exist in a great diversity of domains, including environmental science and materials engineering. For instance, it is an important aspect of the assessment of indoor semi-volatile organic compounds (SVOCs), as highlighted in recent research. The low volatility of SVOCs allows them to persist and spread among indoor compartments, influencing human exposure through inhalation, ingestion, or dermal permeation.

This theoretical framework of thermodynamic interactions not only informs near-term health and safety protocols but also serves as the basis for modeling approaches anticipating chemical activity and exposure risk (Eftekhari et al., 2020). Raoult's Law thus provides a theoretical basis for considering the implications of chemical behavior in real-world contexts, enabling researchers to create better consumer goods and effective protection from dangerous chemicals in the domestic environment. Aside from its applications in environmental science, Raoult's Law is also a basic principle in energy transfer and storage. For example, recent studies have investigated the photophysical processes of molecular systems with the aim of optimizing the efficiency of optoelectronic devices and photovoltaic cells. The thermochemical coupling among different components of these systems needs to be determined for optimizing charge transfer mechanisms and energy storage. With the application of Raoult's Law, scientists can account for how different molecular configurations influence energy efficiency and stability (Scardino et al., 2012). Such an experiment-theory collaboration not only advances the development of new materials but also verifies the relevance of Raoult's Law in current investigations, exemplifying its cross-disciplinary applicability (Figure 9.12).

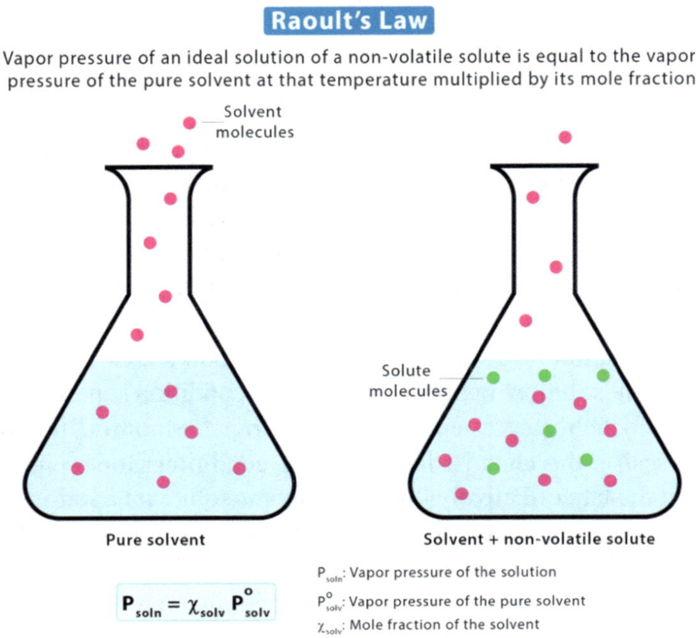

Figure 9.12. *Raoult's law.*

Source: https://www.chemistrylearner.com/wp-content/uploads/2022/11/Raoults-Law.jpg.

9.5. CALORIMETRY IN THERMOCHEMISTRY

Calorimetry is a key component of thermochemistry, more so in connection with the energy change in chemical reactions involving mixtures and solutions. Calorimetry as an analytical technique quantifies the flow of heat that occurs during these reactions and enables chemists to determine enthalpy change and evaluate the thermodynamic stability of various compounds. By measuring temperature changes in a calorimetric system, researchers can calculate the heat absorbed or evolved and thus obtain information regarding reaction mechanism. Techniques such as differential scanning calorimetry (DSC) have achieved prominence due to their ability to provide high-resolution data regarding phase transitions, heats of fusion, and other thermodynamic properties, and thus contribute towards the characterization of simple and complex mixtures.

The application of calorimetry extends outside the walls of the laboratory to practical use, including in materials and energy storage. For example, it is used in characterizing multicomponent fluoride salt mixtures to be used in solar dynamic space power systems. Melting points have been confirmed and new eutectic compositions that are critical for optimizing the heat storage material have been found, demonstrating real-world impact of calorimetric analysis on product development. System design-thermal behavior correlation is critical because accurate measurements of fusion heats and heat capacities ensure compatibility of materials with the elevated working temperatures typical of such systems (Misra et al., 2025). Calorimetry in this case not only determines material selection but also determines the design of advanced technology in energy systems. Calorimetry also offers a means of investigating the thermodynamic interactions in solutions and mixtures, broadening our understanding of solvation and reaction kinetics. From heat capacities, heats of solution, and the like, scientists can tell whether solute-solvent and solute-solute interactions influence overall energetics. This is particularly the case in transition metal compound systems, where phase stability and reactivity may be highly dependent on the thermodynamic environment (Hong et al., 2019). The outcomes that are realized through calorimetric measurements serve to build the foundation for data used to optimize theoretical models and enhance predictive power in thermochemistry. Thus, the use of calorimetric techniques in the overall approach to thermochemical research enhances our information regarding energy processes, thereby leading to an increase in basic chemistry and applied fields in many areas (Figure 9.13).

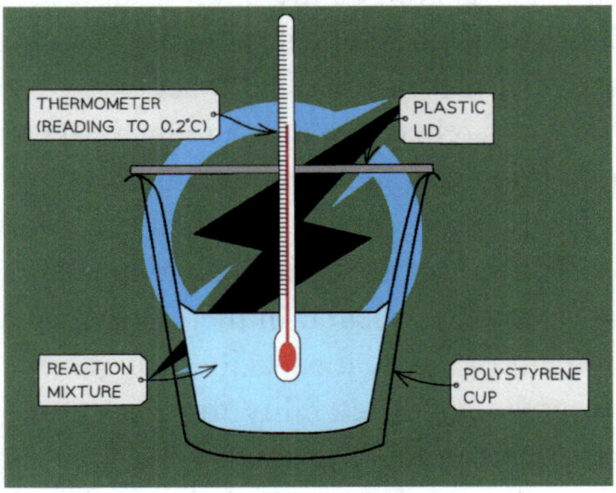

Figure 9.13. *Calorimetry in thermochemistry.*

Source: https://cdn.savemyexams.com/cdn-cgi/image/f=auto,width=3840/https://cdn.savemyexams.com/uploads/2023/07/carlorimetry-equipment-set-up.webp.

9.5.1. Types of Calorimeters Used in Studies

Calorimetry is a significant analytical tool in thermochemistry, namely in the study of solutions and mixtures. Various calorimeters are employed, each for quantifying heat transfer in various environments. The most common calorimeter is the coffee cup calorimeter, commonly for the determination of the heat of reaction at constant pressure. Its simplicity and low price make it an ideal instrument for learning and minor research where precision is not so necessary. On the other hand, bomb calorimeters find application in reactions that occur at constant volume and are utilized for combustion studies. The more advanced instruments provide accurate determinations of the heat released from reactions, which is crucial in energy value calculations for fuels and biochemical substances. Hence, calorimeter selection has a direct bearing on accuracy and utility of thermochemical data obtained in investigations. Another key type is the differential scanning calorimeter (DSC), which detects heat flows in material phase transitions. This type is most significant in the examination of mixtures, as it allows scientists to study melting points, crystallization, and glass transitions with accuracy. DSC is a precious commodity in the pharmaceutical industry, where drug polymorphism and formulation stability must be studied. It provides details regarding the interactions between various solvents and solutes based on their thermal properties. Researchers are increasingly utilizing DSC in combination with traditional calorimetry methods to obtain a wider overall overview of the thermodynamic performance of solutions and mixtures. Combination of DSC data with other analytical data can make a significant contribution towards the formation of new materials as well as the optimization of the existing ones for purposes. Besides, isothermal titration calorimetry (ITC) has also emerged as a useful tool for the study of thermochemical reactions, particularly biochemical interactions. From ITC, researchers can measure the heat change associated with binding events in a solution, e.g., enzyme-substrate complexation or antigen-antibody complexes. The real-time analysis can provide useful information on the enthalpy and entropy changes occurring in such binding events, which can help understand molecular interactions better. In thermochemistry research, ITC also assists in the measurement of the heat capacity changes of various solvents and solutes at varying temperatures. The ability to obtain multiple thermodynamic parameters from a single experiment positions ITC as an important tool in the development of our knowledge of molecular behavior in solutions and mixtures, adding significantly to the area of thermochemistry and its applications.

9.5.2. Measuring Heat Changes in Solutions

Solution heat changes measurement is a critical component of thermochemistry that gives insight into solute and solvent energetic behavior as chemical reactions take place. It involves the application of calorimetry, which is a technique of measuring heat transferred in a chemical reaction or phase change. Differential scanning calorimetry (DSC) is particularly valuable since it quantitates heat flow pertaining to material transitions by temperature. For instance, when analyzing solute dissolution in water, DSC will reveal endothermic or exothermic reactions, detecting energy absorbed or released in the process of solvation. Detection of such heat changes is basic, as they do not only describe solute interactions but also predict a variety of applications in biology, materials science, and industrial processes, for the successful design of systems based upon precise thermal control (Carlo et al., 2025). In addition, the heat transformations that take place during the measurement of solutions can significantly affect the stability and activity of biochemical compounds in aqueous solutions. The thermodynamic characteristics of the compounds have a direct connection with their activity in solution, and this may be explained in terms of calorimetric methods. For example, dissolution would influence molecular interactions and stability, and therefore alter the efficacy of drugs and biochemical reactions.

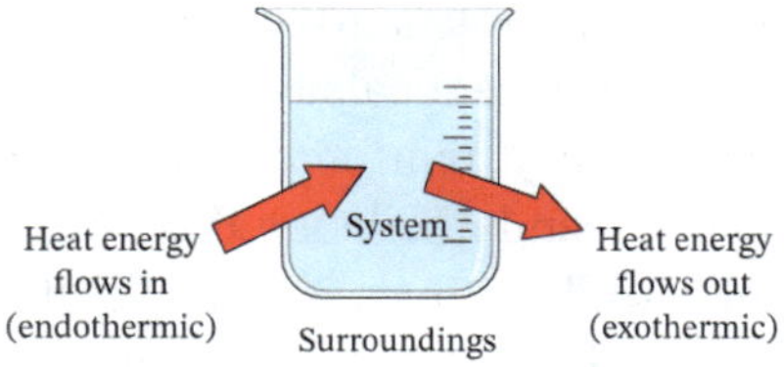

Figure 9.14. *Measuring heat changes in solutions.*

Source: https://images.nagwa.com/figures/explainers/259104205403/3.svg.

As suggested, calorimetric techniques are instrumental in uncovering how energy relationships control the solubility, reaction rates, and position of equilibria of various compounds within a solution, and shed light on the underlying principles of biochemical reactions on a thermodynamic foundation (Carlo et al., 2025). The ability to measure and interpret such heat changes has profound implications to the design of more efficient chemical systems. In considering the broader relevance of heat measurement in solutions, the emergence of innovative solid-state battery technology underscores the value of thermochemical principles in contemporary research. Solid-state electrolytes (SSEs) employed in lithium batteries, for instance, must be efficient at managing heat transitions to optimize performance and safety. The interaction between SSEs and solvents governs key thermodynamic properties that influence ionic conductivity and chemical stability within the battery system. An understanding of these interactions from the quantification of heat changes determines the composition of composite materials to achieve high energy density and safety needs. It illustrates the necessity for thermochemical research, bridging fundamental scientific inquiry with real-world applications in cutting-edge technologies, setting the importance of precise thermal modeling in creating scalable and effective energy storage devices (Figure 9.14).

9.5.3. Applications of Calorimetry in Thermochemistry

Calorimetry in thermochemistry is an important tool for understanding the heat characteristics of mixtures and solutions. With the help of measurements on heat passed in chemical reactions or physical changes, calorimetric techniques provide data on enthalpy change, specific heat capacity, and phase transition characteristics. These measurements are critical in making forecasts regarding the behavior of substances under varying conditions, especially in solutions where solute-solvent interactions play a critical role in affecting the behavior. For instance, calorimetry can give insights into the dissolution process energetics, thereby allowing chemists to have optimized formulations for applications ranging from pharmaceuticals to materials science. Consequently, the unique merits of calorimetry reinforce our fundamental understanding of chemical thermodynamics and facilitate the development of innovative technologies in many industrial settings. The application of calorimetry also reaches systems of nanoalloy bases, as demonstrated in studies aimed at Cu-Ni nanoparticles. In such experiments, differential scanning calorimetry (DSC) provides scientists with important data regarding the melting temperatures and phase behavior of nanomaterials. The application of the CALPHAD method to determine phase diagrams proves that nanoparticle size affects the melting temperatures and that the theoretical model accordingly agrees with experimental findings. By combining such methods as dynamic light scattering (DLS) and transmission electron microscopy (TEM), scientists are in a position to completely characterize the thermal behavior of these nanoalloys and thus bridge the gap between experimental data and theoretical computations. Such data supports the optimization of materials for catalytic or electronic applications, emphasizing the decisive role of calorimetry in the creation of nanotechnology. Calorimetry also plays a key role in research on phase equilibria within complex systems like the Hf-Zr-C-N-O system. This field of thermochemistry is a telling demonstration of how calorimetric data can advance our understanding of phase stability and phase transitions in transition metal compounds. Through extensive experimental studies and ab initio calculations, researchers have explored the thermodynamics of these materials in a systematic manner in recent decades. New methods such as electrical pulse annealing and laser melting are set to revolutionize experimental approaches by generating novel data points validating theoretical predictions. The ability to forecast phase diagrams and melting temperatures with high accuracy provides new opportunities for advanced material design, so calorimetry still occupies a central place in the science of thermochemistry in mixtures and solutions.

9.6. REAL-WORLD APPLICATIONS OF THERMOCHEMISTRY

Applications of thermochemistry extend to many fields where thermochemical information is essential to know about energy changes in chemical reactions. In the pharmaceutical industry, for instance, thermodynamic properties of compounds in aqueous solutions are crucial in drug design. This involves the prediction of solubility, stability, and reactivity under pressure and temperature influence. Computational thermochemistry calculations allow chemists to predict whether a drug will properly dissolve in body fluids, hence influencing its functionality. Further, the ability to replicate these interactions makes the creation of new drugs more robust in that scientists can measure optimum drug performance conditions. Thus, thermochemistry not only facilitates the creation of effective pharmaceuticals but also assists in determining drug behavior under a variety of physiological conditions as well as being directly related to clinical results from lab research. In environmental science, thermochemistry assists in comprehending and reversing the influence of pollutants. The thermodynamic energy conversions during chemical degradation processes provide the foundation upon which effective remediation technologies are designed. Thermodynamic data facilitate predictions regarding the behavior of the contaminants when interacting with different matrices in the environment, i.e., soil and water bodies. For instance, the solvation structure of a contaminant can be modeled to determine its mobility and reactivity in different ecological environments, which is significant in assessing its environmental persistence. Advanced approaches, such as statistical mechanics, enable researchers to establish significant conclusions about contaminant behavior through meticulous analysis of molecular interactions. Approaches based on principles provided in thermochemistry therefore offer effective tools for planning interventions with minimal ecological impact and maximum safeguarding of public health. There is another significant application of thermochemistry in the production and storage of energy. Improvements in technology for batteries rest upon understanding the energetic processes involved with electrochemical reactions. Scientists learn to optimize battery efficiency and longevity by comprehending the thermochemical properties of electrode materials and electrolyte solutions. Efficient storage technology is particularly valuable when designing renewable energy technologies, where supply and demand must be matched. Moreover, greater understanding of heat management through thermochemical cycles can make fuel cells and other alternative energy systems more efficient. In this regard, thermochemistry is one of the prime drivers of innovation, guiding engineers and scientists to solutions that promise cleaner energy solutions.

9.6.1. Industrial Processes Involving Solutions

Manipulation and utilization of solutions figure centrally in industrial processes, particularly in the synthesis and production of high-tech materials. One of the best examples is the development of all-solid-state batteries (ASSBs), where the emphasis on ionic conductivity and stability in their electrolyte systems is especially critical. Researchers are becoming more focused on the intricate interactions between solid-state electrolytes (SSEs), polymers, and solvents because they have a profound influence on the overall performance of the batteries. Notably, the successful combination of a $Li_7P_3S_{11}$ SSE and a SEBS polymer and xylene solvent has been promising with a decrease in the thickness of the SSE without losing conductivity up to 0.7 mS cm-1. The above example is a proof of the industrial benefit of having in-depth knowledge of molecular interaction in solution having a direct impact on material property as well as scaling-up for industrial commercialization (Banerjee et al., 2019). In addition, the application of transition metal carbides and nitrides in industry is another example of solution-based technologies in material synthesis and thermal management. The high melting points and chemical stability of zirconium and hafnium compounds make them good candidate materials for applications ranging from cutting tools to high-temperature insulation. Recent research addresses phase equilibria in the Hf-Zr-C-N-O system based on previous data and modern computational methods. These findings encapsulate the role solutions play in augmenting the synthesis of complex compounds, such as oxynitrides and solid solutions of Zr and Hf. Based on the knowledge of thermodynamics and experimental conditions necessary for the synthesis of these compounds, industries can produce and optimize their manufacture process to enhance material performance and operational efficiency (Hong et al., 2019). Also, the thermochemical process in industrial solutions is at the center of operation for most uses, particularly in materials design and energy storage. While competing to excel in high-performance material production, industrial competition highlights the importance of solution chemistry through optimizing the thermochemical process. The addition of advanced characterization tools to explore the molecular level behavior in various electrolyte systems presents new possibilities for enhancing the properties of materials used in energy devices. This evolving understanding of phase relationships and merging of materials at a molecular level affects the operating properties and durability of products like batteries. Lastly, the recent research points out that solutions are not merely a medium but a core part of industrial processes that drive technological advancement and economic growth via greater material solutions (Banerjee et al., 2019; Hong et al., 2019) (Figure 9.15).

Figure 9.15. *Industrial processes involving solutions in thermochemistry.*

Source: https://www.endress.com/__image/a/9236974/k/01720a96d48ef17fb3a56f38698ccecb8a5d8756/ar/1-1/w/928/t/jpg/b/ffffff/fn/process_analyzer_solutions.jpg.

9.6.2. Environmental Implications of Thermochemical Studies

The environmental impact of thermochemical studies is significant since it provides greater insight into the behavior and alteration of materials in various combinations and solutions. By means of thermodynamics in such processes, scientists are able to predict with greater accuracy how compounds will react under other conditions, and this is critical in identifying their environmental implications. Thermochemical modeling aids in the determination of the volatility and condensation behavior of elements, particularly in the atmospheric condition and potential pollution contexts. The data can be applied to evaluate risks related to industrial emissions, thereby guiding the development of counter-measures against the negative effects on ecosystems as well as human health. Hence, extensive thermochemical analysis is critical in organizing sustainable processes in the production of chemicals and waste disposal. Besides, the environmental importance even extends to planetary research, where thermochemical knowledge explains the development of planetary atmospheres and surface conditions. Investigations of planetary atmospheres, as highlighted by recent research, revolve around the transport and distribution of volatile elements at the hands of thermochemical processes. These researches help to mimic the effects of extraterrestrial events, including cometary and asteroidal impacts, on the atmospheres of planets, as well as how such events can impact atmospheric stability and chemistry. The integration of thermochemical data allows researchers to assess previous atmospheric conditions and predict future changes, hence enhancing our understanding of planetary habitability and the potential of life beyond Earth. Such studies are important because they not only enhance our knowledge of Earth's ecosystems but also give feedback to planetary exploration missions. Besides planetary atmospheres, thermochemical investigation has significant inputs for climate change and its reasons. By the analysis of the thermochemical processes that lead to greenhouse gas emissions and interactions, researchers can identify the reasons behind global warming. For instance, knowledge of the thermodynamics of carbon dioxide and methane under varying conditions enables the quantification of their effect on warming the atmosphere. Further, the examination of outcomes of chemical reactions and phase transitions within mixtures can give rise to feedback processes that increase or reduce climatic impacts. As the rate of climate change mitigation accelerates, thermochemical science is a critical foundation for the development of effective policy frameworks, technologies, and strategies for enabling environmental sustainability and resilience in light of evolving global conditions.

9.6.3. Role in Pharmaceuticals and Drug Formulation

In pharmaceuticals, solution and mixture thermochemistry makes a strong contribution to drug formulation and activity. Drug compounds' thermodynamic properties, such as solubility, stability, and reactivity, need to be known to develop effective medication. Each of these qualities is established by a variety of analytical techniques by which scientists can predict the behavior of drugs under different conditions. Computational tools are sophisticated enough to analyze the thermodynamic properties of drug molecules, focusing on the liquid phase that is very hard to study. Research in this area typically seeks to enhance the accuracy of the thermochemical evaluations, thereby contributing to the production of effective and safe medicines, as stated in existing texts regarding the theoretical evaluation of thermochemistry (Alibakhshi et al., 2021). The process ensures that drugs achieve the expected bioavailability and therapeutic effect upon administration. Drug synthesis generally involves the employment of deep eutectic solvents (DESs), which are gaining recognition for their positive attributes in drug solubilization and delivery. DESs have the potential to profoundly increase the solubility of poorly soluble drugs, solving a significant limitation in drug development. By combining low-cost and biocompatible compounds, such as choline chloride, and renewable bioproducts, researchers have created therapeutic DESs (THEDES) with better characteristics, including their safety and efficacy in the biological system (Rahman et al., 2022). The solvents not only increase the drug molecule's solubility but also possess low toxicity and are environmentally friendly, establishing a new benchmark in drug production. Incorporation of DESs into pharmacy practice indicates the potential of new solvent systems in revolutionizing drug delivery systems with a focus on maximizing thermodynamic properties influencing solubility and activity. In addition, the use of principles of thermochemistry extends beyond drug solubility to encompass stability and interaction between drugs and biological systems. The understanding of thermodynamic interactions can help in the prediction of how drugs will behave in physiological conditions, which is of utmost importance for ensuring their efficacy and safety. Techniques such as molecular dynamics simulations help in grasping the mechanism of stabilization and interaction forces between drug molecules and their target molecules within the body. Present studies have proven that sound thermochemical modeling can lead to better predictions of the mechanisms of how drugs behave in various biological environments, including their metabolic pathways and potential side effects. Blending outcomes from thermochemistry allows for pharmaceutical scientists to design products that not only meet the necessary regulatory needs but also possess better therapeutic profiles essential to modern medicine (Alibakhshi et al., 2021). As science moves forward, thermochemistry's place in the pharmaceutical industry remains at the forefront in ensuring the development of new medications.

SUMMARY

An excursion into the thermochemistry of solutions and mixtures, integrating the key discoveries which reveal the subtle dynamics underlying thermodynamic principles and chemical behavior is an utmost priority. Mastering enthalpy, entropy, and free energy is crucial in comprehending the ways in which solutions and mixtures react during chemical reactions. The use of thermodynamic principles provides useful information regarding the energy changes that occur upon solvation and mixing, which have a direct bearing upon the stability and reactivity of the resulting solutions. Through the study of these principles, we are better able to foresee the spontaneity of chemical reactions and the energy efficiency of a range of processes, all of which serve to highlight the underlying themes of this work focused on thermochemical behavior. In addition, the value of learning the thermodynamic properties of different mixtures cannot be underestimated. The unique properties exhibited by atomic, diatomic, and polyatomic species, as highlighted in recent literature, demonstrate the significance of grasping how molecular structure affects thermochemical behavior. The use of free energy and enthalpy calculations based on partition functions, such as given in Horton et al. (2025), illustrates how increased knowledge regarding molecular interactions enriches our understanding of solution dynamics. Further, the characterization of various forms of combustion systems, such as those studied with the stagnation Hugoniot method in Shepherd et al. (2006), gives the level of energy transformations involved in combustion processes. Together, these results even further increase our understanding of thermochemistry in engineering. In the end, the comprehensive investigation of thermochemistry is a never-ending debate between theory and real-world applications, particularly those related to energy production and material technology. By combining the principles governing energy conversion in solutions and mixtures, scientists and engineers can develop more efficient processes and produce new materials with desired properties. The investigation of combustion reactions and their related thermodynamic implications is a grand example of the utility of these principles in actual engineering application, emphasizing the evolution of propulsion systems and other technologies. Current research in the field of thermochemical approaches is critical to future development models of sustainable development and world energy issues, emphasizing that perfect understanding of thermochemical concepts is a necessity for further scientific and engineering progress.

The field of thermochemistry is also pivotal in the study of mixtures and solutions, where the fields of entropy, enthalpy, and Gibbs free energy gain more relevance. Among the key concepts addressed is the thermodynamic properties that attend solution formation, particularly the absorption or release of heat in solvation reactions. Dissolution of substances by the solvent is endothermic or exothermic and affects the temperature of the system overall. In addition, the interaction process among solute and solvent molecules is regulated by intermolecular forces that also account for the variations in energy during these processes. From an understanding of such reactions, one is sensitized to the reactivity and stability of different mixtures, thereby creating a basis for more sophisticated research in chemical thermodynamics (Marble et al., 1956). One of the main focuses of the discussion is the principles of calorimetry in mixtures and solutions. Calorimetry is a reliable means to assess the heat changes that constitute the basis in determining the specific heat capacities as well as enthalpy changes of the constituents in a mixture. This examination is particularly important in practical applications, such as determining reaction yields in industrial chemistry and understanding biological processes where biochemical thermochemistry plays a controlling influence. The understanding of some heat contributions has further potential implications, such as environmental research and energy requirements for chemical processes. Through the analysis of such applications, students and researchers can better appreciate how thermodynamic principles relate to real applications (Woods et al., 2025). Furthermore, consideration should be taken for the concept of non-ideal behavior in solutions, which manifests in the application of Raoult's Law and Henrys Law. The two laws account for the departure from ideal behavior in vapor pressures and solubilities due to interactions between solute and solvent. The discussion capsulates the role of activity coefficients in capturing an enhanced perception of solution behavior at elevated concentration. Understanding the character of such deviation is important to forecast the thermodynamic property of more complex mixtures and solutions typically encountered in fields such as pharmaceuticals and environmental science. Examining such non-ideal behavior allows for opportunities to enhance chemical process engineering, especially those that involve thermodynamic stability and efficiency in practical applications (Marble et al., 1956).

REFERENCES

1. Alibakhshi, A. (2021). Developing advanced theoretical methods for enhanced study of thermochemistry. https://core.ac.uk/download/621260608.pdf (accessed on 4th October 2025).
2. Allen, J. E., Nava, D. F., Payne, W. A., et al. (2025). Upper atmosphere research: Reaction rate and optical measurements. https://core.ac.uk/download/pdf/42814758.pdf (accessed on 4th October 2025).
3. Anderson, P. G., Pearson, M. L., Penny, M. M., Smith, R., et al. (2025). Supersonic flow of chemically reacting gas-particle mixtures. Volume 1: A Theoretical Analysis and Development of the Numerical Solution. https://core.ac.uk/download/pdf/42883856.pdf (accessed on 4th October 2025).
4. Austin, J. M., Edney, B. E., Edney, B. E., Hornung, H. G., Hornung, H. G., Konnov, A., Oppenheim, A., et al. (2009). Reactant jetting in unstable detonation. https://core.ac.uk/download/4884467.pdf (accessed on 4th October 2025).
5. Banerjee, A., Chen, Z., Cheng, J. H., Deng, Y., et al. (2019). Enabling thin and flexible solid-state composite electrolytes by the scalable solution process. https://core.ac.uk/download/323308721.pdf (accessed on 4th October 2025).
6. Carlo, M., Chao, J., & Wilhoit, R. C. (2025). Thermodynamic properties of compounds of biochemical interest in aqueous solution. A survey of thermodynamic properties of the compounds of the elements CHNOPS: Progress report. https://core.ac.uk/download/pdf/85236995.pdf (accessed on 4th October 2025).
7. Dingwell, D. B., & Virgo, D. (1988). Viscosities of melts in the Na_2O–FeO–Fe_2O_3–SiO_2 system and factors controlling relative viscosities of fully polymerized silicate melts. https://core.ac.uk/download/12166831.pdf (accessed on 4th October 2025).
8. Eftekhari, A. (2020). Advancing prediction of transdermal uptake of SVOCs from clothing and lotion. https://core.ac.uk/download/534501082.pdf (accessed on 4th October 2025).
9. Gilbert, L. M., & Scala, S. M. (2025). Theory of hypersonic laminar stagnation region heat transfer in dissociating gases. https://core.ac.uk/download/pdf/80680365.pdf (accessed on 4th October 2025).
10. Gregory, D. H., Gu, Q., & Mao, J. (2015). Revisiting the hydrogen storage behavior of the Na–O–H system. https://core.ac.uk/download/96883253.pdf (accessed on 4th October 2025).
11. Hong, Q. J., Navrotsky, A., Ushakov, S. V., van de Walle, A., et al. (2019).

Carbides and nitrides of zirconium and hafnium. https://core.ac.uk/download/323067876.pdf (accessed on 4th October 2025).

12. Horton, T. E. (2025). The computation of partition functions and thermochemistry data for atomic, ionic, diatomic, and polyatomic species. https://core.ac.uk/download/pdf/85238937.pdf (accessed on 4th October 2025).
13. Keller, R. B., Jr., & Twichell, S. E. (2025). Solid rocket motor internal insulation. https://core.ac.uk/download/pdf/42877658.pdf (accessed on 4th October 2025).
14. Koop, T., Pfrang, C., Pöschl, U., Shiraiwa, M., et al. (2011). Kinetic multi-layer model of gas-particle interactions in aerosols and clouds (KM-GAP): Linking condensation, evaporation and chemical reactions of organics, oxidants and water. https://core.ac.uk/download/360073.pdf (accessed on 4th October 2025).
15. Kovalenko, A. (2015). Molecular theory of solvation: Methodology summary and illustrations. http://arxiv.org/abs/1510.06520 (accessed on 4th October 2025).
16. Lewis, J. S. (2025). Origin and evolution of planetary atmospheres. https://core.ac.uk/download/pdf/42811921.pdf (accessed on 4th October 2025).
17. Marble, F. E. (1956). Flame theory and combustion technology. https://core.ac.uk/download/4887597.pdf (accessed on 4th October 2025).
18. Middelburg, J. J. (2025). Thermodynamics and Equilibria in Earth System Sciences: An Introduction. https://core.ac.uk/download/637928404.pdf (accessed on 4th October 2025).
19. Misra, A. K., & Whittenberger, J. D. (2025). Fluoride salts and container materials for thermal energy storage applications in the temperature range 973 to 1400 K. https://core.ac.uk/download/pdf/42836449.pdf (accessed on 4th October 2025).
20. Mohammadi, A. H., & Richon, D. (2008). The Scott-Magat polymer theory for determining onset of precipitation of dissolved asphaltene in the solvent + precipitant solution.
21. Moyer, C. B., Powars, C. A., Rindal, R. A., Wool, D., et al. (2025). Ablative response of a silica phenolic to simulated liquid propellant rocket engine operating conditions. https://core.ac.uk/download/pdf/85234166.pdf (accessed on 4th October 2025).
22. Rahman, M. S. (2022). Formulation, structure, and applications of therapeutic, amino acid, and water-based deep eutectic solvents. https://core.ac.uk/download/519731643.pdf (accessed on 4th October 2025).

23. Sandström, H. (2022). Nitriles in prebiotic chemistry and astrobiology. https://core.ac.uk/download/521283174.pdf (accessed on 4th October 2025).
24. Scardino, D. (2012). Theoretical and spectroscopic studies of molecular systems for charge transfer and energy storage and transfer. https://core.ac.uk/download/288063203.pdf (accessed on 4th October 2025).
25. Šesták, J., & Holba, P. (2012). Czechoslovak footprints in the development of methods of thermometry, calorimetry and thermal analysis. https://core.ac.uk/download/pdf/26946170.pdf (accessed on 4th October 2025).
26. Shepherd, J. E., & Wintenberger, E. (2006). Stagnation Hugoniot analysis for steady combustion waves in propulsion systems. https://core.ac.uk/download/4878852.pdf (accessed on 4th October 2025).
27. Styskalik, A., Anderson, P., Bale, C., Brillo, J., Brooks, R., Butler, T., Chen, L., et al. (2014). Cu–Ni nanoalloy phase diagram–Prediction and experiment. https://core.ac.uk/download/51296963.pdf (accessed on 4th October 2025).
28. Woods, R. L. (2025). Publications of the Jet Propulsion Laboratory, July 1964 through June 1965. https://core.ac.uk/download/pdf/85253247.pdf (accessed on 4th October 2025).

CHAPTER 10

THERMOCHEMISTRY IN INDUSTRIAL AND BIOLOGICAL SYSTEMS

Contents

Introduction 361
10.1. Fundamental Principles of Thermochemistry 365
10.2. Thermochemical Processes in Industrial Applications 370
10.3. Thermochemistry in Biological Systems 375
10.4. Thermochemistry and Environmental Impact 381
10.5. Advances in Thermochemical Research 386
10.6. Case Studies of Thermochemistry in Action 391
Summary 395
References 397

INTRODUCTION

Thermochemistry, the study of the connection between heat and chemical reactions, finds industrial and biological applications. An understanding of energy transfer in chemical reactions is crucial in maximizing reaction efficiency and optimizing reactions. Applications of the concepts of thermochemistry assist industries in designing chemical reactors, thereby enabling industries to effectively use energy during production. Temperature and pressure control have direct impacts on reaction routes, yield, and process sustainability. This interface of thermochemistry and engineering principles demonstrates the importance of thermodynamic parameters in optimizing output at minimizing environmental impact, propelling the quest for cleaner technologies ultimately. In biological usage, thermochemistry is an integral component of metabolic processes that support life. Organisms break down macromolecules to release energy, which they use to perform necessary functions. Successful conversion of chemical to biological energy is of most importance since it fuels processes like cellular respiration and biosynthesis.

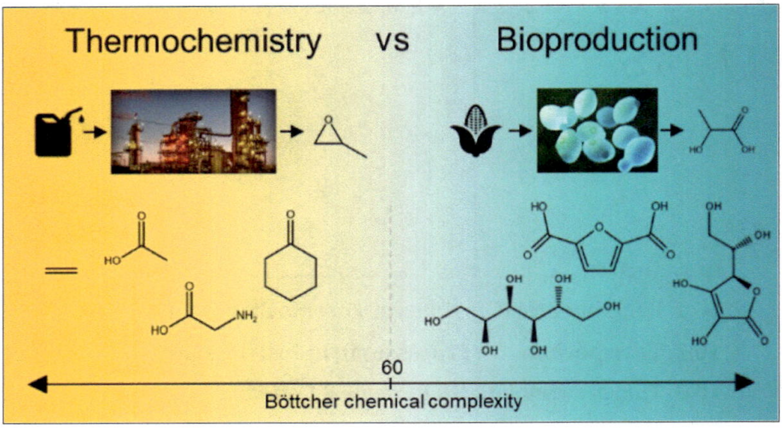

Figure 10.1. *Thermochemistry in industrial and biological systems.*

Source: https://external-content.duckduckgo.com/iu/?u=https%3A%2F%2Fpubs.acs.org%2Fcms%2F10.1021%2Facs.iecr.4c01958%2Fasset%2Fimages%2Facs.iecr.4c01958.social.jpeg_v03&f=1&nofb=1&ipt=2c1251dcdc7eeb372e39fb05b5bb6311013442ea770a1db00992a21b2e67574b.

Biochemical reactions will often involve enzymes that lower activation energies, thus making thermodynamically favorable routes more attainable. Enzyme-catalyzed reaction analysis, such as amine transaminases, is a clear demonstration of the sensitive balance of thermodynamics and kinetics in biological systems. For instance, the rational engineering of enzymes like those disclosed by Allen et al. (2016) highlights the way in which thermochemical expertise can lead to biocatalysis advances enhancing the production of value-added products. Moreover, the domain of thermochemistry encompasses the discovery of novel synthetic pathways in organic and medicinal chemistry, highlighting the significance of enzyme-mediated transformations towards the synthesis of natural products. The revelation of new enzymatic reactions, as uncovered in Garcia-Borràs et al. (2017), shows the growing potential of thermochemistry to spearhead innovation in the realm of biosynthesis. Through the revelation of new ways of building complex molecules, researchers can shed light on the path to sustainable method that diminishes reliance on traditional chemical synthesis methods. Lastly, thermochemistry is a fundamental principle that not only directs our understanding of energy processes but also drives innovation across various fields, from industrial engineering to biology (Figure 10.1).

Definition of Thermochemistry

Thermochemistry, a fundamental branch of physical chemistry, is primarily concerned with the examination of energy changes in chemical reactions. This science investigates the way heat energy is transferred as materials undergo molecular-scale changes. These thermal properties are worth knowing since they control spontaneous reaction and reaction rates. For example, in industry, thermochemical factors are employed to maximize processes such as combustion and material production for optimal efficiency and energy content. This efficiency is crucial in a variety of industries, including manufacturing and the generation of energy, where an understanding of thermal dynamics can mean immense savings and productivity. Aside from industry, thermochemistry has a significant role in biological systems, that is, the comprehension of metabolic processes. Biological reactions also possess complex pathways wherein energy transfer is a matter of utmost concern to ensure life. For instance, the heat released during metabolic reactions is vital to maintaining the body's temperature and for facilitating enzymatic action.

How essential heat energy is to living things is illustrated by the interconnectivity between the concepts of thermochemistry and biological processes. Moreover, knowledge gained from the research of thermochemistry can inform medical research and drug design, through illustrating how drug interaction affects metabolic rates and body energy homeostasis. With the evolution of thermochemistry, its use extends very far. Modern-day scientists employ thermochemical data to design new materials and improve existing processes, particularly in energy storage and conversion. This work has implications on the existing need for sustainable processes in which heat transfer and energy efficiency are known to be realized and result in innovations like better batteries and fuel cells. Besides, the production of polymeric materials, as discussed within the context of fire protection and aerospace applications, highlights the applied significance of thermochemistry in enhancing safety and performance in systems of interest (Heimbuch et al., 2025). Additionally, knowledge of chemical formulation and strategic utilization of thermochemical data can assist technologists in creating efficient and creative products, highlighting the significance of thermochemistry in applied contexts (Ortiz-Cantú et al., 2011).

Importance of Thermochemistry in Various Fields

Thermochemistry forms an essential part in a wide variety of industrial processes, particularly in materials science and manufacturing. It assists engineers in making processes energy-efficient and product-improved by knowing the heat changes in chemical reactions. Polymer synthesis and other material synthesis, for instance, is typically based on close temperature control to manage reaction kinetics and product properties. Applying the concepts of thermochemistry, industries can minimize energy needs and environmental impacts while enhancing sustainability. Moreover, advancements in the area of thermochemistry have also assisted in the manufacture of advanced materials, like light-weight composites used in air transportation, in which a comprehensive understanding of thermal characteristics is needed for application under extreme conditions. In biological systems, the contribution of thermochemistry is equally significant, particularly in metabolic processes and catalysis.

The thermodynamic processes associated with biochemical reactions facilitate a deeper understanding of cell functions and organism behavior. Enzyme-catalyzed processes often involve advanced thermodynamic principles that dictate reaction spontaneity and velocity. For example, the identification of Brønsted acid enzymes that catalyze cationic epoxide rearrangements points to the influence of thermochemistry on biosynthesis, leading to complex natural product formation (Garcia-Borràs et al., 2017). These results not only enrich the understanding of metabolic pathways but also offer means of biotechnological application, for instance, in the construction of biosynthetic processes that simulate natural processes but maximize synthetic efficiency and specificity. Utilization of the thermochemical information in product formulation strategies is also needed in several fields, such as the pharmaceutical and material science industries.

The availability of thermochemical data encourages the development of effective, focused formulations capable of promoting drug delivery and stability. With adequate identification and utilization of relevant databases and information sources of thermochemistry, technologists can utilize more rational formulation approaches (Ortiz-Cantú et al., 2011). Such strategic application of thermochemistry therefore leads to the manufacture of innovative products meeting consumer needs and regulatory needs. By and large, the applicability of thermochemistry does cross departmental boundaries, highlighting its value in catalyzing innovation across industrial and biological platforms (Figure 10.2).

Figure 10.2. *Thermochemistry in various fields.*

Source: https://external-content.duckduckgo.com/iu/?u=https%3A%2F%2Fassets.quizgecko.com%2Fcdn-cgi%2Fimage%2Fwidth%3D800%2Cquality%3D90%2Cformat%3Dwebp%2Fquiz%2F74a4518ceaff86182a01f3ffcba0208d.jpg&f=1&nofb=1&ipt=09feff18b686bc3f5aa984ae48d2e1ab4e838a64c2ac143eb683c82bb732b532.

10.1. FUNDAMENTAL PRINCIPLES OF THERMOCHEMISTRY

Basic principles of thermochemistry lead to a broad framework of understanding energy change in chemical reactions with emphasis on the interdependence of heat and work. At the foundation of thermochemistry is the principle of enthalpy, a thermodynamic property that accounts for the heat content of a system under constant pressure. It is a measure used to make predictions about the spontaneity and equilibrium of reactions. When systems move from reactants to products, the change in enthalpy (ΔH) can be measured, which tells us whether the reaction is exothermic or endothermic. Consequences of these principles are vast in industrial applications such as chemical production, where reaction condition accuracies are paramount to optimizing product yield and minimizing energy consumption. It is information that makes industry processes not only cost-effective but also sustainable environmentally, fulfilling both the goals of profitability as well as the environment. Another fundamental principle of thermochemistry is the concept of Gibbs free energy, which unifies enthalpy, entropy, and temperature into a unified criterion for spontaneity. Chemists are able to ascertain the direction of reaction and assess the feasibility under varied conditions by examining the change in Gibbs free energy (ΔG). A negative ΔG indicates a spontaneous process, particularly applicable in industrial applications as well as biology. For instance, the use of Gibbs free energy calculations in catalysis has helped researchers create more efficient catalysts, allowing reactions to be done at lower energy cost. Moreover, in biological processes, the regulation of metabolic pathways typically relies on the variation of Gibbs free energy of biochemical reactions, illustrating the central significance of thermochemistry in sustaining life processes. Hence, biological as well as industrial systems utilize these principles to optimize efficiency in reactions. The role of catalysis in thermochemistry cannot be underestimated as catalysts take a dominant role in the rate of chemical reactions without altering the general thermodynamic properties. Acceptance of catalysis with the viewpoint of thermochemistry allows for the development of more stable catalytic systems that can operate with milder conditions while being extremely efficient. Recent advancements, such as the application of iron catalysts in the synthesis of ammonia, demonstrate the importance of thermochemical principles to maximize reaction rates. The use of such catalysts has yielded more effective turnover numbers and efficiencies, consequently improving industrial processes (Chalkley et al., 2017). As stressed in the science of formulation chemistry, effective utilization of thermochemical information and tools could lead to novel product ideas, for which various chemical industries are under development (Ortiz-Cantú et al., 2011). In this context, the combination of thermochemistry and catalysis is an essential function in enhancing industrial efficiency as well as biological activity, proving interplay and utility of these fundamental principles.

10.1.1. Laws of Thermodynamics

Principles of thermodynamics are fundamental directing principles that guide the functioning of energy in industrial and biological processes. The law of conservation of energy is the first, which holds that energy cannot be created or destroyed but merely changed from one form to another. This principle is significant, particularly for industrial processes where energy efficiency has implications on operating costs and the sustainability of the environment. For example, in chemical processing, knowledge of energy conversions enables the designing of reaction conditions to achieve the optimum yield with minimal waste. Examples of this principle enable innovations in energy-requiring sectors through the design of systems that convert energy more efficiently, thereby aiding sustainable industrial operations. The second law of thermodynamics introduces entropy, a measure of disorder within a system. With processes occurring, entropy will always increase as an indication of the universal tendency for energy dispersal. The second law has very important ramifications in both thermochemistry and engineered system design. In biological systems, processes such as cellular metabolism obey the second law as organisms transfer energy but unavoidably increase overall entropy. In addition, Johnson et al. (2016) indicates that understanding of entropy in engineered systems can reveal emergent behaviors that render system performance unpredictable. Understanding how entropy affects energy use enables scientists and engineers to create systems that resist inefficiencies, thus advancing biological understanding and industrial engineering.

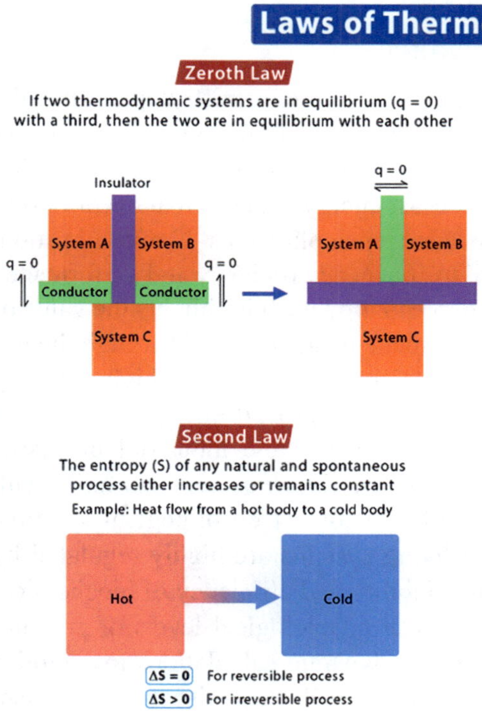

Figure 10.3. *Laws of thermodynamics.*

Source: https://external-content.duckduckgo.com/iu/?u=https%3A%2F%2Fwww.chemistrylearner.com%2Fwp-content%2Fuploads%2F2022%2F08%2FLaws-of-Thermodynamics.jpg&f=1&nofb=1&ipt=79d376f16899f7377918c5e8c3e30f96bb2b27619b40f8b30f8af70354af2c3b.

Finally, the third law of thermodynamics provides that as the temperature of a system approaches absolute zero, its entropy approaches a definite minimum. This law has serious consequences in low-temperature chemistry, primarily in industrial and biological applications. For instance, in cryogenic engineering, the application of third law principles helps to improve the preservation of biological samples and the efficiency of certain industrial processes involving low temperature. Moreover, the knowledge gained from this law complements the applications of statistical mechanics, as proposed by Kovalenko et al. (2015), to lead to innovative techniques in molecular simulations and solvation interactions. Ultimately, laws of thermodynamics serve to enhance our understanding of energy transformations in other applications, pointing to their significant usefulness in thermochemistry in industrial and biological applications (Figure 10.3).

10.1.2. Enthalpy and Its Significance

Enthalpy, a significant term in the subject of thermochemistry, quantifies the total heat content of a system at constant pressure. Enthalpy is a significant factor in numerous processes of both industrial and biological systems. Having knowledge of enthalpy enables us to forecast energy changes that occur during chemical reactions, hence its significance in applications like combustion, where accurate values of enthalpy lead to enhanced efficiency and performance of fuels. The measurement of combustion enthalpy itself highlights the value of sophisticated theoretical methods in thermochemical analysis. Through the use of sophisticated computational methods, scientists can execute high-accuracy measurements that are important in designing improved energy sources and combustion engines and thus contribute to sustainable industrial processes (Alibakhshi et al., 2021). The use of enthalpy does not cease to exist with industrial processes but also has a significant impact on biological systems. For instance, biochemical reactions in living systems are highly regulated by enthalpy changes, which determine the viability and efficiency of biochemical reactions. These reactions primarily occur at physiological levels of pressure and temperature and involve precise thermodynamic calculations to establish energy conversions involved. In addition, enthalpy also explains the mechanism of enzyme activity, whereby substrate binding triggers enthalpic changes resulting in biological processes. The ability to model these energy conversions is crucial, particularly in the pharmaceutical industry, where an understanding of the enthalpic interactions between prospective drug molecules and biological targets may lead to more therapeutic strategies (Woods et al., 2025). Overall, the significance of enthalpy in the realm of thermochemistry can be seen in numerous areas, indicating its central role within industrial and biological reactions. By allowing quantification and prediction of energy change, enthalpy allows scientists and engineers to optimize processes, increase energy efficiency, and understand complex biological interactions. The addition of advanced theoretical methods for estimating enthalpy brings extra accuracy to these calculations and thereby increases the scope of application of the principles of thermochemistry in real-world situations. In light of the continued search for sustainable processes and effective drug therapies, enthalpy continues to be in the limelight as its critical contribution in both these fields of exploration continues to emphasize its intrinsic position (Alibakhshi et al., 2021; Woods et al., 2025).

10.1.3. Heat Capacity and Calorimetry

Knowledge of calorimetry and heat capacity is critical in the study of thermochemistry, particularly in industrial and biological systems. Heat capacity, as the quantity of heat required to raise the temperature of a material by a °C, plays a critical role in the definition of thermal response between processes. Heat capacity, in real-world applications, dictates the efficiency of reactions at high temperature, where precise control of heat is needed for optimal yield with minimal loss of energy. Measurement of the heat capacity of substances allows scientists and engineers to forecast the behavior of systems in conditions of thermal stress, an essential component in design and operating procedures. Calorimetry is the practical science of measurement of heat transfer, providing invaluable information in processes thermochemical. Calorimeters allow scientists to quantify the transferred heat in chemical reactions, phase changes, and other thermal transitions.

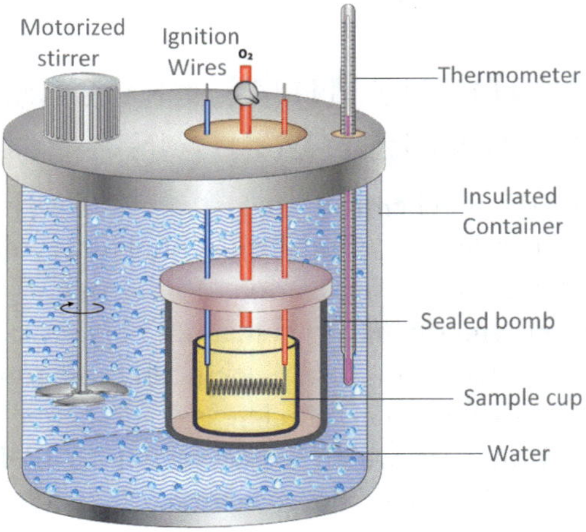

Figure 10.4. *Heat capacity and calorimetry.*

Source: https://external-content.duckduckgo.com/iu/?u=https%3A%2F%2Fwww.sciencеabc.com%2Fwp-content%2Fuploads%2F2019%2F10%2Fcalorimeter.jpg&f=1&nofb=1&ipt=6a602217879e94934b9c741a5442be9de566811d5a57ca23f9463ff7afbc4e2a.

Precision calorimetry is essential in a wide range of applications, from the pharmaceuticals sector, where thermal properties determine drug formulation, to materials science based on phase behavior under varying conditions. Precise determination of the changes in heat enables better predictive models of reaction paths and stability, which is important in the development of new materials or for the optimal production process. Generally speaking, calorimetry provides a fundamental toolbox to understanding thermal response under chemical and biological transformations. New developments in calorimetry techniques, particularly for emerging materials such as graphene oxide, have broadened the science's scope. As a case in point, GO-amended fly ash pervious concrete development illustrates the combination of material property and thermal response that conditions construction practice and environmental sustainability. Studies indicate that graphene oxide enhances the thermal stability and freeze/thaw resistance of concrete—a crucial factor in regions with extreme climate. The new use of nuclear magnetic resonance and X-ray diffraction to characterizing these materials not only brings clarity to their microstructure but also brings to light their superior thermal performance in varying environmental conditions (Shi et al., 2018). These findings underline the interconnectivity of calorimetry, heat capacity, and high-technology materials science in its constant pursuit of sustainability and efficiency in industrial processes (Figure 10.4).

10.2. THERMOCHEMICAL PROCESSES IN INDUSTRIAL APPLICATIONS

Incorporation of thermochemical processes in industrial applications over the last several years has demonstrated the potential for revolutionary breakthroughs in materials science and engineering. Pioneering this development is the development of polymeric materials that not only have improved performance but also better safety profiles. Specifically, thermochemistry has made it possible to fabricate materials that are far better in fire resistance and mechanical performance, which are essential to high-risk sectors such as aerospace, construction, and transportation. The safety of fires within these sectors is paramount because it directly relates to human survivability during emergencies. The research on fire protection systems has led to the application of new materials to provide greater resistance to hostile conditions, illustrating the tangible benefits of thermochemical processes for the improvement of industrial standards and practices (Heimbuch et al., 2025). Outside the material development aspects, thermochemistry carries significant relevance in the design of chemical products in various industrial sectors. The use of the scientific approach to using chemical formulations allows technologists to navigate the intricacy embedded in product design and optimization. Technologists, through the use of scientific information databases and tools for industries, are empowered to make the correct decision on material thermochemical properties. This not only enhances the effectiveness of the product but also minimizes waste in the process of developing resources. Technological mapping strategies and technological integration strategies are of utmost significance in this context, as these allow for innovative solutions and green measures with strict compliance with regulatory standards (Ortiz-Cantú et al., 2011). Finally, application of thermochemical principles in industrial processes assures energy efficiency and reduced environmental impact. Industries can enhance reaction pathways via governed thermochemical processes, allowing maximum energy consumption with minimal emissions and waste. This is particularly important for sectors such as manufacturing and energy generation, where traditional processes are likely to result in extreme resource exhaustion and environmental degradation. The ability to innovate with the implementation of thermochemical processes not only drives profitability but also facilitates a move towards more eco-friendly industrial procedures. With the environmental issues confronting industries increasingly on the rise, incorporation of thermochemical procedures will be inevitable in the pursuit of long-term sustainability targets at the expense of efficiency in operations.

10.2.1. Role of Thermochemistry in Chemical Manufacturing

Thermochemistry plays a critical role in chemical manufacturing by aiding in understanding energy changes during chemical reactions. This data assists manufacturers in optimizing processes in a way that maintains energy costs as low as achievable with product quality maintained. For instance, thermochemical data aids in determining the viability of reaction under various conditions, allowing for better heat control and energy recovery practices. With business more and more oriented towards sustainability, thermochemistry plays a vital role in achieving environment-friendly processes through the integration of energy-efficient methods. The expertise that originates from thermochemical theory not only propels productivity but also aids in regulatory abidance with eco-friendly standards, thus balancing corporate goals with environmental responsibility. Thermochemistry also becomes important in formulating improved materials in the production of chemicals. The composite formation and polymerization, in turn, might necessitate advanced comprehension of heat transfer, reaction kinetics, and changes in enthalpy. Employing thermochemical principles, manufacturers are now capable of formulating processes to yield the best material properties, such as flame retardance, strength, and durability. Specifically, materials science progress is represented through applications such as aerospace and construction building fire safety, where the products are specifically tailored to minimize risks and enhance survivability of human beings (Heimbuch et al., 2025). These strategic applications point towards the broad impact that thermochemistry has on overall product function in the industrial sector. Furthermore, the introduction of thermochemical data into the design cycles backfeeds into technology strategies that facilitate innovation in chemical manufacturing. Using thorough databases and sophisticated search methods, technologists are able to accurately chart the chemical formulation landscape, thus making research easier. Such analyses not only improve new product development effectiveness but also promote inter-disciplinary collaboration within the sector (Ortiz-Cantú et al., 2011). The findings obtained from thermochemical research inform decision-making, opening up routes for the development of new formulations that meet changing market needs. As technology evolves, the application of thermochemistry continues to play increasingly vital roles in driving chemical production and keeping pace with industry norms (Figure 10.5).

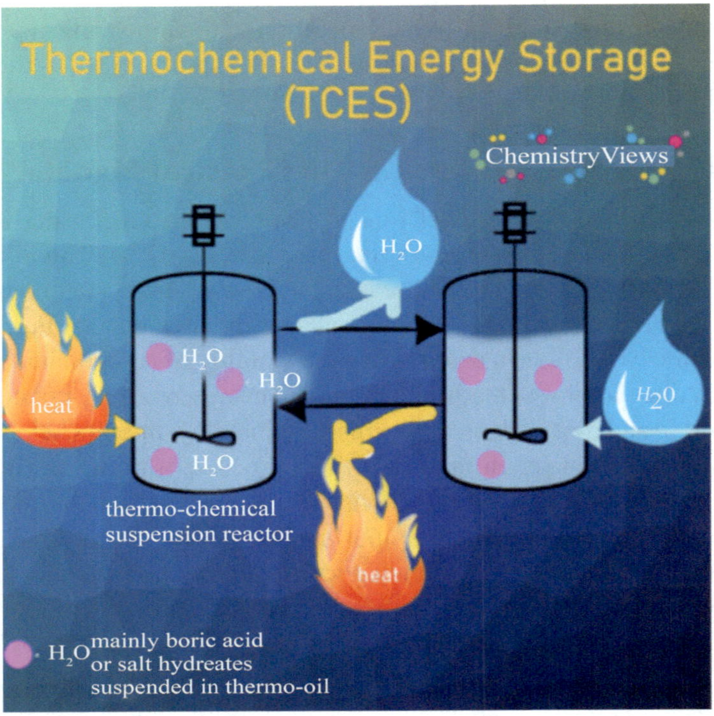

Figure 10.5. *Role of thermochemistry in chemical manufacturing.*

Source: https://external-content.duckduckgo.com/iu/?u=https%3A%2F%2Fwww.chemistryviews.org%2Fwp-content%2Fuploads%2F2022%2F12%2F2023_TCES_CleverPicture_ChemistryViews-1-400x400.png&f=1&nofb=1&ipt=36b8d326b5d28f5947003aac1350e80a9b1f1c7f7857eceb6532adfdd4433d55.

10.2.2. Energy Efficiency in Industrial Processes

Energy efficiency remains a dominant topic in industrial applications, particularly as there is heightened demand for sustainable methods globally. With industries seeking to cut the consumption of energy while boosting production, thermochemistry principles come to the forefront. Various industries such as chemical manufacturing, metallurgy, and food processing increasingly rely on new thermal control methods that enhance the efficiency of energy usage. With the incorporation of advanced heat recovery technology and optimizing reaction conditions, thermal losses can be significantly reduced by industries. Moreover, Energy Management Systems (EnMS) can be utilized to continuously monitor and optimize energy performance. Not only is it a method of resource conservation, but even with regulation systems that promote environmental sustainability, energy efficiency is an integral component of modern industrial operations. The incorporation of energy-saving technologies not only optimizes performance but also enhances the general sustainability of industry operations. Advances in technologies, such as chiral ionic liquids (CILs), have provided groundbreaking applications in chemical separations with significant implications on energy consumption. For instance, development of an efficient CIL-based liquid-liquid extraction system for racemic amlodipine separation has demonstrated higher enantioselectivity and lower energy requirement compared to the conventional method. Such technologies promote minimizing energy use in chemical reactions, finally reducing industrial manufacturing's carbon footprint. By embracing innovative solutions integrating thermochemistry with operational efficiency, industries can maximize production levels while promoting ecological health, hence illustrating the interdependence of scientific innovation and real-world application. In addition, the application of integral equation theories in molecular liquids also highlights the potential for increased energy efficiency through a greater understanding of chemical interactions in solution. Through statistical mechanics and diagrammatic techniques, such theories enable industries to produce precise correlation functions describing solvation structures and interactions. Through this process, it is possible to design processes for maximum energy transfer with minimum waste, thereby increasing overall chemical system efficiency. Moreover, the ability to compute effects over broad temporal and spatial scales, as exemplified by approaches such as three-dimensional reference interaction site model (3D-RISM), highlights the intrinsic role played by theoretical models in minimizing industrial process waste. As industries deploy these advanced methods, energy efficiency transcends mere cost minimization and becomes a building block of sustainable industrial practice in harmony with broader thermochemical principles (Figure 10.6).

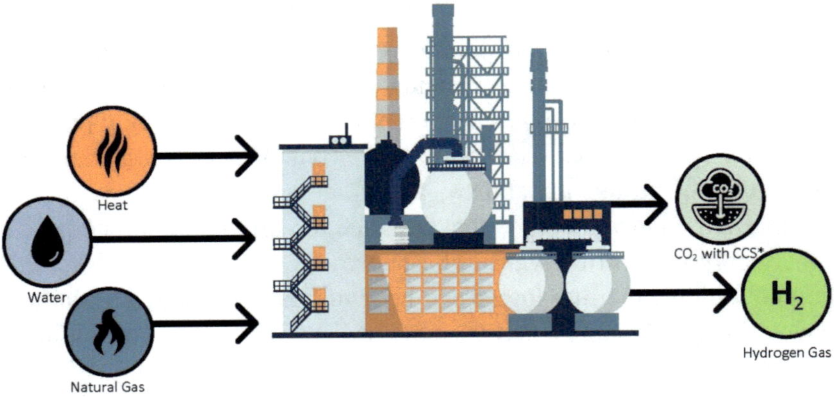

Figure 10.6. *Energy efficiency in industrial processes in chemical.*

Source: https://external-content.duckduckgo.com/iu/?u=https%3A%2F%2Fwww.energy.gov%2Fsites%2Fdefault%2Ffiles%2F2023-03%2Ffaq-steam-methane-reforming.jpg&f=1&nofb=1&ipt=be6e3742dbebb2c1be223df0986c64733398196be59fc3fb4b5c9c284a8f664d.

10.2.3. Case Studies of Thermochemical Applications in Industry

Applications of thermochemistry in industrial processes are increasingly receiving attention as a method to enhance efficiency and product quality. A case study is the production of advanced composite materials that are aimed at ensuring improved fire safety for the aviation and high-rise buildings industries. By applying intense thermochemical testing, researchers were able to create materials that not only provide enhanced thermal stability but are also self-extinguishing in nature. These are even more important in commercial and military aircraft applications, where survivability for the human is paramount. Bench-scale and pilot-scale testing demonstrated that these novel materials, derived from aerospace engineering applications, were effective at reducing fire hazard and assisting in enhancing safety controls in industries, demonstrating the successful synergy of thermochemistry and engineering innovation (Heimbuch et al., 2025). Another very interesting application of thermochemistry in an industrial setting is provided by the synthesis of chiral amines using engineered amine transaminases. Steric hindrance historically limited the synthesis of these compounds, especially in attempting to catalyze reaction with large substrates like aromatic ketones. Recent research has demonstrated a rationale design approach to enhancing the catalytic effectiveness of the *Vibrio fluvialis* amine transaminase. By introducing specific mutations, scientists made striking improvement in reaction rates by enhancing catalytic performance more than 1716-fold for the challenging substrate. This optimization not only demonstrates the potential of thermochemical ideas in enzyme engineering but also showcases its potential to produce enantiomerically pure compounds necessary in drugs (Allen et al., 2016). Aside from the straightforward necessity for efficiency increments, the implications of these case studies reflect a broader trend towards sustainable manufacturing processes. By leveraging the use of thermochemical reactions, industries are able to spare the environment while advancing the boundaries of technology. The aerospace industries development of fire-retardant materials and the pharmaceutical industries enhanced enzyme design illustrate how thermochemistry improves safety and sustainability objectives. Such developments indicate mounting attention to environmental stewardship in industrial systems, making thermochemistry a prominent field for future research and application. With industries evolving continuously, these case studies identify the significant contribution of thermochemistry towards developing safer and more efficient applications in various industries.

10.3. THERMOCHEMISTRY IN BIOLOGICAL SYSTEMS

Thermochemistry also has a crucial role to play in biological systems, as it aids in explaining the energy transitions that occur in the process of metabolism. This branch of chemistry examines the manner in which organisms acquire energy from the surroundings and convert it into forms that can be utilized for growth, reproduction, and upkeep of homeostasis. Dominant to all these processes is the concept of enthalpy, a quantitative measure of the heat content of biological reactions. As an example, in cellular respiration, the hydrolysis of glucose generates adenosine triphosphate (ATP), the energy currency of the cell, by a series of exothermic reactions. The fine regulation of such thermodynamic operations allows organisms to harness energy efficiently with a minimum wastage, demonstrating the profound synergy between thermochemical principles and biological process. Besides, thermochemistry is used to explain enzyme catalysis, where reaction rates and equilibrium are influenced extensively by temperature and energy changes. Enzymes, as biological catalysts, lower the activation energy necessary for metabolic reactions in order to take place under the soft conditions present within living things. Temperature-reacting kinetics is crucial since it may set the rate by which enzymes convert substrates to produce necessary biomolecules. In enzyme design examples, such as amine transaminases, thermochemical data could be employed to guide alterations towards higher efficiency and specificity. Rational design strategies, for example, demonstrated an impressive increase in catalytic activity—and thus a favorable thermodynamic profile—by engineering specific mutations in the enzyme structure, thus demonstrating thermochemistry is crucial to bioengineering breakthroughs (Allen et al., 2016). The use of thermochemistry in biological systems also has echoes in ecological systems, whose energy circulation and nutrient cycling are regulated by the rules of thermodynamics. Each trophic level within an ecosystem is a sign of significant energy loss, primarily as heat, due to metabolic processes. This energy transfer is the foundation of food web research and energy efficiency of utilization in organisms. The rules of thermochemistry not only clarify the metabolic strategies employed by different species but also inform conservation mechanisms for the sustenance of ecosystems. By examining how organisms adapt their energy-saving methods in the light of climatic fluctuation, researchers become able to comprehend the strength of ecosystems and the probable impacts of climate change. Thus, thermochemistry provides a systems view for the examination of biological systems, with interfaces into broader environmental and industrial settings (Heimbuch et al., 2025) (Figure 10.7).

Figure 10.7. *Thermochemistry in biological systems.*

Source: https://external-content.duckduckgo.com/iu/?u=http%3A%2F%2Fwww.globetrotterscience.com%2Fuploads%2F2%2F1%2F8%2F4%2F21849914%2Fpublished%2Fthermo.png%3F1528991148&f=1&nofb=1&ipt=10dfc1c9a4cc32bf558c7a86d03629583dd6249e4e60838c3a84df82ab4b2cc1.

10.3.1. Metabolic Pathways and Energy Transformations

Metabolic pathways consist of intricate networks of enzymatic reactions that make the conversion of substrates to energy and cell building blocks possible. These processes are crucial to biological systems and industrial processes, where an understanding of energy conversions can optimize processes such as fermentation and biosynthesis. Cellular respiration and photosynthesis are large-scale metabolic processes in biological systems that convert biochemical energy from the sun or foodstuffs into usable forms, i.e., ATP (adenosine triphosphate). In the context of thermochemistry, these energy transformations also illuminate how living things maintain their biochemical homeostasis, demonstrating the fine balance between generating and expending energy for growth and functioning. By an understanding of these processes, researchers can ascertain how energy efficiency can be optimized in microbial and plant systems, with consequent implications for productivity in farm and industrial processes. The integration of metabolic engineering into metabolic pathway research allows targeted modification, which biases processes of energy transformation in favor. Biotechnology exploits the engineered pathways to produce valuable chemicals and biofuels from industrial microbes. Various approaches including native pathway optimization, foreign gene pathway integration, and synthetic pathway construction can lead to highly efficient microbial factories that can convert low-value substrates into high-value products (Hadadi et al., 2015). This is in alignment with the laws of thermochemistry by concentrating on the energetic character of metabolic processes, such as enthalpic and entropic change involved in biochemical conversions. Understanding of how to manage these pathways not only allows sustainable production within industrial networks but also is in harmony with environmental sustainability, placing additional value on metabolic investigations in both contexts. Additionally, the application of nanotechnology in the context of metabolic pathways reveals new environmental and biological issues. The introduction of nanoparticles (NPs) into biological systems, while promising in applications within areas like agriculture and remediation, poses a problem in terms of their effect on metabolic processes and energy flow in living organisms. Experiments have shown that certain NPs can interact negatively with plant root systems and hence may alter nutrient intake and energy availability (Kumar et al., 2012). This interaction helps to illustrate the indirect ways in which nanoparticles disrupt metabolic processes, with implications for both ecosystem health and crop yield. Therefore, while advances in nanotechnology have the aim of facilitating more efficient production, it is necessary to develop integrated strategies that account for the complex interactions of these materials with biological metabolic processes, thereby ensuring ecological integrity while expanding industrial use.

10.3.2. Role of Enzymes in Thermochemical Reactions

In thermochemistry, enzymes are also key catalysts catalyzing key biochemical reactions by lowering the energy barriers associated with thermochemical reactions. The biological catalysts accelerate reactions without being permanently changed in the process, a feature that enhances reaction rates and selectivity. By stabilizing transition states and intermediates, enzymes can catalyze reactions efficiently under mild conditions, critical in biological environments where harsh temperatures and pressures would be dangerous. This catalytic ability is particularly evident in complex conversions, such as epoxide rearrangements, where the enzyme PenF has been shown to catalyze new pathways that create quaternary carbon centers in physiological conditions. The ability of enzymes to direct reaction pathways reflects their fundamental role in driving biologically relevant thermochemical processes, essential for cell function and metabolic efficiency (Garcia-Borràs et al., 2017). The significance of enzymes is also seen in industrial operations, where their role to catalyze thermochemical reactions can uncover new ways of enhancing production in most industries, including pharmaceutical and biofuels. Enzymes such as DMSP lyase are significant in the conversion of substrates into products through mechanisms that often mimic or even transcend nature.

These enzymatic reactions can be engineered to work at lower pressures and temperatures, conserving energy and thereby the industrial carbon footprint as a whole. Sulfur compounds research has also helped foster the versatility of enzyme action, showcasing the ability of enzymes for the production of vital reactants from otherwise unreactive substrates. Enzymes do not only facilitate streamlined production processes but also open the way for sustainable production by using renewable resources, thus enabling alignment with the modern ecological and economic objectives (Nikoo et al., 2019). In addition to their core catalytic functions, enzymes also control thermochemical reactions by responding to variability in conditions of the external environment such as temperature and pH. This function is critical to the maintenance of metabolic homeostasis in living things, where fluctuating conditions in the external environment can influence the metabolic process. For instance, enzymes can exhibit altered kinetics with altered thermal conditions, optimizing the reaction rates to enable optimal metabolic demands with efficacy. The fine understanding of enzyme activity towards thermochemical alterations may induce new biotechnological applications, for instance, enzyme engineering for greater stability and functionality under industrial conditions. This versatility emphasizes the pivotal role of enzymes in thermochemical dynamics, further proving their importance in biological and industrial systems (Garcia-Borràs et al., 2017; Nikoo et al., 2019) (Figure 10.8).

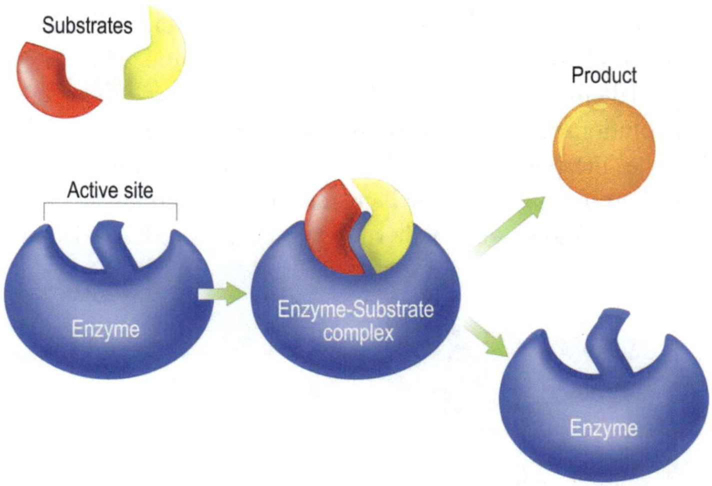

Figure 10.8. *Role of enzymes in thermochemical reactions.*

Source: https://external-content.duckduckgo.com/iu/?u=http%3A%2F%2Fibiologia.com%2Fwp-content%2Fuploads%2F2018%2F11%2Fenzyme-lock-and-key-model.jpg&f=1&nofb=1&ipt=180805a2dc2e0e3519fb089632c36edf07997fc35ed70a62f064ea6574a2c008.

10.3.3. Thermoregulation in Living Organisms

Thermoregulation is a fundamental physiological process that enables living organisms to maintain their internal body temperature within a normal range despite external fluctuation. This ability is not only vital for optimal metabolic function but also for homeostasis maintenance. For instance, endothermic animals such as mammals and birds utilize a sequence of complex systems like behavioral adaptation, physiological changes, and metabolic mechanisms to regulate their core temperatures. For instance, mammals are able to shiver, increase perspiration, or adjust their activity levels with regard to temperature changes. These processes are indicative of high coordination between thermodynamics and biological systems, illustrating the way living organisms utilize thermochemistry to efficiently navigate their environment. Knowing these processes cast lights on the fine balance supporting life as well as the energy transformations involved during thermoregulation. Unlike endotherms, ectothermic animals such as reptiles and amphibians largely rely on external environmental temperatures for regulating body temperature. Through the adoption of behaviors such as basking in sunlight or shade seeking, ectotherms regulate their thermal circumstances through passive processes and not by active metabolic processes. This dependence on external sources of heat allows ectotherms to be energy conservers, especially under conditions of low food availability. Ectothermy, however, is fraught with huge risks, as fluctuation in environmental temperature can impose huge physiological stress or even kill them. The thermoregulatory strategies in ectotherms demonstrate the diverse evolutionary solutions to environmental requirements that emerge, showing the fine interplay between thermochemistry and survival in biological systems. The thermoregulatory mechanisms in living organisms reveal bigger implications in the framework of thermochemistry in industrial and biological systems.

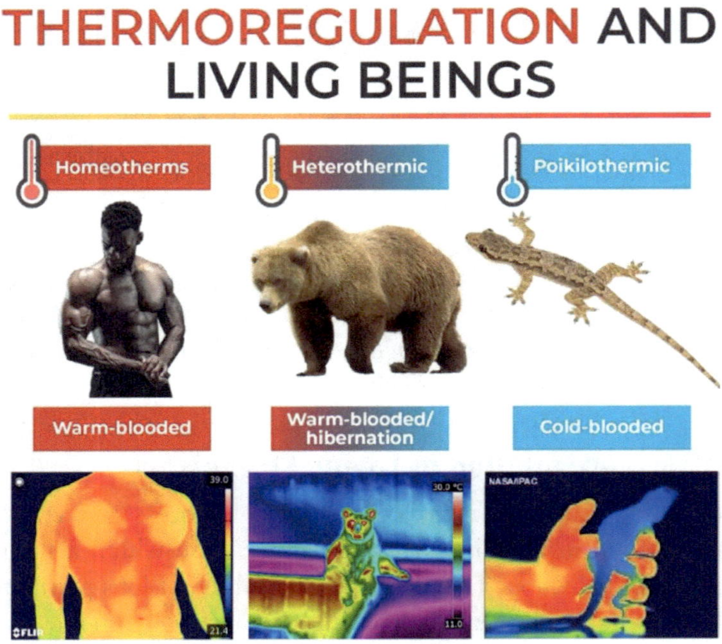

Figure 10.9. *Thermoregulation in living organisms.*

Source: https://thermohuman.com/_next/image?url=https%3A%2F%2Fcdn.sanity.io%2Fimages%2Fh1t24h0t%2Fproduction%2F52e1f9679fa072f8593d392bf1ca3191135297c5-1628x1632.jpg%3Fw%3D1000%26auto%3Dformat&w=3840&q=75.

For example, understanding the principles of heat transfer can inform the design of technologies that simulate or expand biological thermoregulation. Likewise, understanding how organisms adapt to elevated temperatures can trigger innovation in materials and processes requiring robust heat management systems. Moreover, descriptions of thermoregulation strategies can also teach about the impact of climate change on species survival and biodiversity and about the need for cross-disciplinary understanding. With the integration of biological thermoregulation principles into thermochemistry, environmentally friendlier industrial applications can be created which are in harmony with ecological laws, thereby making possible development that respects natural ecosystems as well as human technological developments (N/A, 2025a, 2025b) (Figure 10.9).

10.4. THERMOCHEMISTRYANDENVIRONMENTAL IMPACT

Understanding the thermochemical processes involved in industrial processes will help determine their environmental impact. Energy transformation in chemical reactions would result in the emission of greenhouse gases and other poisons, which cause global warming and air pollution. Chemical reactions that either utilize or release significant amounts of energy are the backbone of manufacturing and energy production industries. The operations involved here have implications beyond immediate industrial products; they have long-range implications on climate and ecosystems. Therefore, a comprehensive study of the thermochemical processes employed by these industries can assist in developing strategies that minimize environmental degradation. Detection of alternative routes or materials of lower harmfulness can significantly promote sustainability and reduce the carbon footprint, bringing industry practice into line with global moves towards controlling climate change and promoting environmental stewardship. In biological systems, thermochemistry plays a central role in metabolism and energy transfer. The biochemical reactions responsible for energy production and utilization in organisms are fundamentally thermochemical in nature. These processes determine how energy is extracted from food and made available to cells in the form it can be used, and they determine growth, reproduction, and survival. In addition, external environmental factors control these thermochemical processes strongly, particularly temperature and pressure, which may influence enzyme activity and metabolic rates. Knowing these interactions is critical to the ability to predict how organisms could cope with climate change, pollution, and habitat loss. Lastly, an awareness of the thermochemical processes within living organisms not only enhances our knowledge regarding life sciences but also offers insights into ecological management because it depicts the delicate balance that ensures biodiversity in dynamic ecosystems. Utilization of thermochemistry towards the creation of materials and chemical compositions is fundamental in enabling industrial development as well as protection of the environment. Polymer chemistry and biophysical chemistry research, for instance, has yielded products that improve fire safety and human survival in varying settings from skyscrapers to transport infrastructure (Heimbuch et al., 2025). This milestone underscores the importance of material innovation with enhanced performance and lower ecological impacts. In addition, the establishment of technological maps with various approaches to chemical formulation allows for more logical material choice suitable for specific applications. This translates into a very good knowledge of relative databases and information organization to maximize research efficiency and product innovation (Ortiz-Cantú et al., 2011). By focusing on sustainable processes in chemical and material design, companies are able to save the planet while still meeting productivity and efficiency goals.

10.4.1. Green Chemistry and Sustainable Practices

Green chemistry is a radical philosophy of chemical research and utilization that seeks to minimize the adverse environmental consequences commonly generated by industrial processes. Through its emphasis on renewable resources, energy efficiency, and elimination of harmful substances, green chemistry endeavors to foster environmentally and public health-friendly processes that are sustainable. Green chemistry encourages the design of reactions that produce fewer by-products and use more benign solvents. Advances within this topic area include the use of catalysis techniques and the use of biocompatible materials as devices to increase chemical process efficiency and sustainability. Therefore, the incorporation of green chemistry principles into industry practice not only enhances production effectiveness but also imposes economic sustainability, outlining a synergistic partnership between business prosperity and environmental stewardship. The quintessential demonstration of the practice of green chemistry is the production of graphene oxide-modified pervious concrete utilizing coal fly ash as a binding agent. This new material was characterized for its permeability of stormwater infiltration and environmental stability with respect to freeze/thaw and salt exposure. Analytical techniques, X-ray diffraction and nuclear magnetic resonance, were used to give knowledge about the chemical structure of these materials and its ability to reduce pollution levels in stormwater. The study highlighted that the addition of graphene oxide greatly enhances the materials' resistance and promotes effective management of water resources. The development complements the strength of green chemistry protocols in addressing real-world environmental challenges as well as in supporting innovation in eco-friendly building materials (Shi et al., 2018). The emphasis on sustainable methods also reaches beyond material science to cover industrial and biological applications. The concepts of green chemistry can be implemented across industries like pharmaceutical, agriculture, and biotechnology, reducing waste and efficient use of energy in various processes. These practices induce an industry-scale paradigm shift towards sustainability, where the development of economy is no longer incompatible with the preservation of the environment. Moreover, initiatives such as public awareness and policy regulations ease the implementation of these sustainable practices by both industries and scientists. As thermochemistry evolves and incorporates these new techniques, a wide understanding of industrial processes and biological systems will be necessary to guarantee a sustainably healthy future (N/A, 2014) (Figure 10.10).

Figure 10.10. *Green chemistry and sustainability.*

Source: https://webfiles.amrita.edu/2018/08/g27yINQC-green-chemistry-blog.jpg.

10.4.2. Thermochemical Processes in Waste Management

Thermochemical waste management is one exciting sector for the maximization of resource recovery with minimized environmental impact. Techniques such as pyrolysis, gasification, and partial oxidation are the cutting-edge of these technologies, efficiently converting complex and dirty wastes into valuable chemicals, fuels, and energy carriers. These processes use thermal energy to break down organic compounds, hence enabling the conversion of waste into more convenient forms and less landfill dependency. Kinetic modeling of such thermochemical reactions is essential, as it supports reactor design and optimization for industrial applications, thus maximizing efficiency and effectiveness towards waste valorization. In this regard, techniques based on the use of Computational Fluid Dynamics (CFD) are especially helpful, primarily to define the interaction between chemical kinetics of thermal degradation processes and system performance, and in making waste treatment economically viable and environmentally friendly (Locaspi et al., 2024). In addition to the maximization of traditional waste treatment processes, innovative uses of industrial by-products such as sewage sludge and red mud in thermochemical energy storage prove the versatility of thermochemical processes. Their inclusion in thermal energy recovery systems not only maximizes utilization of resources but also aids carbon emission reduction. Red mud and sewage sludge, when properly treated, can be transformed into energy storage materials, which are specifically designed to be used in solar energy harvesting and the enhancement of building energy efficiency. This is in accordance with a circular economy approach, where waste materials are utilized for alternative applications rather than towards environmental degradation. However, challenges still remain, particularly with respect to the stability of such energy storage devices and potential environmental toxins released during treatment, yet emphasizing the importance of ongoing research and development work in this field (Ding et al., 2024). The achievement of thermochemical technologies for waste management is contingent on the convergence of researchers, engineers, and policymakers to solve the multifaceted issue of sustainability. These operations will have to be optimized constantly through rigorous experimental evidence and kinetic modeling to be effectively applied to various forms of wastes. As environmental awareness intensifies and pressures rise for sustainable waste treatment technology, scale-up of use of thermochemical processes and integration into mainstream infrastructure will become a necessity. A series of case studies and data from experiments support the promise of achieving successful waste transformation, thus outlining a strategic path for minimized landfill waste and optimized resource recovery. Lastly, the advances in thermochemistry not only have the potential to revolutionize industrial waste management but also contribute to grander efforts toward carbon neutrality and environmental sustainability.

10.4.3. Impact of Thermochemistry on Climate Change

The relationship between thermochemistry and climate change is a significant area of study, particularly in the context of increased dependency on industrial processes that are based on thermodynamic principles. Thermochemistry, which is the scientific study of heat energy in chemical reactions, plays a significant role in energy production methods that impact the environment. Examples are the combustion reactions that occur in industries, which emit greenhouse gases, leading to high atmospheric concentrations of carbon dioxide and other pollutants. The thermochemical reactions release a lot of energy but toxic by-products. A better grasp of the thermochemistry principles facilitates the development of new technologies and strategies for minimizing the environmental impact, further justifying the importance of sustainable processes in energy production and utilization. In the industrial sector, the challenge of achieving energy requirements without compromising environmental sustainability is at the core. Application of thermochemical knowledge promotes energy efficiency by means of conversion, which can actually reduce carbon footprints. Renewable energy production processes, including alternative sources like biofuels and others, are primarily reliant on thermochemical processes to ensure maximum utilization of resources and avoid wastage. For instance, gasification and pyrolysis process improvements show the potential for thermochemistry to lead to cleaner manufacturing processes that achieve significant greenhouse gas emissions savings.

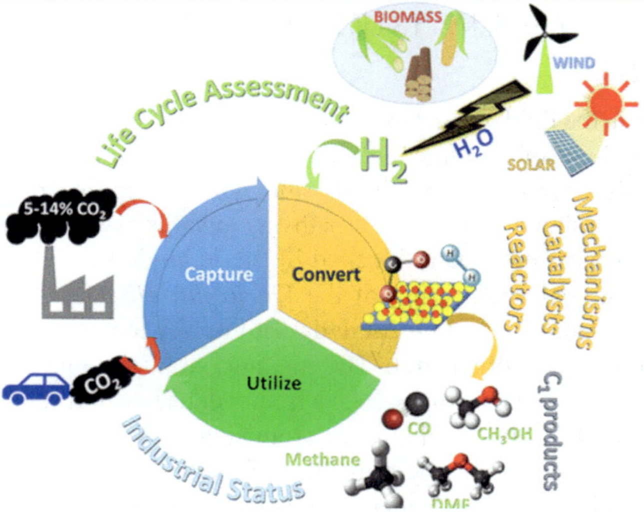

Figure 10.11. *Impact of thermochemistry on climate change.*

Source: https://pubs.acs.org/cms/10.1021/acsenergylett.8b00740/asset/images/medium/nz-2018-00740w_0018.gif.

This synergy of industrial processes and thermochemical principles is reflective of an evolutionary direction towards sustainability, wherein prudent energy production systems are engineered to put climate change into balance while also enabling environmental cure, as hinted at by recent policy suggestions for addressing global energy concerns (Barbero et al., 2011). Moreover, the authority of thermochemistry extends beyond industry into the bioprocesses that are part of the larger climatic narrative. Biological processes, such as respiration and fermentation, are unavoidably thermochemical in character and are foundational to carbon cycling. An understanding of these processes permits insight into their contribution to greenhouse gas dynamics. For instance, agriculture methods based on thermochemical knowledge can modify food production systems to reduce methane emissions from animals or improve soil carbon capture techniques. This intersection of climate science and thermochemistry assists in policy framing towards promoting sustainable agricultural practices and mitigation of the effects of climate change. Resources such as NASAs extensive investigations of the upper atmosphere emphasize the interdependence of thermochemical reactions and climatic consequences, with the necessity of systems thinking for addressing both industrial and biological systems for the purposes of global climate change (Kurylo et al., 2025) (Figure 10.11).

10.5. ADVANCES IN THERMOCHEMICAL RESEARCH

New findings in the field of thermochemical science have significantly enhanced our understanding of enzyme catalysis for biotechnological applications, particularly the synthesis of chiral amines. The engineering of amine transaminase has shown a sustainable pathway towards synthesizing the critical molecules that were previously blocked by steric hindrances. A pivotal paper revealed the rational engineering of the *Vibrio fluvialis* amine transaminase, which initially demonstrated no catalytic activity for the conversion of bulky substrates like 2-acetylbiphenyl. By making strategic mutations, W57G/R415A, scientists were able to produce a tremendous boost in catalytic activity, which surpassed 1716-fold without compromising enantioselectivity. The huge improvement illustrates how the principles of thermochemistry can be utilized effectively to enhance the performance of enzymes, thereby paving the way for their incorporation into industrial synthesis routes that focus on sustainability and efficiency (Allen et al., 2016). Parallel to this, research on nanomaterials in thermochemistry has established a new frontier in industrial uses, primarily in terms of environmental sustainability. Nanoparticles (NPs) possess various properties that facilitate various uses, including remediation and agriculture as well as catalysis. They exhibit high specific surface areas and reactivity, which offer possibilities for enhancing chemical processes in industry. However, the release of NPs into nature and living organisms is generating more and more significant worries regarding their toxicity. Research has established that certain nanoparticles do not only affect plant growth indirectly but can also damage root tissues once accumulated. Understanding the thermochemical interactions within such systems is significant, since it brings out both the benefits and risks of nano-enabled technologies and therefore calls for ongoing research into their effects on the environment (Kumar et al., 2012). The intersection of nanotechnology, biochemistry, and thermochemistry presents a chance to innovate at both the industrial manufacturing and environmental safeguarding levels. As researchers delve further into the processes of enzyme catalysis and nanoparticle action, the potential to achieve peak chemical reaction potential by leveraging thermochemical principles becomes increasingly evident. These findings are not merely theoretical; they have the potential to revolutionize the manner in which industries tackle production efficiency, minimizing waste and optimizing resources. By embracing this cross-disciplinary approach, scientists are working toward a future where thermochemistry not only enhances traditional chemical processes but also enables sustainable practices that can moderate environmental impacts, resulting in a balance coexistence of industry and nature.

10.5.1. Innovations in Thermochemical Technologies

Developments in thermochemical technologies have played crucial roles in the development of industrial and biological systems. One of the significant areas of development is the production of hydrogen by thermochemical cycles, wherein thermal energy is utilized to enhance chemical reactions. The ongoing research shows the potential of solar energy to be utilized efficiently in these systems and can lead to reducing the cost of hydrogen production and enhancing its market positioning in the energy sector. The examination of a few candidate systems has indicated that market prices correlate closely with product prices within niche markets for the hydrogen market. Such developments are not only economically viable but also encourage practices of sustainable generation of energy, thus contributing to reducing greenhouse gas emissions and promoting transition towards cleaner energy (Escher et al., 2025). In addition, the use of innovation in thermochemical technology extends beyond hydrogen production to material processing and resource development. Recent workshops have explored the potential for collaborative public and private sector endeavors to encourage technologies that can support space habitation. These workshops have facilitated an understanding of the synergies between thermochemical methods and other technological development toward space travel. Working groups have come to realize essential needs in areas such as automation, biosystems, and life support, representing a growing awareness of the interconnectedness of thermochemical processes with various supporting technologies. This cross-disciplinary process can lead to significant breakthroughs that can not only encourage space exploration but also enable processes in present use on the planet in industrial applications (Fairchild et al., 2025). Moreover, the use of thermochemical technologies in biological systems presents a vast innovation frontier. Such technologies have the ability to impact various bioprocesses, where chemical reactions are optimized for enhanced efficiency, minimized wastes, and increased product yield. Utilization of thermochemical reactions in the manufacture of biofuels presents a very promising area of research and application, for instance. As the industry shifts towards green modes of operation, the development of efficient thermochemical processes will be the pillar of biomass to energy-dense fuel conversion. Developments in this realm, in fact, not only reflect a technological leap but also a step towards global environmental goals, positioning thermochemical technology at the forefront of green industrial processes (Escher et al., 2025) (Figure 10.12).

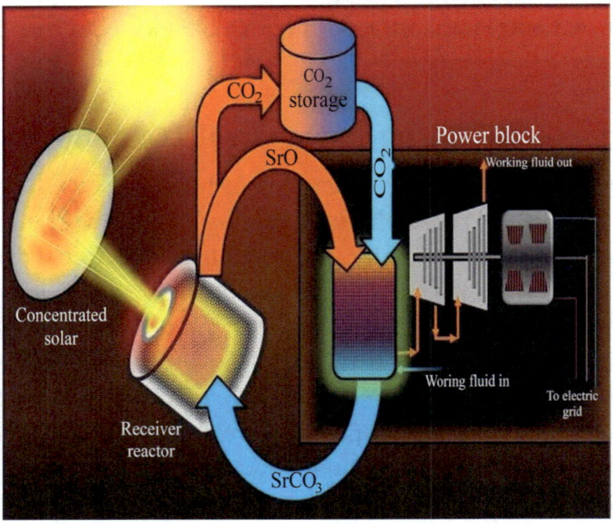

Figure 10.12. *Innovative thermochemical energy storage technology.*

Source: https://external-content.duckduckgo.com/iu/?u=https%3A%2F%2 Feepower. com%2Fuploads%2Fthumbnails%2F33360__.jpg&f=1&nofb=1&ipt= c693e48b19ec7049dec517eb73f4b547b3bc193b1c490cf97911f2577b5e4a72.

10.5.2. Role of Computational Chemistry in Thermochemistry

The fusion of computational chemistry and thermochemistry has significantly impacted the way scientists understand and manage thermodynamic properties in industrial as well as biological processes. High-level computational methods allow the modeling of molecular interactions, providing insight into reaction pathways and energy landscapes that are crucial to the prediction of thermochemical behavior. Techniques such as density functional theory (DFT) and molecular dynamics simulations enable researchers to assess stability, reactivity, and kinetics of chemical systems without the need for large-scale experimental testing. The computational approach not only accelerates the discovery of new materials and reactions but also optimizes current processes, thus enhancing processes in areas such as catalysis, materials science, and biochemistry where thermochemical principles are of utmost significance. Its impact cannot be overstated in industrial application. Computational chemistry allows the design and optimization of chemical processes via prediction of thermodynamic properties like enthalpy, entropy, and Gibbs free energy. Accurate thermochemical calculations make it easier to create more efficient, less environmentally taxing processes that waste less and use less energy. For instance, computational simulations have the ability to predict numerous reaction conditions, which allow engineers to select optimum routes and catalysts. In addition to that, the models can also be applied to scale up processes from laboratory to industrial scales, hence positively impacting production efficiency. With industries increasingly turning towards sustainability, the application of computational chemistry to gain economically viable thermodynamic solutions is increasingly becoming a key method. In biological organisms, the use of computational chemistry also involves the comprehension of thermodynamics at a molecular level, which is pivotal in unraveling biochemical reactions. Modeling of macromolecules such as enzymes and proteins allows researchers to investigate their thermodynamic stability and substrate binding. This information is pivotal in drug development and design, where thermodynamic favorability of the binding interaction has a substantial bearing on therapeutic effectiveness. The developments in machine learning algorithms have also complemented traditional computational methods such that examination of large data sets from experimental thermochemical studies becomes accessible. By the union of these novel approaches, the scientific fraternity hopes to advance knowledge on thermodynamic activity and characteristics in biological systems, thus furthering innovations in biotechnology advancements and health technologies (Owida et al., 2024) (Figure 10.13).

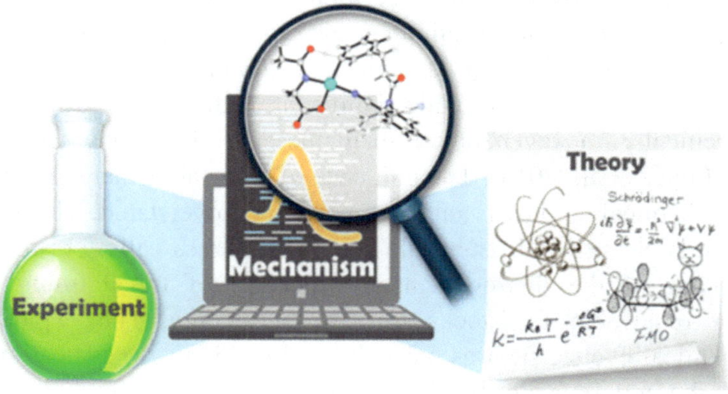

Figure 10.13. *Role of computational chemistry in thermochemistry.*

Source: https://pubs.acs.org/cms/10.1021/ja5112749/asset/images/medium/ja-2014-112749_0024.gif.

10.5.3. Future Directions in Thermochemical Research

One such bright direction for the future of thermochemistry is the marriage of sophisticated computational strategies with experimental techniques. Sophisticated simulation enables scientists to make more accurate predictions of reaction kinetics and thermochemical characteristics. Molecular dynamics and quantum mechanics-based models, for instance, can model challenging reaction pathways that are hard to detect under normal laboratory conditions. These innovations do not only push our understanding of simple thermochemical processes forward but also help to develop industrial applications, particularly in energy production and storage systems. The marriage of computational predictions with experimental testing has the potential to accelerate innovation in thermochemistry by far, leading to more efficient energy conversion technology and more efficient industrial reactions that minimize waste and energy consumption, ultimately addressing global sustainability challenges. Another highly prospective field of thermochemical research deals with biomimetic materials, particularly in biological systems. The recent trend has indicated the potential of nanocrystalline calcium phosphate apatites, which constitute a critical component of hard tissues. Researchers are now more inclined toward creating synthetic analogs that mimic the structural and compositional characteristics of biological apatite compounds. This field can provide significant advancement in applications such as bone tissue engineering and drug delivery systems.

Better understanding of crystallization effects and interfacial behavior of these materials will facilitate the development of new, innovative bioactive materials capable of meeting requirements in biologic environments. Therefore, the marriage of thermochemistry and biochemistry not only promises to further material design, but also to provide significant contributions towards medical uses, specifically in regenerative medicine and therapeutic interventions (Delgado-López et al., 2013). The influence of renewable energy topographies on thermochemical studies is one that cannot be overstated. The increasing trend towards cleaner and renewable energy resources calls for an extensive understanding of the thermochemical processes of transforming biomass and other renewable feedstocks to valuable energy. The exploration of thermochemical cycles, such as solar thermal or biomass gasification cycles, highlights the importance of breakthrough approaches with higher efficiencies and reduced carbon footprints. By optimizing these processes, researchers can significantly contribute to the development of cleaner energy systems that continue to become more sustainable substitutes for the traditional fossil fuels. Scalability and economics need to be the bywords for research in the future while maintaining a keen sensitivity towards sustainability of the environment.

Thus, thermochemistry will be a key force in shaping the future path of energy systems in industrial and biological processes, to construct a complete solution to energy problems globally (Woods et al., 2025).

10.6. CASE STUDIES OF THERMOCHEMISTRY IN ACTION

Industrial and biological applications of thermochemistry help to demonstrate its vital role in creating safety and efficiency. A good example case is the development of fire-resistant materials for high-risk uses such as aerospace and construction. There has been great research in the area of polymer chemistry and composite technology that has resulted in great improvement in fire performance and survivability of humans. They are not just on paper but the result of careful bench-scale and full-scale testing, indicating the integration of thermochemical principles into practical safety solutions. The materials systems included in these examples describe the ways in which an understanding of heat transfer, combustion, and degradation of materials can be applied to create innovations saving lives and property during extreme situations (Heimbuch et al., 2025). These case studies demonstrate the importance of applying thermochemical principles to minimize threats in dynamic environments. In biological processes, thermochemistry has a significant role to play in enzyme design and biocatalysis. One of the most fascinating case studies on amine transaminases demonstrates how understanding the thermodynamic and kinetic properties of enzyme reactions can be a solid foundation for innovation in synthetic chemistry. In this particular study, the aim was to enhance catalytic activity by the enzyme for large-sized ketone substrates, which commonly have been problematic by way of steric repulsion. By means of a rational design approach that combined computational and experimental approaches, researchers achieved an astonishing increase in enzyme performance, with the rates of reaction being more than a 1716-fold improvement. The engineered variant was not only characterized by enhanced catalytic activity but also high enantioselectivity, which produced the target chiral amines with unparalleled purity (Allen et al., 2016). This is an example of how thermochemistry has the revolutionary potential to develop sustainable and efficient biotechnological substitutes. Besides, integration of thermochemistry into industrial use broadens its multi-dimensional applications across various industries. For instance, optimization of endothermic and exothermic reactions in the chemical production industry enhances effective production of major chemicals. By sustaining thermal control and reaction conditions, companies can improve yield as well as conserve energy, describing a picture of sustainability and affordability. Secondly, thermochemistry's role in the development of alternative fuels and energy storage devices further solidifies its significance in relieving the world energy crisis. Through research in novel thermochemical reactions, such as those designed to make access to biofuels possible or enhance battery efficiency, industries can reduce their carbon footprint and build a more sustainable future. These case studies collectively show the pivotal role of thermochemistry in driving innovation in industrial and biological systems.

10.6.1. Industrial Case Study: Thermochemical Energy Storage

Thermochemical energy storage is an industrial technology that has been a breakthrough in advancing sustainable energy solutions. It depends on chemical reactions that absorb and release heat energy so that the storage of energy is both efficient and enduring. When renewable energy sources such as wind or solar generate more power than they are currently able to utilize, thermochemical systems store this excess in the form of endothermic processes that leave power in chemical bonds. This not only makes full use of available power but also addresses the issue of intermittency from renewable resources, offering a systematic solution to stabilizing the energy supply, particularly industrial applications. Significantly, such wastes as red mud and sewage sludge find prominence in research on thermochemical energy storage owing to their dual benefit of waste treatment and recovery of energy.

What are otherwise referred to as environmental liabilities, these substances can be employed as innovative solutions for energy storage if treated accordingly. These by-products can be made liable for the application of thermochemical principles to enhance the efficiency of energy use while concurrently lessening the emissions of carbon that are allied with their disposal. Through the energy storage media conversion of these by-products, industry can convert towards carbon peaking and neutrality strategies as it aligns with crucial environmental concerns inherent in traditional strategies for waste management (Ding et al., 2024). Though industries are still focused towards integration of the thermochemical storage systems, problems still remain in the long-term stability and viability of waste materials. The research on the application of sewage sludge and red mud indicates hopeful trends towards the development of composite materials with superior energy storage ability. Environmental pollution threats through treatment processes and economic feasibility for wider adoption remain significant challenges. These challenges demand prolonged research into the optimization of treatment methods and low-cost measures that enable the transition towards sustainable thermochemical energy storage in industries (Barbero et al., 2011). Studying these measures not only advances the field of thermochemistry but redefines the parameters of waste materials in energy applications.

10.6.2. Biological Case Study: Thermochemistry in Cellular Respiration

During the process of studying thermochemistry in the biological context, cellular respiration is an example of how organisms release stored chemical energy in nutrients in forms usable to them. All this is achieved through several metabolic pathways, primarily glycolysis, the citric acid cycle, and oxidative phosphorylation. All these processes are important for the breakdown of glucose in order to generate adenosine triphosphate (ATP), the cellular currency. Understanding of such biochemical changes from a thermochemical viewpoint explains how energy transformations regulate cellular activities. Energy discharge in electron transport chain and substrate-level phosphorylation indicate the energy conservation and transfer rules that are the foundation of biological as well as industrial thermodynamic processes. These discoveries highlight how energy derivation is optimized in organisms, which is what equivalent measures in industrial applications of thermochemistry seek to achieve. The efficiencies of cellular respiration are also aided by the activity of enzymes and coenzymes, which catalyze chemical reactions at a lower activation energy. The activity of enzymes significantly disrupts the equilibrium of the metabolic processes towards synthesizing ATP, providing a classic example of thermodynamic principles in action.

The process of catalysis decreases energy loss in the form of heat so that biological systems can maintain their metabolic activities at relatively low temperatures. The enzyme-substrate binding can be explained using Gibbs free energy that provides a deeper insight into biochemical reaction energetics. In addition, determining the role of coenzymes such as NAD+ and FAD in the transfer of energy through cellular respiration shows the interconnectedness of thermochemical pathways, similar to similar mechanisms found in industrial chemical reactions where catalysis enables higher efficiency and product formation. Furthermore, the utilization of thermochemistry extends into environmental spheres, namely within the scope of understanding the role that byproducts of cellular respiration have in determining ecological phenomena. Carbon dioxide, as a prime product of aerobic respiration, demonstrates how biological processes influence atmospheric phenomena and global climatic cycles. The increased concentration of the gas, both of natural and human origin, affects the maintenance of temperature within the biosphere. Notably, thermochemistry research in cellulair respiration can enhance the understanding of the larger picture in environmental systems by relating energy production and consumption with ecosystem health. The cellulose nature of the application of nanoparticles as discussed in ecological studies justifies the use of an integrated approach that is sensitive to both biological processes and environmental impact, highlighting the importance of investigating thermochemical interactions in detail (Kumar et al., 2012; Majid et al., 2025).

10.6.3. Comparative Analysis of Industrial and Biological Thermochemical Processes

Comparative analysis between industrial and biological thermochemical processes indicates immense difference in efficiency, scalability, and impact on the environment. Industrial thermochemical processes, often exemplified by high-energy needs and chemical harshness, are designed to maximize yield through explicit control of reaction conditions. These thermochemical reactions, commonly applied in the production of fuel, chemicals, and materials, tend to rely on fossil fuels, which release greenhouse gases. Conversely, biological thermochemical reactions leverage enzymatic pathways and microbial activities, which operate under milder conditions and prefer renewable materials. This inherent disparity in operating conditions not only highlights the sustainability potential intrinsic in biological systems but also tests conventional industrial practice in its efforts to lessen its environmental footprint. In terms of the operating processes, industrial procedure depends predominantly on high temperature and pressure operating conditions to disintegrate compound substances into reduced components. This is in direct contrast to biological responses, where enzymes are used as catalysts to enable reactions at room temperature. As noted in some uses, such as in the oxidation of arylamines, laccases have been proposed as biocatalysts for substituting conventional chemical oxidants, providing a cleaner alternative as highlighted in Acebes et al. (2016). The effectiveness of such enzymic reactions relies on the optimization of substrate binding and electron transfer processes, and on the delicate interaction of thermodynamics and kinetics. This ability to tailor reactions in biological systems is an important milestone towards the further purification of biocatalytic applications throughout industries and demonstrates the potential for bioprocesses as an exemplar for industrial progress. Additionally, due to progress in biotechnological techniques, great improvement in the efficiency of biological thermochemical processes has been witnessed. Scientists can predict and control enzymatic functions using computer tools such that they can balance redox potentials of desired substrates for effective reactions, including engineered laccases optimizing oxidation efficacies in the synthesis of aniline (Acebes et al., 2016). These findings not only provide a glimpse into maximizing enzyme engineering but also indicate that biological systems can become as efficient as their industrial counterparts. As industrial pressures to be more sustainable grow, moving towards biologically inspired thermochemical processes may be an avenue through which industries can incorporate environmental concerns into their operational models and thus cultivate an integrated model of chemical production that fosters a balanced approach towards chemical manufacturing that meets both economic requirements and ecological responsibility.

SUMMARY

In brief, the thermochemistry science carries profound importance for various industries, especially industrial and biological processes. The laws governing heat transfer and energy changes are at the core of optimizing chemical production to biological metabolism processes. Implementing the principles of thermochemical concepts not only facilitates streamlined production processes but also facilitates energy conservation, balancing industrial procedures with sustainability objectives. The mechanisms and applications discovered direct innovations that have a significant contribution to greener technology innovation. To conclude, taking up these principles encourages a scientific platform on which the effectiveness of processes can be enhanced, leading to more sustainable industrial processes eventually. Moreover, recent advances in enzyme catalysis go to show the dynamic position that thermochemistry plays in biological processes. Identification of new enzymatic pathways is a prime example of how thermochemical reactions are capable of initiating complex biochemical conversion, as demonstrated by the novel behaviors of epoxide-altering enzymes in biosynthesis. These findings not only underscore the complexity of biological thermodynamics but are also an inspiration for synthetic methodologies within industrial chemistry. The correlation between thermal dynamics of such enzymes and their substrates is a key factor in revealing how biological systems operate at an optimal level regardless of changing environmental conditions. Such findings, thus, bridge the gap that exists between biological processes and industrial applications, portraying a synergistic nature that benefits both. The use of theoretical models like integral equation theory further deepens our understanding of molecular interactions in thermochemical processes. This approach allows for accurate description of solvation architectures and free energy modifications influencing a broad range of systems from synthetic materials to biomacromolecules. Statistical mechanics by researchers allows them to obtain valuable information about the interaction and stability of molecules within a solution, which serves to advance materials science and biochemistry. These methods enable broader examination of thermochemical processes across scales, culminating ultimately in breakthroughs in both research and application. By taking a multiscale methodology to thermochemistry, researchers enclose their inquiry within a robust theoretical framework that is not confined by single disciplines, paving the way for future breakthroughs and uses.

The intricate relationship between thermochemistry and industrial and biological systems forms the foundation for many processes fundamental to modern society. Thermochemistry essentially means comprehending how the exchange of energy in the form of heat influences chemical reactions. In an industrial process, awareness of such changes in energy can optimize production efficiency, minimize energy consumption, and optimize product yields. This basic knowledge benefits industries from pharmaceuticals to manufacturing, where controlling temperature and reaction kinetics is critical to obtain desired results. The understanding of thermochemistry is also integral to the formulation of sustainable processes reducing the environmental effects linked with conventional processes, thus harmonizing economic interests with ecological responsibilities. For biological systems, thermochemistry provides insight into energetic transactions that form the basis of metabolic pathways and cellular functions. Organisms constantly experience chemical reactions in order to maintain homeostasis, with the energy input or output, and this is described by thermochemical principles. For example, thermodynamic principles play an important role in studying biochemical processes such as glycolysis, with the enthalpy changes providing the means to explain how organisms convert nutrients to energy that can be harnessed. This understanding is pivotal to advancements in medicine and healthcare fields, in which the management of thermodynamic variables can lead to new treatments and drug development strategies, such as through processes handling chiral separation techniques in medication, with emphasis on efficiency and efficacy of drug synthesis (Becke et al., 2018). Further, the application of thermochemistry is relevant in environmental science, in which thermodynamics informs studies on the observation of pollution and the energy cycle. For instance, thermochemical energy transfer processes are significant when the sustainability of different technologies and materials is being considered. In renewable energy technologies like solar energy and biofuels, the use and optimization of thermochemical processes can lead to less environmentally degrading alternatives that use fewer fossil fuels. By merging concepts of thermochemistry, industries can establish efficient systems that reduce waste while maximizing energy use in numerous applications, thereby finding a sustainable future (Woods et al., 2025). All these areas demonstrate how much influence thermochemical knowledge has for propelling industrial innovation as well as biological research.

REFERENCES

1. Abu Owida, H., Abuowaida, S., Al Sharah, A., Alhaj, M., et al. (2024). Application of machine learning in chemical engineering: Outlook and perspectives. https://core.ac.uk/download/642649312.pdf (accessed on 4th October 2025).
2. Acebes, S., Camarero, S., De Salas, F., Guallar, V., et al. (2016). Computer-aided laccase engineering: Toward biological oxidation of arylamines. https://core.ac.uk/download/46606171.pdf (accessed on 4th October 2025).
3. Alibakhshi, A. (2021). Developing advanced theoretical methods for enhanced study of thermochemistry. https://core.ac.uk/download/621260608.pdf (accessed on 4th October 2025).
4. Allen, C. C. R., Brown, R. T., Carvalho, A. T. P., Caswell, J. M., et al. (2016). Rational design of a (S)-selective-transaminase for asymmetric synthesis of (1S)-1-(1,1′-biphenyl-2-yl) ethanamine. https://core.ac.uk/download/74229950.pdf (accessed on 4th October 2025).
5. Barbero, S., & Fassio, F. (2011). Energy and food production with a systemic approach. https://core.ac.uk/download/pdf/11423953.pdf (accessed on 4th October 2025).
6. Becke, A. D., Bhattacharyya, R., Boys, S. F., Cao, Z., Cui, Y., Dong, S., Fileti, E. E., et al. (2018). Quantum chemistry calculation aided design of chiral ionic liquid-based extraction system for amlodipine separation. https://core.ac.uk/download/160606542.pdf (accessed on 4th October 2025).
7. Chalkley, M. J., Del Castillo, T. J., Matson, B. D., Peters, J. C., et al. (2017). Catalytic N_2-to-NH_3 conversion by Fe at lower driving force: A proposed role for metallocene-mediated PCET. https://core.ac.uk/download/77930176.pdf (accessed on 4th October 2025).
8. Delgado-López, J. M., Drouet, C., Gómez-Morales, J., Iafisco, M., et al. (2013). Progress on the preparation of nanocrystalline apatite's and surface characterization: Overview of fundamental and applied aspects. https://core.ac.uk/download/12044395.pdf (accessed on 4th October 2025).
9. Ding, Y., Xiong, Y., Xu, Q., Zhang, T., et al. (2024). A mini review on sewage sludge and red mud recycling for thermal energy storage. https://core.ac.uk/download/604053661.pdf (accessed on 4th October 2025).
10. Escher, W. J. D., Foster, R. W., Hanson, J. A., Tison, R. R., et al. (2025). Solar/Hydrogen Systems Assessment. Volume 1: Solar/Hydrogen Systems for the 1985–2000-Time Frame. https://core.ac.uk/download/pdf/42864063.pdf (accessed on 4th October 2025).

11. Fairchild, K., & Mendell, W. W. (2025). Report of the *in-situ* resources utilization workshop. https://core.ac.uk/download/pdf/42830419.pdf (accessed on 4th October 2025).
12. Garcia-Borràs, M., Hirayama, Y., Houk, K. N., Li, X., et al. (2017). Enzyme-catalyzed cationic epoxide rearrangements in quinolone alkaloid biosynthesis. https://core.ac.uk/download/323080002.pdf (accessed on 4th October 2025).
13. Gilman, J. A., Paige, J. B., & Schwenk, F. C. (2025). The NASA SBIR product catalog. https://core.ac.uk/download/pdf/42806099.pdf (accessed on 4th October 2025).
14. Hadadi, N. (2015). Computational studies on cellular metabolism: From biochemical pathways to complex metabolic networks. https://core.ac.uk/download/148015164.pdf (accessed on 4th October 2025).
15. Heimbuch, A. H., & Parker, J. A. (2025). Chemical research projects office functions accomplishments programs. https://core.ac.uk/download/pdf/42889885.pdf (accessed on 4th October 2025).
16. Johnson, J. J., IV. (2016). A general theory of emergence in engineered systems. https://core.ac.uk/download/217293973.pdf (accessed on 4th October 2025).
17. Kovalenko, A. (2015). Molecular theory of solvation: Methodology summary and illustrations. http://arxiv.org/abs/1510.06520 (accessed on 4th October 2025).
18. Kumar, R., Mishra, A. K., & Rawat, K. S. (2012). Nanoparticles in the soil environment and their behaviour: An overview. https://core.ac.uk/download/158352526.pdf (accessed on 4th October 2025).
19. Kurylo, M. J., Ormond, F. M., Prather, M. J., Watson, R. T., et al. (2025). Present state of knowledge of the upper atmosphere 1990: An assessment report. https://core.ac.uk/download/pdf/42821958.pdf (accessed on 4th October 2025).
20. Locaspi, A. (2024). A lumped condensed-phase kinetic model for plastic waste thermochemical recycling. https://core.ac.uk/download/650095391.pdf (accessed on 4th October 2025).
21. Majid, N. A. M. N. A., Osman, K., & Yee, T. S. (2025). Energy education: Adapting to learning preferences and pedagogical strategies. https://core.ac.uk/download/665327678.pdf (accessed on 4th October 2025).
22. N/A. (2014). SciTech News, 68(4), 2014. https://core.ac.uk/download/46973459.pdf (accessed on 4th October 2025).

23. N/A. (2025). Aerospace Medicine and Biology: A Continuing Bibliography with Indexes, Supplement 267, January 1985. https://core.ac.uk/download/pdf/42845954.pdf (accessed on 4th October 2025).
24. N/A. (2025). Aerospace Medicine and Biology: A Continuing Bibliography with Indexes, Supplement 132. https://core.ac.uk/download/pdf/42891625.pdf (accessed on 4th October 2025).
25. Nikoo, S. (2019). Computational insights into bio-relevant sulfur chemistry. https://core.ac.uk/download/250624180.pdf (accessed on 4th October 2025).
26. Ortiz-Cantú, S. J., & Pedroza-Zapata, Á. R. (2011). Sources of information for technology intelligence in chemical formulation. https://core.ac.uk/download/47247144.pdf (accessed on 4th October 2025).
27. Pratihar, S. (2023). Physical Chemistry I Laboratory Manual. https://core.ac.uk/download/552592691.pdf (accessed on 4th October 2025).
28. Shi, X., & Xu, G. (2018). Environmentally friendly pervious concrete for treating deicer-laden stormwater: Phase II. https://core.ac.uk/download/185623460.pdf (accessed on 4th October 2025).
29. Woods, R. L. (2025). Publications of the Jet Propulsion Laboratory, July 1964 through June 1965. https://core.ac.uk/download/pdf/85253247.pdf (accessed on 4th October 2025).

INDEX

A

Absolute zero 15, 67, 70, 128, 209, 367
Acclimatization mechanisms 136
Actual physical behavior 145
Adiabatic compression 53
Advanced computational methods 3
Advanced radiation-hydrodynamics codes 31
Alternating current (AC) 135
Amino acid-based ionic liquids 28
Ammonium polyphosphate (APP) 154
Angular impulse-momentum equation 59
Atmospheric reactions 17, 333
Atomic diffusion 145
Augmented reality (AR) 48

B

Binding mechanisms 151
Biochemical systems 15, 344
Biochemistry 29, 30, 85, 87, 88, 125, 136, 145, 150, 152, 253, 260, 264, 270, 279, 297, 300, 323, 330, 334, 340, 344, 386, 388, 390, 395
Black hole 54, 55, 168, 203, 206, 207, 211, 213, 216, 230, 232, 240, 284, 319
Bomb calorimeter 23, 134
Bond Dissociation Energy (BDE) 123
Bond formation 18, 102, 249, 292
Brownian particle 61

C

Calibration procedures 139
Calorimetric analysis 135, 138, 142, 143, 144, 145, 147, 149, 347
Calorimetric behavior 125, 128
Calorimetric data 94, 123, 125, 126, 127, 137, 139, 141, 144, 146, 147, 149, 150, 154, 350
Calorimetric measurement precision 142
Calorimetric measurements 6, 85, 116, 123, 129, 138, 139, 141, 142, 144, 146, 153, 154, 178, 347
Calorimetric techniques 22, 99, 125, 127, 128, 135, 139, 140, 146, 147, 149, 153, 154, 155, 156, 157, 347, 349, 350
Calorimetry 6, 20, 21, 97, 98, 121, 123, 125, 126, 136, 137, 146, 147, 148, 149, 151, 152, 154, 177, 178, 346, 347, 348, 350, 356, 369
Calorimetry devices 135
Calorimetry principles 23
Calvet microcalorimetry 24
Chemical bonds 11, 16, 102, 286, 292, 294, 297, 392
Chemical change 89, 153, 253, 257, 290, 299
Chemical disposal 168
Chemical energy 43, 45, 46, 48, 49, 52, 53, 57, 88, 91, 92, 119, 218, 283,

284, 302, 305, 306, 307, 308, 317, 393
Chemical engineering 6, 14, 27, 32, 93, 115, 116, 117, 166, 171, 174, 182, 305, 310, 335, 397
Chemical equilibrium 241, 243, 245, 248, 252, 255, 259, 263, 274, 275, 332
Chemical kinetics 4, 26, 31, 248, 263, 300, 316, 383
Chemical process 12, 34, 96, 99, 187, 237, 277, 285, 305, 309, 344, 357, 382
Chemical reaction 1, 2, 5, 7, 15, 16, 17, 36, 87, 88, 98, 99, 100, 104, 115, 116, 163, 165, 166, 167, 168, 170, 173, 176, 187, 188, 189, 193, 226, 227, 228, 243, 251, 252, 253, 256, 257, 259, 263, 265, 269, 275, 286, 301, 302, 324, 349, 386
Chemical reactions 1, 3, 4, 5, 6, 7, 8, 9, 10, 11, 12, 13, 15, 16, 20, 21, 22, 23, 25, 28, 30, 31, 33, 34, 35, 48, 52, 83, 84, 85, 86, 87, 88, 89, 90, 91, 92, 93, 94, 97, 98, 99, 100, 101, 102, 103, 104, 105, 106, 107, 109, 111, 112, 113, 114, 115, 116, 121, 123, 124, 128, 130, 136, 143, 156, 165, 168, 169, 170, 172, 173, 174, 176, 177, 178, 181, 182, 183, 184, 187, 189, 190, 192, 193, 194, 195, 222, 224, 226, 227, 234, 241, 242, 243, 244, 245, 247, 248, 249, 252, 254, 255, 256, 257, 258, 260, 264, 268, 270, 272, 273, 274, 275, 281, 284, 285, 286, 287, 288, 289, 290, 292, 296, 297, 299, 300, 301, 303, 304, 305, 306, 308, 313, 315, 316, 321, 323, 324, 325, 326, 329, 330, 331, 333, 336, 338, 339, 345, 346, 349, 350, 351, 354, 355, 359, 361, 363, 364, 365, 366, 368, 369, 371, 373, 381, 384, 387, 392, 393, 396
Chemical synthesis 32, 94, 97, 114, 176, 299, 304, 363
Chemical system 12, 105, 226, 276, 373
Chemical transformation 84, 108
Chemical vapor deposition (CVD) 31, 117
Classic thermodynamics 203
Closed system 57, 67, 77, 103, 128, 215, 222, 224, 233, 234, 253, 254, 263, 265, 330
Combustible materials 99, 126
Combustion chemistry 4, 6, 7, 25, 52
Combustion systems 2, 6, 95, 356
Combustion theory 85, 95
Composite materials 125
Computational models 3, 29, 31, 36, 96, 117, 125, 272, 275
Conduction 47, 62, 63, 129, 294
Cone calorimetry 137
Conformational change 87
Conservation of Energy Law 67
Constant pressure 12, 20, 85, 87, 89, 96, 101, 107, 132, 133, 136, 169, 184, 213, 241, 244, 246, 251, 253, 257, 260, 266, 348, 365, 368
Constant pressure calorimeters 132, 133
Constant rate thermal analysis (CRTA) 122, 129
Convection 12, 47, 62, 63, 64, 65, 76, 79, 129, 130, 294, 295, 296
Cooling systems 46, 295

D

Density functional theory (DFT) 26, 35, 388
Density Functional Theory (DFT) 100, 170, 173, 175
Dielectric analysis (DEA) 147
Differential scanning calorimetry (DSC) 98, 125, 126, 130, 144, 145, 147, 153, 346, 350
Differential Scanning Calorimetry (DSC) 20, 97
Digital signal processing techniques 136, 146

Dipropylene glycol propyl ether (DPnP) 144
Direct current (DC) 135
Dissipation forces 60
Dual-readout systems 131
Dynamic mechanical analysis (DMA) 147

E

Ecological equilibrium 187
Electrical energies 48
Electrical work 61
Electromagnetic calorimeters (ECAL) 131
Electromagnetic conditions 66
Electrostatic forces 50
Endergonic reactions 261
Endothermic 2, 4, 5, 11, 14, 16, 18, 19, 20, 33, 84, 85, 87, 88, 89, 91, 93, 94, 96, 99, 101, 102, 108, 116, 169, 170, 177, 187, 265, 290, 292, 296, 298, 299, 301, 310, 315, 325, 326, 328, 338, 340, 341, 349, 356, 365, 379, 391, 392
Endothermic reactions 18, 19, 298, 299
Energy conversion 8, 11, 15, 27, 35, 40, 67, 68, 69, 70, 74, 75, 76, 77, 85, 107, 115, 117, 127, 135, 166, 167, 170, 177, 188, 189, 193, 206, 208, 211, 214, 217, 222, 223, 235, 255, 262, 284, 287, 288, 290, 294, 296, 297, 298, 299, 300, 302, 303, 305, 306, 307, 308, 310, 311, 313, 314, 315, 356, 390
Energy conversion principle 11
Energy conversions 4, 284, 310
Energy conversion technology 15, 70, 117, 314, 390
Energy delivery consistency 177
Energy dissipation 63, 258, 331
Energy efficiency 13, 15, 27, 31, 34, 35, 47, 50, 56, 57, 64, 66, 71, 72, 73, 74, 88, 108, 166, 167, 171, 176, 185, 187, 189, 214, 230, 235, 296, 299, 308, 309, 312, 328, 343, 345, 355, 363, 366, 368, 371, 373, 376, 377, 382, 384
Energy management 3, 57, 61, 65, 72, 164, 171, 172, 176, 182, 189, 193, 195, 197, 223, 231, 284, 290, 297, 302, 314
Energy Management Systems (EMS) 196
Energy production 2, 3, 7, 8, 15, 23, 28, 29, 34, 55, 76, 77, 95, 112, 117, 127, 134, 168, 171, 175, 184, 188, 189, 191, 284, 308, 328, 356, 381, 384, 385, 390, 393
Energy storage mechanisms 164
Energy Storage Systems (ESSs) 195
Energy technology development 164, 188
Energy topology 60
Energy transfer 7, 13, 17, 25, 27, 45, 46, 47, 48, 53, 55, 61, 62, 63, 64, 65, 66, 67, 69, 77, 90, 98, 127, 128, 164, 186, 189, 209, 213, 214, 215, 218, 226, 244, 256, 262, 291, 295, 296, 297, 305, 306, 328, 333, 334, 345, 361, 363, 373, 376, 381, 396
Energy transformation 6, 36, 52, 58, 89, 100, 135, 195, 204, 213, 223, 234, 285, 292, 293, 294, 297, 298, 301, 302, 303, 304, 306, 307, 310, 311, 313, 315, 377
Energy transformations 11, 16, 28, 33, 60, 77, 84, 101, 124, 126, 165, 168, 190, 193, 205, 207, 209, 217, 279, 283, 284, 285, 290, 292, 294, 296, 299, 302, 303, 304, 306, 307, 308, 309, 310, 311, 314, 327, 356, 368, 377, 379, 393
Energy transition process 20
Energy transport analysis 45
Enthalpy 5, 12, 13, 14, 20, 22, 32, 34, 83, 84, 85, 86, 87, 88, 89, 90, 92, 93, 94, 95, 96, 97, 101, 106, 107, 109, 110, 111, 116, 169, 173, 175, 177, 183, 184, 249, 327, 337, 368

Enthalpy change 2, 6, 12, 34, 84, 85, 86, 87, 88, 89, 92, 93, 94, 97, 98, 99, 104, 105, 106, 109, 112, 113, 114, 115, 116, 117, 163, 165, 166, 167, 168, 169, 170, 172, 173, 174, 175, 176, 177, 178, 179, 182, 183, 184, 185, 187, 188, 189, 190, 193, 194, 326, 333, 337, 338, 346, 350
Enthalpy changes 3, 5, 10, 14, 16, 21, 24, 32, 34, 84, 85, 86, 87, 88, 89, 92, 93, 96, 97, 98, 99, 100, 101, 102, 104, 105, 106, 107, 108, 109, 110, 111, 112, 113, 115, 116, 117, 126, 128, 141, 165, 167, 170, 171, 172, 173, 174, 175, 176, 177, 178, 179, 181, 182, 183, 184, 185, 190, 193, 194, 323, 328, 332, 337, 338, 356, 368, 396
Entropy principle 201
Environmental chemistry 4, 5, 10, 30, 32, 33, 88, 96, 97, 184, 189, 190, 194, 243, 321, 323, 345
Environmental protection 28, 65, 312
Environmental sustainability 56, 74, 109, 176, 181, 185, 191, 274, 314, 354, 370, 373, 377, 384, 386
Equilibrium behavior 9
Equilibrium chemistry 3
Exothermic 2, 4, 5, 11, 14, 16, 17, 18, 19, 33, 84, 85, 86, 87, 88, 89, 91, 93, 94, 96, 97, 99, 101, 102, 104, 107, 112, 116, 169, 177, 187, 249, 265, 290, 296, 297, 298, 301, 310, 315, 325, 326, 328, 338, 340, 341, 349, 356, 365, 375, 391
Exothermic processes 17
Experimental chemistry 9
Experimental techniques 136
Experimental validation 177

F

Feedback control 135, 229
Finite Element Method (FEM) 64
Fluid dynamics 3, 49, 63, 64, 65, 72, 117, 129, 293
Fluid viscosity 62
Fuel synthesis 84

G

Gas-phase reactions 6
Gas turbine combustion chambers 2
Geochemical state 88
Gibbs free energy 5, 29, 32, 107, 225, 226, 242, 243, 244, 245, 246, 247, 249, 250, 251, 253, 254, 256, 257, 258, 260, 263, 265, 266, 267, 268, 269, 270, 272, 273, 326, 336, 338, 356, 365, 389, 393
Gibbs Free Energy 241, 243, 244, 245, 246, 248, 249, 251, 252, 255, 256, 258, 259, 260, 261, 263, 266, 267, 268, 269, 270, 271, 274, 275, 338
Gibbs Free Energy equation 252, 264
Gibbs paradox 203, 218, 237
Gibbs-von Neumann entropy 53, 58
Gibbs-von Neumann entropy formalism 53
Ginzburg-Landau equations 202, 204, 206, 207
Glycolysis process 260
Gravitational heating 50, 51
Gravitational interaction 50
Guggenheim-Anderson-de Boer (GAB) 147

H

Hadronic calorimeters (HCAL) 131
Hamiltonian dynamics 58, 205
Heat behavior 12, 128
Heat conduction 45, 62, 82
Heat engines 74, 75
Heat transfer 3, 5, 8, 9, 12, 13, 21, 27, 31, 39, 46, 47, 48, 52, 57, 60, 62, 63, 64, 65, 66, 69, 76, 80, 82, 97, 112, 121, 124, 126, 129, 138, 140, 141, 147, 148, 187, 206, 213, 271, 272, 277, 295, 301, 322, 323, 330, 331, 332, 345, 348, 358, 363, 369, 371, 380, 391, 395
Helmholtz free energy 256, 276

Hesss Law 94, 99, 163, 165, 167, 168, 169, 172, 173, 174, 176, 178, 181, 182, 183, 185, 186, 188, 189, 191, 192, 193
High-energy hadrons 132
High-energy particle physics 131
Hybrid Energy Storage Systems (HESS) 176
Hybrid Energy Storage Systems (HESSs) 195
Hypothetical thermodynamic 244

I

Ignition resistance 137, 149
Industrial synthesis 86, 96, 386
Internal energy 5, 12, 13, 16, 53, 83, 85, 88, 89, 90, 91, 102, 212, 213, 327, 330
International Linear Collider (ILC) 125
Intra-cluster medium (ICM) 50
Ionic liquids (ILs) 28
Ionization conditions 16
Isocaloric test 23
Isolated systems 222, 223
Isothermal titration calorimetry (ITC) 99, 151, 348

K

Key performance indicators (KPIs) 57
Kinetic energy 12, 15, 48, 49, 50, 52, 53, 55, 58, 60, 67, 101, 286, 287, 288, 289, 290, 292, 295, 300, 301, 302, 310, 335, 337, 340

L

Laminar flames 15
Lattice Boltzmann (LB) method 49
Laws of thermodynamics 10, 15, 48, 54, 55, 57, 61, 85, 223, 228, 252, 255, 283, 322, 368
Load flexibility 2
Lower particle density 62
Lower thermal mass 135
Low-temperature physics 70
Lysozyme stability 144

M

Magnesium-aluminum layered double hydroxides (MgAl-LDH) 154
Magnetic energies 16
Magnetic resonance imaging (MRI) 125
Magnetohydrodynamic simulations 16
Material conductivity 129
Maximum Entropy Production (MEP) 219, 228
Mechanical homogenization 138
Metal-organic frameworks (MOFs) 84
Meteorite analysis 10
Microcalorimetry 123, 144
Microelectronics 46
Microfluidic calorimeters 110, 118, 149, 150, 159
Microfluidic calorimetry 150
Micro-phase transitions 87
Microscale behavior 234
Micro-scale calorimeters 154
Microscopic particle behavior 49
Modern thermochemical research 9
Molecular behavior 27, 85, 87, 244, 258, 348
Molecular bonding 12
Molecular interaction 7, 12, 93, 113, 329, 334, 352
Molecular reactivity 16
Monte Carlo simulations 96, 252, 275, 280
Multi-energy systems 164

N

Nanomagnetic hyperthermia (NMH) 125
Nickel cadmium batteries 14
Non-cool core (NCC) clusters 51
Non-equilibrium conditions 45, 214, 297, 302
Nonequilibrium dynamics 223
Nonequilibrium states 60
Non-equilibrium systems 58, 235
Nonequilibrium systems 60, 222, 234

Non-equilibrium thermodynamics 58, 204
Nonequilibrium thermodynamics 69, 79, 81, 210, 225, 240, 243, 247, 263
Nuclear fission 52
Nuclear magnetic resonance (NMR) 153

O

Organic Rankine Cycle (ORC) 75
Oxygen evolution reaction (OER) 35

P

Particle flow calorimetry 152
Phase change reactions 114
Phase equilibria 21, 22, 33, 140, 141, 142, 267, 336, 350, 352
Phase stability 21, 347, 350
Phospholipid synthesis 113
Photoionization mechanisms 16
Physical chemistry 1, 244, 254, 297, 363
Physico-chemical properties 9
Pile-up correction methods 137
Plasma ionization dynamics 27
Plasma mass separation 28
Poly(1,4-butylene terephthalate) (PBT) 146
Polytetrafluoroethylene (PTFE) 122, 146
Polyunsaturated fatty acids (PUFAs) 113
Potential energy 16, 48, 50, 51, 55, 57, 58, 67, 105, 195, 218, 261, 268, 286, 287, 288, 289, 290, 291, 294, 301, 302, 310
Pseudomonas aeruginosa l-arginine sensor 151

Q

Quantitative approach 3, 243
Quantum chemical methods 30
Quantum mechanics/molecular mechanics (QM/MM) 29

R

Radiation 12, 31, 41, 47, 62, 63, 65, 66, 78, 79, 129, 206, 207, 230, 240, 284, 295, 319
Radical species-molecule reactions 17
Reaction mechanisms 2, 7, 15, 21, 29, 31, 34, 36, 84, 87, 94, 97, 108, 122, 146, 149, 156, 181, 274, 344
Redox-Flow Batteries (RFBs) 195
Renewable energy technology 164, 188
Reversible solid oxide fuel cells (RSOCs) 188

S

Second Law of Thermodynamics 45, 69, 201, 203, 204, 205, 206, 207, 210, 212, 213, 215, 217, 218, 219, 220, 222, 223, 225, 227, 230, 231, 232, 234, 235, 254, 326
Smart Energy Hubs (SEHs) 164
Solar System 10
Solid-state reactions 6, 327
Solvent composition 123
Spectroscopy 7, 16, 21, 119, 137, 146, 153, 161
Spin transfer processes 59
Spontaneous reaction 107, 226, 242, 251, 363
Standard enthalpy 32, 33, 96, 101, 174
State of equilibrium 15, 246, 253, 254, 257, 259, 262, 263, 265, 267, 268, 270
Static equilibrium 54
Steady-flow processes 89, 120
Stochastic forces 45
Sustained development 137
Synergies management 192

T

Thermal analysis 72, 79, 122, 125, 129, 132, 135, 138, 145, 147, 153, 159, 327, 360
Thermal conduction 48, 64, 66
Thermal conductivity 63, 138, 272, 295, 333
Thermal control 47, 48, 66, 71, 349, 373, 391

Thermal decomposition 20, 32, 127
Thermal decomposition reactions 32
Thermal degradation 122, 129, 137, 142, 146, 149, 153, 383
Thermal degradation processes 122, 383
Thermal energy 9, 12, 18, 20, 47, 48, 52, 55, 57, 60, 62, 63, 64, 69, 74, 106, 114, 210, 265, 286, 290, 291, 292, 295, 297, 303, 306, 315, 359, 383, 384, 387, 397
Thermal equilibrium 51, 213, 222
Thermal insulation engineering 62
Thermal insulation technology 63
Thermal lattice Boltzmann model 50, 79
Thermal management 63, 126, 295, 352
Thermal Response and Ablation Program (TITAN) 3
Thermal response codes 3
Thermal stress 122, 142, 306, 369
Thermo-charged capacitor 55, 80
Thermochemical analysis 12, 21, 26, 28, 341, 353, 368
Thermochemical cycles 12, 87, 352, 387, 390
Thermochemical data 4, 21, 23, 25, 36, 348, 354, 363, 364, 371, 372, 375
Thermochemical inquiry 11
Thermochemical modeling 9, 11, 355
Thermochemistry 1, 3, 5, 7, 9, 21, 25, 26, 28, 30, 31, 38, 119, 321, 322, 323, 325, 326, 332, 346, 350, 351, 361, 362, 363, 364, 365, 371, 375, 376, 381, 384, 388, 391, 393, 396
Thermodynamic equilibrium 45, 51, 211, 214, 215, 219, 221, 223, 227, 242, 265, 275
Thermodynamic feasibility 21, 83, 339
Thermodynamic laws 45, 55, 61, 113, 127, 170, 205, 208, 216, 221, 228, 252, 274, 341
Thermodynamic principles 1, 2, 14, 45, 54, 67, 76, 80, 85, 96, 98, 105, 109, 110, 124, 128, 132, 174, 185, 189, 191, 192, 193, 202, 206, 207, 208, 210, 212, 213, 219, 221, 225, 227, 228, 230, 232, 252, 255, 256, 258, 262, 264, 268, 276, 285, 298, 314, 327, 328, 332, 334, 340, 341, 355, 357, 364, 384, 393, 396
Thermodynamic property 95, 169, 357, 365
Thermodynamic responses 135
Thermodynamic stability 22, 32, 93, 102, 104, 116, 185, 246, 261, 268, 273, 277, 329, 346, 357, 389
Thermodynamic systems 212
Thermogravimetric analysis 143
Thermogravimetric analysis (TGA) 153
Thermogravimetry (TG) 149
Thymidylate synthase (TSase) 29
Tight calibration standards 140
Titanium carbide (TiC) 114
Transmission electron microscopy (TEM) 154, 350
tThermodynamic system 5, 25, 215, 230, 232, 241, 248, 270, 275, 310, 338

U

Ultrasonic testing 123

V

Velocity 49, 55, 63, 287, 364

W

Water cooling garments (WCG) 139
weathering process 11
Work dynamics 57
Work-energy principle 55, 56
Work-energy relationship 53
Work-energy theorem 53, 60

X

X-ray diffraction (XRD) 154